3 4028 08956 5312
HARRIS COUNTY PUBLIC LIBRARY

D1282195

DISCARD

In the Shadow of
the Chinatis

THE TEXAS EXPERIENCE

*Books made possible by
Sarah '84 and Mark '77 Philpy*

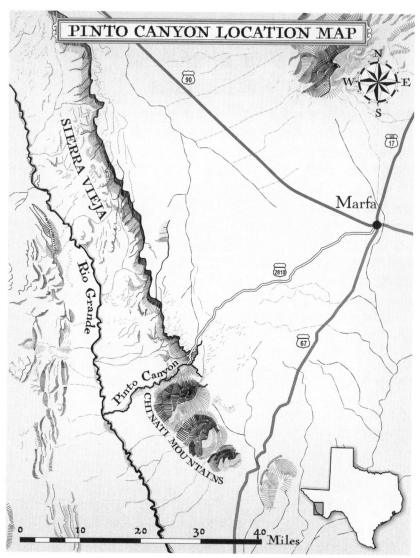

Map by Tish Wetterauer and David Keller.

In the Shadow of the Chinatis

A History of Pinto Canyon in the Big Bend

David W. Keller

TEXAS A&M UNIVERSITY PRESS

COLLEGE STATION

This paper meets the requirements of ANSI/NISO Z39.48–1992
(Permanence of Paper).
Binding materials have been chosen for durability.
Manufactured in the United States of America

Library of Congress Cataloging-in-Publication Data

Names: Keller, David W., author.
Title: In the shadow of the Chinatis : a history of Pinto Canyon in the Big
    Bend / David W. Keller.
Other titles: Texas experience (Texas A & M University. Press).
Description: First edition. | College Station : Texas A&M University Press,
    [2019] | Series: Texas experience | Includes bibliographical references
    and index.
Identifiers: LCCN 2018043502| ISBN 9781623497354 (book/cloth : alk. paper) |
    ISBN 9781623497361 (e-book)
Subjects: LCSH: Pinto Canyon (Tex.)—History. | Canyons—Texas—Presidio
    County—History. | Big Bend Region (Tex.)—History. | Chinati Mountains
    (Tex.)—History. | Natural history—Texas—Big Bend Region. |
    Ranchers—Texas—Pinto Canyon—History.
Classification: LCC F392.P7 K45 2019 | DDC 976.4/93—dc23 LC record
available at https://lccn.loc.gov/2018043502

Contents

Preface

THIS BOOK WAS born by a chance confluence of events that would find an East Coast businessman going native in the outback of Far West Texas. When Jeff Fort bought the Pinto Canyon Ranch, it was a defining moment in the history of the canyon, for he soon found evidence of a far more distant past, which brought him to the offices of the Center for Big Bend Studies, launching a collaboration now sixteen years long. By the time I met Jeff, he had already been working with and providing support for the center for nearly a decade. With me in the mix, the project expanded to include the historical record. My role was to write the history of the Pinto Canyon Ranch and to record the historic sites it contained. For a time I threw my research net over a much broader area—one that eventually embraced the entire Sierra Vieja. But I had gone too far. I discovered there was more history contained within the canyon alone than could fit into a single book. I ended up right back where I started.

Limiting the book to Pinto Canyon would allow me to produce a more cohesive narrative and a more pronounced and meaningful boundary. Unlike ranch or county boundaries, the natural borders of the Rimrock and the surrounding mountains are more geographically meaningful. As such, this book may be the deepest historical look at the smallest piece of the greater Big Bend yet written. However, to assume it is therefore definitive would be wrong. As with any history, the availability of source materials determined the limits of the book, and these sources consisted almost entirely of public records and oral histories once or twice removed. Since nearly the entire cast of characters have long since passed, it fell to descendants to tell the stories of their parents and grandparents. But not every person who owned or leased land in Pinto Canyon over the years appears within these pages. Their absence is less an indicator of significance than a result of the limitations of the extant record. I feel especially fortunate, however, to have had the pleasure of detailing the history of José Prieto, the most enduring resident of Pinto Canyon and the person whose story—bearing the hallmarks of an epic tragedy—eclipses all others.

Because the book relies heavily on narrative storytelling, I have embellished some scenes in an effort to convey stories without sacrificing their descriptive power. Even so, I went to considerable lengths to maintain fidelity to the historical record. So while the details are sometimes conjecture they are informed conjecture and not simply fabricated fancy. For those who prefer a more objective treatment, the remainder of the book—the vast majority of it—is more conventional in approach.

My goal in writing this book was simply to reveal the stories behind the crumbling rock and adobe ruins tucked within this remote canyon. Who lived here? When did they come? What were their lives like? Why did they leave? The answers, as I discovered, were far more complex than I could have known. My hope is that this story offers readers more than simply a case study of ranching in the Big Bend borderlands. I hope that it will also offer context to this dramatic and mysterious landscape, to imbue the land with deeper meaning in a way that fosters a stronger sense of place.

Acknowledgments

THE BEST PART of writing a book is having the opportunity to thank all those who played a role in creating it. I owe a debt of gratitude to a great many but none more than Jeff Fort, who generously funded the project and opened his ranch to my colleagues and me. More than that, Jeff served as a source of information, a personal guide, and—not least of all—a friend. I would also like to thank Bob Mallouf for bringing the two of us together and Andy Cloud, director of the Center for Big Bend Studies, for helpful editorial advice and unflagging support. Office manager Susan Chisholm not only keeps the center functioning, she is also a bastion of common sense, reminding us all that we should relax and eat more chocolate. Of course, she is right.

This work greatly benefited from the able assistance of Mattie Matthaei, who served as my right hand throughout most of the project. She conducted a great many of the interviews, made some of the most important contacts, and discovered some of the richest sources. I would also like to thank Dawnella Petrey, Ashley Baker, and Jim Miller, who waded through mounds of newspaper articles and tax records, which helped immensely in informing the story. The historian Glenn Justice was more than generous with his vast storehouse of knowledge on the history of the Mexican Revolution and remains the preeminent scholar of the Sierra Vieja. In addition to the insights he offered while reviewing my chapter on the Mexican Revolution, he helped shape my understanding of the history and significance of the larger Sierra Vieja region. I would similarly like to thank rancher

Chip Love for reading parts of the manuscript and validating some of the overarching themes in the book.

There are no words to express my gratitude to Ty Smith for graciously sharing his hard-earned, original research in US military records. In response to one of my inquiries, he pored through hundreds of obscure archival documents and then compiled the most comprehensive treatment ever of the operations of the US Cavalry and especially the National Guard in the Big Bend Military District during the Mexican Revolution. With great attention to detail he produced several documents that I relied on extensively. Ty conducted the research, typed up the manuscripts, and then handed them to me in a gesture of academic courtesy that has no parallel in my years of scholarship. For all the work he did, I am very happy that effort resulted in a book of his own, *The Old Army in the Big Bend of Texas, 1911–1921: The Last Cavalry Frontier*.

I also wish to thank the staff of the many archival repositories I consulted, especially Melleta Bell, Jeri Garza, and Michael Howard of the Archives of the Big Bend at the Bryan Wildenthal Memorial Library and Liz Jackson, Mary Bones, and Matt Walter of the Museum of the Big Bend—both at Sul Ross State University—as well as Maggie Marquez and the late Richard Williams of the Marfa and Presidio County Museum, and Mandy Roane of the Marfa Public Library. Thanks also go to the staff of the Judd Foundation, especially Caitlin Murray, who went out of her way to assist me with the foundation archives and photographs, and to Andrea Walsh, who made many helpful corrections and suggestions to the section on Donald Judd. Thanks also to Lisa Sharik of the Texas Military Forces Museum at Camp Mabry in Austin, to Grady O'Connor of the David Zwirner Gallery in New York, and to Judy Baker Elliott on behalf of the Robertson County Historical Survey Committee for facilitating permission to use historical photographs. Thanks also to the staff of the Presidio County Clerk's Office, the Presidio County Tax Office, the Presidio County District Clerk's Office, the Presidio County Abstract and Title Company, the archives of the El Paso Public Library, and the National Archives in Washington DC.

I owe a special debt to Jerry Lujan for sharing his childhood stories of Juan Morales and, more significantly, for introducing me to the extended Prieto family, which, in providing a focal point for the narrative, was pivotal. To that end, I am proud to be able to extend heartfelt thanks to members of the Perez family—especially Oscar, Nancy, and Manny—who enthusiastically shared family history, photographs, heirlooms, and—significantly—their time, including several days spent revisiting their old haunts in Pinto Canyon. Because of their generosity, the story deepened in ways it never could have otherwise. I would like to extend a special note of thanks to Oscar Perez, the keeper of the family history, for his incredible generosity of spirit.

Thanks to Diane Hankins and Dan Burbach, caretakers of the Chinati Hot Springs, for hours of engaging conversation and keeping me up to date on the latest local news and gossip. Thanks also go out to Ashley Baker, Candace Covington, and Jesse Nowak, who provided fieldwork assistance by helping with the archeological recording of a number of complex historic sites in Pinto Canyon. I offer particular gratitude to Sam Cason for sharing his expansive knowledge of Pinto Canyon prehistory and for being an exemplary colleague throughout our years of working together.

Oral histories—more than a hundred—form the backbone of this book. I am honored to have the opportunity to thank the individual interviewees for graciously sharing their time and memories. Although the entire body of interviews helped shape my thinking, only about half of those directly informed this book and are mentioned here. Thanks to Apache Adams, Terry Jean Shely-Allen, Clementine Bales, Benny Benavidez, Ida Benavidez-Taulbee, Steve Benavidez, Teresa Benavidez-Porras, Alfredo Bryalba, Don Cadden, Tony Cano, Jamie Dearing, Kay Ellis, Monroe Elms, Julie Finch, Jean Anne Fort, Jeff Fort, Ann Fuentes, Ted Gray, Jeanette Houston, Flavin Judd, Rainer Judd, Daniel Logan, Chip Love, Jerry Lujan, Benjamin Martinez, Adam Miller, John Newby, Frank O'Dell, Manny Perez, Nancy Perez, Oscar Perez, Ruben Perez, Chon Prieto, Guadalupe Prieto, José Manuel Prieto, Robert John Rod, Bill Rogers, Shirley Rooney, Wayne Seipp, Oliver Smith Jr., Ann Sochat, Janell Stratton, Jason Sullivan, Frederick Dale Sutherlin, Jackie Sutherlin, Martin Sutherlin, Michael Sutherlin, Nancy Sutherlin-Gibson, Antonia Vasquez, Demetrio Vasquez, Maria Flora Vizcaino, and

Ginny Watts. I would also like to thank Christopher Henry, Will Jewett, John Karges, Jesse Kelsch, and A. Michael Powell, all of whom helped clarify my thinking on the natural history of the area.

Many thanks to Jay Dew, Patricia Clabaugh, Shannon Davies, and others at Texas A&M University Press. Thanks also to my anonymous reviewers for making important suggestions and to the freelance copyeditor, Maureen Creamer Bemko, for correcting errors and bringing the book into conformity with house style. Thanks to Letitia Wetterauer for patiently working with me to create the maps and to Susan Chisholm for her attention to detail in creating the book's index. Thanks to my father, Charles W. Keller, for being my first reader and for offering helpful insights and constructive criticism, and to my mother, Judith P. Keller, for being my greatest advocate. I owe a bow of gratitude to writer and friend Anna Badkhen, who offered a fresh perspective and her keen editorial skills. Finally, I would like to thank my partner, Jessica Lutz, for making me laugh when I needed it the most, for refusing to allow me to lose faith, and for sharing her gift for seeing the bigger picture.

In the Shadow of
the Chinatis

Map by Tish Wetterauer and David Keller.

1

The Fault Line

THE LONG MORNING shadow cast by Chinati Peak shrouds a ragged desert lowland as mysterious as the Marfa Plain is vast. Between them runs a rimrock composed of ancient lava flows that heaved skyward late in geological time but fractured along a much older zone of weakness, one that has been a defining feature of the region throughout its storied geological past. For the Rimrock Fault represents the dividing line between two worlds, a feature that is as much a border as the Rio Grande. It is arguably more so, for the lowlands on both sides of the river hold more in common with each other than either do with the mountain peaks and plateaus that surround them. In many ways the Rimrock is a more enduring boundary than the river could ever be.

The Rimrock Fault is a divide that through deep time has been nearly absolute. History unfolded differently in the lowlands than above, and it unfolded differently for reasons that had more to do with the character of the landscape than its proximity to the border with Mexico. That is because the fault line marks a boundary between two distinct bioregions, a division born of geological processes that affected everything that followed: physiography, hydrology, climate, ecology, and—of more interest anthropologically—the culture that developed below its escarpment.

Above the Rimrock, the Marfa Plain tilts gently eastward in a vast expanse of fertile grasslands that eventually lap against the flanks of the verdant Davis Mountains. There are few features that restrict the view, and cattle graze peacefully on the gently sloping hills. Rain that falls here flows to the Salt Flat Bolson, a closed basin where water pools and eventually evaporates on open land. By contrast, below the Rimrock, the land pitches away into a maze of lesser faults and volcanic intrusions as chaotic as the grasslands are serene. Below, it is a forbidding place of precipitous cliffs and rock-strewn slopes, where catclaw thickets and lechuguilla post a savage guard. Where every single thing living or not seems eager for your blood. Where grass is scarce and the scant rain that does fall flows to the Rio Grande and onward to the sea.

And yet, these sun-scorched lowlands of the Sierra Vieja are also places of epic, if rugged, beauty—a landscape forged from heaving anticlines, massive volcanoes, igneous intrusions, and an endless jumble of block faults now weathered from millions of years of erosion. What has been left behind is a geological wonderland as grand as any in the greater Big Bend. Pinto Canyon, the southern-most drainage of the Rimrock Country, showcases these variegated

A view of turtle-backed Chinati Peak from the well-grassed Marfa Plain. Photo by D. Keller.

A view over Pinto Canyon toward the south, with Chinati Peak (*back left*) and Sierra Parda (*back right*). Photo by D. Keller.

lowlands through stunning panoramas rarely encountered outside of a national park. For these reasons, over the last several decades, people have sought to protect it from forces that might destroy its character.

The hostility of the landscape is the principal reason Pinto Canyon and the Sierra Vieja lowlands were among the last areas to be settled in Presidio County. And the first to be abandoned. The earliest settlers naturally chose the luxuriously grassed highlands, regarding the low country as little more than a wasteland. With the state legislature's passage of the 1905 Eight-Section Act, however, a second wave of eager settlers arrived to find nearly all the good country taken, forcing them to scramble for what remained. The lowlands, settled with much greater difficulty, were also less forgiving of failure. Turnovers became routine. A precious few settlers were able to enjoy a handful of wetter years and stronger markets. But the seven-year drought of the 1950s ultimately spelled the beginning of the end of ranching the lowlands. Never productive to begin with, the land had declined past the tipping point of profitability. The ratchet effect of decades of episodic drought and continuous grazing had pared the country down to but a shadow of its former self.

For the first half of the twentieth century, only one man persevered in the face of such adversity. Whatever limitations were imposed upon him through biases against his ethnicity, or his meager education and humble background, José Prieto enjoyed success that was his alone. But tragedy ultimately eclipsed his triumph. The Prietos' tenure in Pinto Canyon would be marred by conflict, some of which would come from the outside but most of which would come from within. The family's own internal strife eventually forced its departure, marking the end of an era.

With the dawn of the twenty-first century, the demographic, economic, and cultural profile of Pinto Canyon and the surrounding lowlands had changed fundamentally from that of its past. As stockraising grew increasingly challenging and the old ranching families left, those who replaced them no longer depended on the land as a source of income. Instead, wealth made elsewhere allowed newcomers the luxury of owning land without requiring it to turn a profit. In its transition away from agriculture, Pinto Canyon narrowly missed the kind of subdivision and development that easily could have marred its majesty. The minimalist artist Donald Judd introduced a set of values that ultimately made its preservation a reality. By accidents of history, the gnarled splendor of Pinto Canyon has been afforded a rare protection that has safeguarded its ecological and historical integrity.

As the raven flies, Pinto Canyon is about thirty-two miles southwest of Marfa—the seat of Presidio County in Far West Texas. Meandering within the recesses of the canyon, Pinto Creek marks the northern edge of the Chinati Mountains and the southern edge of the Sierra Vieja. Pinto's waters rise in the Cuesta del Burro, a cluster of rolling hills above the canyon's rim. As the creek descends, it winds around volcanic intrusions and faulted Cretaceous mountains before cutting through bolson deposits that erode into caves in conglomerate and create vertical walls in sandstone. Finally, south of the hamlet of Ruidosa, Pinto Creek debouches into the Rio Grande in a delta, as if a river to the sea. This swath of broken country, in the morning shadow of the Chinati Mountains, is the setting for our story.

Today, Pinto Canyon is best known as the scenic route to the Chinati Hot Springs, where a loose gravel road descends precipitously from the high plain, like a passageway to some distant past. The canyon and the mountains around it seem ancient and inscrutable, an enigmatic land as full of secrets as it is empty of people. Only the melting adobe ruins, the fallen fence posts, the crumbling stone corrals remind us that humans were ever here at all. It is a place that seems frozen in a time that has been lost. A past that has been forgotten. Much of it has, but a few stories, a few threads survive. This book weaves them into a loose tapestry to reveal a history as complex as the churned topography of the land itself.

A landscape is never simply a passive backdrop for the play of history, but here aridity and toppled topography conspire to force it center stage. The land assumes a certain primacy, defined not by the ephemera of plants and animals but by the very bedrock that rises far above the silvery flow of Pinto Creek, in looming masses that break the horizon into a hundred different vistas. The history of Pinto Canyon, like the landscape itself, reflects a violent geological past. If the general outline of that history is applicable to much of the borderlands of the Big Bend, the details reveal its singularity—a story as grand as the scenery.

2

The Terrible Mountains

The Natural and the Early History
of Pinto Canyon

[Pinto Canyon] is a very rough country . . . being surrounded
by terrible mountains[;] the State might do well to let them
have it.

—State school land agent William M. Baines, 1885

AROUND THIRTY-THREE million years ago, magma began to
rise through ancient faults along the southern edge of the North
American continental plate. Blocked by overlying bedrock, the magma formed a huge pool. As the molten rock churned and bubbled,
the gases within it accumulated until the crust of the earth could no
longer hold. When it broke, the pent fury of the largest and most
destructive volcano in the history of the Big Bend burst forth.

It is known as a Plinian eruption. A deafening explosion rocked
the earth as the crust cracked open, releasing a vertical column of
molten magma that shot skyward for twenty miles, penetrating
the stratosphere. There it spread laterally, growing into a massive,
roiling cloud before streaming back to the earth. The ash and rock
rained to the ground, building into a pyroclastic surge that massed
and began to flow outward, racing across the landscape at more
than 100 miles per hour, incinerating everything in its path. As the
molten mudslide engulfed landforms hundreds of feet tall, vigor-

ously convecting clouds of fine ash rose ominously above its leading edge. The dense ash obscured the sun until no light could be seen aside from the glow of the molten rock and the flash of trees as they exploded into flame. The deafening roar, the fiery column, the molten flow, the darkness lasted for weeks. When it finally quieted, some six thousand square miles of land—an area larger than the state of Connecticut—lay beneath a thick blanket of searing volcanic ash.[1]

The Tambora volcano in Indonesia killed tens of thousands of people in 1815. The thunderous roar could be heard from twelve hundred miles away. That volcanic eruption stands as the largest in recorded history. "The whole mountain," an eyewitness wrote, "appeared like a body of liquid fire, extending itself in every direction. The fire and columns of flame continued to rage with unabated fury, until the darkness caused by the quantity of falling matter obscured it." The Chinati eruption was at least twenty times larger. For fifty miles in any direction, all vestiges of life vanished, leaving only a scorched earth, lifeless beneath the ash flow. For at least a hundred miles beyond that, the rainfall of dry ash suffocated anything that breathed. The landscape, as far as the eye could see, was as barren as it would ever be again.[2]

The Chinati volcano erupted intermittently for several hundred thousand years, each event sterilizing and redefining the landscape anew. With every eruption, a fresh mantle of volcanic rock hardened around the vent, creating a vast volcanic alluvial apron. Finally, about thirty-two million years ago, the Chinati eruptions ceased. Although other eruptions continued across the region for another five million years, none compared in scale with Chinati. Producing some 240 cubic miles of rock, it was the most expansive flow in the history of the Trans-Pecos. When the eruptions finally subsided and the active vent collapsed, what remained was a caldera some 12 by 19 miles in area—more than 3,200 feet of rhyolite and trachyte flows. These apron rocks were later intruded by vertical granite stocks, dikes, and horizontal sills, the largest of which is just west of Chinati Peak—the massif known as Sierra Parda. This granite stock stands today in bold relief as the northwestern rampart of the Chinati Mountains.[3]

Chinati Peak (*right*) from above Horse Creek. Photo by D. Keller.

Ridgeline Creek

Other than by foot or horseback, there are only two ways to get to Pinto Canyon. One is to drive north from Presidio following the Rio Grande along the winding River Road before turning east on Pinto Canyon Road. The better way is to head west from Marfa on Ranch-to-Market 2810, which crosses the grassy plain of the Marfa Plateau and winds through the Cuesta del Burro before turning to dirt and descending nearly a thousand feet into Pinto Canyon. After following Pinto Creek for about six miles, the road climbs out to roll westward across ancient bolson deposits to the Chinati Hot Springs or the river hamlet of Ruidosa. Locally famous, it is one of the most spectacular drives in the region, offering an array of Permian, Cretaceous, and Tertiary deposits, including a prominent field of volcanic intrusions set against the rocky hulk of the Chinati Mountains. To the west, across the Rio Grande, rise lofty Mesozoic mountains, sometimes collectively known as the Sierra Quemada, a series of ranges that parallel the river and provide the Mexican analog to the Rimrock escarpment—highlands that bracket and contain the low country between.[4]

Largely serving as a sort of natural boundary between the Chinati Mountains and the Sierra Vieja, Pinto Canyon has elements of both but belongs entirely to neither. It is a place where the cooler, moister slopes and canyons constituting the northern edge of the Chinatis grade into the desert lowlands of the Sierra Vieja. As such, Pinto Canyon is not a canyon in the typical rock-walled sense but is instead defined by Pinto Creek, which winds through a confluence of varied landforms—the flanks of the Chinatis, the volcanic intrusions of the Allen field, faulted Cretaceous limestone. The array of geological structures lends Pinto Canyon a measure of visual complexity that defies easy description. But the complexity doesn't end there.

Physiographically, Pinto Creek flows in seeming defiance of the underlying structure that supports it, running along an anticline of the same name and intersecting a second at a right angle near where the canyon leaves the mountain front. These Mesozoic folds are much older than the volcanic deposits that dominate the surrounding landscape. As a result of subsequent deposition and faulting, today they have less bearing on topography than they once did—so little that a stream now runs along an ancient ridgeline.[5]

The End of the Earth

It's all about edges, for Pinto Canyon and the Sierra Vieja reside on the edge of ancient geological structures and vast geographical provinces. It is a region marked by instability reflecting its geological underpinnings. Grasping it requires an understanding of one of the most fundamental of geological processes: plate tectonics. Located along the far southwestern margin of the North American tectonic plate and the edge of Paleozoic structures such as the Diablo Plateau, the Viejas also border the northeastern margin of the Chihuahua trough—a narrow Mesozoic depression—and the northeastern margin of the expansive Basin and Range physiographic province. Politically they constitute part of the far southwestern edge of the state of Texas and the northeastern edge of the Mexican state of Chihuahua. Throughout time the area has been marginal to everything, central to nothing. It is a place peripheral to the solid mass of continents, to the geography of stability, and to political and cultural centers. As such, the human history of the region becomes inextricable from the land's protean past.

Culturally, the region is a melting pot. For millennia it was under the sway of desert-based human adaptations—small groups of hunters and gatherers—who were joined by farmers in later centuries. Economics hinged on the meager desert pantry of sotol and cactus or the fertile floodplain soils and seasonal overflows of the Rio Grande. Various Indian dialects lost to time and, later, Spanish were the principal languages spoken. Intermarriage between the Spaniards and Indians was common, and, just as blood mixed, so did customs and traditions. Mestizos often embraced Indian adaptations forged in the desert, and Anglos in turn frequently took up the mestizos' traditions. Only in the last century and a half have ranching and irrigated floodplain agriculture predominated. Vernacular architecture evolved from wikiups and pithouses to *jacales* to more permanent mud-mortared rock and adobe. Prehistorically, clothing was hardly worn at all, but regional apparel has historically been loose fitting and bracketed by huaraches and sombreros. Following the US-Mexico War, when the Rio Grande officially became an international boundary, it suddenly split what had long been a geographically and culturally unified region. Yet, despite the increasing disparity between the two countries, something of a border culture persisted, resisting the division the river was supposed to represent.

The Great Rimrock of the Sierra Vieja

As the volcanoes that dominated the region for millions of years gradually calmed, the land lay nearly flat from successive lava flows that filled in the valleys and smothered the low peaks, reducing the contours and raising the base level of the landscape. The Laramide orogeny that had squeezed the earth, causing volcanoes to spew lava, began to ease and slowly reverse. Bedrock long compressed now relaxed and began to be pulled apart, causing a riot of faulting that marked the beginning of basin and range deformation. As it was stretched, rock blocks rose and fell and tilted along fault lines, some newly created, others following ancient structural grains that dated back at least to the Late Paleozoic more than 220

The Great Rimrock of the Sierra Vieja. Photo by D. Keller.

million years earlier. Among the many sections of moving ground, a long northward-trending array of bedrock blocks tilted slightly eastward at the same time that adjacent blocks to the west began to sink. The result was the Great Rim of the Sierra Vieja, which begins just south of Chispa summit about twenty miles northwest of Valentine and extends south-southeastward for some forty-five miles until it grades into the Chinati Mountains.[6]

The rim extends southward from Chispa as a singular towering rock wall as far as Vieja Pass, where it breaks into a zone of fault wedges and narrow parallel fault blocks. Formed along en-echelon faults that bifurcate and coalesce, the main rim becomes obscured among the array of lower rims. These lesser rims ultimately give way to the prominent Capote Rim as the chief rim of the southern third of the Sierra Vieja. It first appears as the capping rim of Capote Mountain before winding southwestward, where it divides, with one branch terminating in a massif known as El Macho. The other branch trends southward, maintaining a prominent rimrock until just shy of Pinto Canyon, where it becomes more subdued, so that where the Pinto Canyon road descends, there is hardly a rim at all. South of Pinto Creek, the rim emerges again as Love Mesa, which extends to the shoulder of Chinati Peak.[7]

Westward from the uppermost rim, the land tumbles away stepwise down to the Rio Grande, forming a valley that is a large downdropped block containing the Sierra Vieja lowlands. Across the Rio Grande, a loose chain of mountains in Mexico—the Sierra Pilares, Sierra los Fresnos, Cerro la Ventana, Cerros Colorados, Sierra Ojo Caliente, and Sierra la Chiva—form the western border of the lowlands. This broken mountain chain—a northern outlier of the Sierra Madre Oriental—is sometimes collectively known as the Sierra Quemada. Composed primarily of lower Cretaceous limestone that was sharply folded and thrust-faulted during Laramide compression and further deformed during normal extensional faulting, the Mexican mountains contrast markedly with the mostly volcanic features on the US side.[8]

In the southern third of the Sierra Vieja, the major down-dropped block accompanying the Rimrock Fault created the Presidio Bolson— an inwardly draining topographical depression that extends from Candelaria southward to Presidio. Some twelve by forty-five miles

The Sierra Quemada in Mexico in evening twilight. Photo by D. Keller.

in extent, the bolson is deepest (around forty-nine hundred feet) just south of the town of Ruidosa. Coincident with the rising Rimrock and sinking basin, rainwater flowed into the depression to create a large saline lake. For millions of years the surrounding mountains eroded and filled the lake with sediment. Tributary streams deposited coarse loads on alluvial fans and pediment slopes near the mountain front, with finer material carried farther downslope to create the bulk of the bolson fill. Over time, three distinct rock bodies of similar composition flowed into the bolson—the lowest a silty clay, followed by sandstone, and topped by a conglomerate containing interbedded sandstones. By the time the depositional phase ended, the basin had filled with some fourteen hundred feet of clay, silt, and rock above the present level of the Rio Grande.[9]

Around 2.5 million years ago, with the onset of the Pleistocene, wetter conditions caused the lake to overflow the lower lip of the basin by a trickle that would ultimately become the Rio Grande. As basins upstream also overflowed their containers, the waters eventually connected one basin to the next, giving birth to the ancestral river. As the base level of the main channel eroded downward, side streams began to excavate the basin, the main stream carrying debris toward the Gulf of Mexico. At intervals, likely during periods of low rainfall, downcutting slowed, allowing tributaries to form nearly flat terraces. This "lateral planation," punctuated with renewed periods of erosion, produced four major terrace surfaces that step downward toward the river. Further erosion has dissected the terraces so that only islands of these tablelands remain. Below these terrace islands, the surrounding eroded bolson deposits form a barren bench-and-slope topography wherever the deposits are comprised of sandstone and rolling hills wherever the deposits are comprised of conglomerate.[10]

Faulting and erosion are the primary forces that have sculpted the present landscape and that continue apace today. Floods carve out canyons and arroyos, uncover volcanic intrusions, and redeposit sediment to form broad alluvial fans. Overlying rock weathers down to uncover more resistant igneous intrusions, creating mountains, mesas, and stone-walled dikes. Meanwhile, mountains continue to rise and basins continue to drop in their inexorable push toward tectonic balance. This is why the Sierra Vieja has the highest earthquake risk in Texas and registered the worst in state history, near Valentine in 1931. Accompanied by rumbling subterranean sounds heard for hundreds of miles from the epicenter, the Valentine earthquake reached a magnitude of 6.0. No deaths were reported, but many structures were damaged and tombstones in the cemetery rotated as much as 45°. Ranchers reported landslides across the Sierra Vieja, and water wells were muddied as far as fifty miles away. At Candelaria, residents recounted how the "ground seemed to rise up in waves."[11]

As the Pinto Canyon road descends from the Marfa Plain, it passes over Quaternary gravel terraces, Tertiary volcanic rocks, and Cretaceous limestones before entering Permian deposits of siltstone, chert, and marl. At the lowest elevation of the canyon, just before the road climbs out below Cerro de la Cruz, the very earliest rocks are exposed. Heavily faulted, dark gray, thin-bedded mudstone—deposited during the Permian some 260 million years ago—is a remnant of the ancient Ouachita Mountains deposited just before the largest extinction in history marked the end of the Paleozoic era.[12]

Geologists sometimes refer to such landscapes as "youthful." Erosion has not had enough time to soften sharp edges or flatten folds. The result is the dramatic topography we see today—one largely caused by erosion along a remarkably steep elevational gradient. From where the road crosses Pinto Creek at 4,600 feet above sea level, the mountain flanks rise steeply to the south for more than 3,000 feet until topping out on Chinati Peak—at 7,728 feet above sea level, the tallest landform for fifty miles in any direction. A line from the peak down to the Rio Grande defines the second-steepest gradient in the state—plunging downward more than 5,000 feet in a distance of only eleven miles, or almost 450 feet per mile. Farther north, the gradient diminishes. From 6,549 feet at Cleveland Peak on the Capote Rim, the land descends in fits and starts some 3,800 feet across an array of smaller peaks and successive plateaus until arriving finally at the Rio Grande near Ruidosa. There, if a flow exists at all, the water courses at 2,760 feet above its final destination—the Gulf of Mexico.[13]

The Way the Waters Flow

Although the Rio Grande is the only perennial stream in the Sierra Vieja, some fourteen major tributaries head in the southern third of the lowlands alone and trend generally westward to meet the river. All but one of these can be considered intermittent, meaning they are often dry but carry a flow along some portion during most seasons. Pinto Creek stands out as one of the two strongest-flowing intermittent streams of the southern Viejas (the other being Capote Creek). Pinto flows westward some sixteen miles along the northern flanks of the Chinati Mountains before emptying into the Rio Grande about five miles south of Ruidosa. There is typically a surface flow along some portion of the streambed year round, even if the water travels only a short distance before disappearing into the sand, only to emerge again farther downstream. This steady subsurface flow supports robust riparian vegetative communities and has made Pinto Creek a focal point for camps and settlements through time. Archeological evidence shows it was used extensively in prehistoric times as a location for temporary campsites, and, most likely, it served as a major travel corridor. More recently, the waters of

The clear water of Pinto Creek has long attracted humans to its banks. Photo by D. Keller.

Pinto Creek attracted settlers and anchored a small ranching community for the first few decades of the twentieth century.[14]

Ojos de agua and Tinajas

The most precious of desert resources, springs allow life to exist in an otherwise hostile land. Springs were especially critical to the spread of livestock ranching in the early 1900s, before surface tanks and wells rendered unwatered land usable. In fact, the abundance of springs in the Sierra Vieja lowlands was likely the most compelling reason people settled in the area at all. Even though the lower country is hotter, drier, and more marginal rangeland than the Marfa Plain, in one of the great hydrogeological ironies it also has far more surface water.

You can thank the faults. Without these geological conduits water might remain trapped below the surface. Of course, water must also be present to tap, and for that you can thank the bolson. Most of the water in the area is contained within the ancient Presidio Bolson that functions as a large underground reservoir. Over the millions of years as the bolson sank, it filled with sediment that is largely porous, allowing it to hold a vast amount of water. The importance of the bolson to the distribution of springs locally is self-evident. Of the thirty-nine major springs across the Sierra Vieja lowlands, two-thirds are in the southern third of the range, the majority located within the bolson. The contrast is striking, and the reason is that the northern two-thirds of the range has no equivalent water reservoir.[15]

The Presidio Bolson is not a closed system but one that has both inputs and outputs. It is recharged through rainfall in bordering mountains—the Chinatis and Sierra Vieja—and through coarse, permeable alluvium in upper tributaries. This water flows underground toward the Rio Grande, but it does so slowly—at an estimated rate of about eight feet per day. The journey of this flowing water is often interrupted. The zones through which it travels are commonly blocked by less permeable strata because as the Presidio Bolson sank, it created a stepwise series of blocks down-dropped toward the river. Such vertical movement of geological strata placed more recent strata on the downslope side and older strata on the upslope side. In cases where the younger strata were less permeable, water pooled, forcing it to the surface.[16]

Aside from streams and springs, the most important water sources are *tinajas*—water-bearing bedrock depressions that develop below waterfalls or else are carved out by minor spring flows, seepage, or other geohydrological processes. *Tinajas* (meaning "large earthen jars") are often found in dry streambeds and even in bedrock exposures on higher landforms, areas that otherwise have no water, making them especially important sources for wildlife. In more protected areas, *tinajas* can also support small riparian communities and create mesic (wet) microenvironments that allow a much more diverse flora and fauna.[17]

The Bounty of the Sky

Pinto Canyon and the Sierra Vieja are part of a much larger climatic province that spans the entire northern Chihuahuan Desert and is distinguished by hot summers and cool, dry winters. But due to the broken topography and stark elevational differences, the mountains and basins are further distinguished by their own regimes. Higher elevation brings cooler summers and topographic relief that triggers orographic precipitation—rainfall spawned by sharp updrafts from the steep, sudden rise of the mountains. By contrast, the lowland basins are much hotter, with less precipitation and much drier air.[18]

The closest reliable climate data come from two weather stations—one at Candelaria along the Rio Grande (elevation 2,875 feet), about fourteen miles northwest of Pinto Canyon—and one above the Rimrock near Viejo Pass (elevation 4,420 feet), about forty miles to the north-northwest. The 1,545-foot elevation difference is instructive. Average high temperatures recorded during June, typically the hottest month, vary by almost ten degrees Fahrenheit between the two stations—Candelaria reporting an average of 102 degrees and Viejo Pass an average of 93. But averages compress the extremes, which are often more illuminating. In Candelaria, temperatures exceeding 100 degrees have been recorded in every month from April to October, with the highest on record being 115. By contrast, Viejo Pass reports only four months exceeding 100 degrees, with an all-time high of 108.[19]

Wide temperature swings are a hallmark of desert climates. Clear skies, dry air, and bare ground cause the spread between daytime highs and nighttime lows to diverge significantly—as much as sixty degrees or more, especially during the spring. Lowlands, like Pinto Canyon, tend to have the greatest divergence, such as when a March afternoon has heat waves shimmering over ground that bore a heavy frost at sunrise. Temperatures sometimes drop below freezing and snow sometimes falls, but both are typically the result of strong cold fronts that rarely stay for long.[20]

Rainfall in the desert is as precious as it is scarce. From year to year the amount and season in which it falls largely determines the biotic productivity of the land. Although precipitation can occur in any month, most comes during summer monsoonal rains—a pattern of afternoon convective thunderstorms that brings the desert to life. Throughout summer, land temperatures steadily rise in Mexico and the southwestern United States, causing a shift in atmospheric circulation that pulls in moist air from the Gulf of California and the eastern Pacific Ocean. The arrival of this moist air, coupled with the instability created by differential heating, brings intense but short-lived thunderstorms. Many storms are local in nature, but the heavi-est, most widespread rains—and those responsible for most of the large deviations from year to year—are typically remnants of pow-erful tropical storms from the Gulf of Mexico or the eastern Pacific Ocean that lift moisture high into the troposphere.[21]

Many factors conspire to cause the dryness of the region, but the mountainous masses of the Sierra Madre Oriental to the southeast and the Sierra Madre Occidental to the southwest are fundamental, blocking most of the moisture-laden air from entering the region. As air lifts above the peaks, the associated cooling causes rainfall on the mountains. The air that descends on the leeward side is now drier, wrung of moisture. The resulting zone of dryness on the lee side casts what is known as a "rainshadow." Localized topography plays a similar role on a smaller scale. Here, the western bounding mountains in Mexico create orographic lift for prevailing westerlies, producing a dry zone across the lowlands. The eastward bound-ing Rimrock has a similar effect—as a result of both elevation and sharp updrafts along the vertical wall of the Rimrock, which forces air upward. Annual precipitation at Candelaria, in the rainshadow of Sierra Ojo Caliente, averages about twelve inches, whereas at Viejo Pass, about twenty miles from the mountains, rainfall averages almost fourteen.[22]

For most of the year, prevailing winds blow from the west and southwest—bringing air that is typically dry. During July and August, however, a more mixed pattern develops, adding easterly surges of air moistened by the Gulf of Mexico. Although gale force winds are rare, dust storms are not, especially during the windier months of spring. A wind as light as 15 miles per hour can begin to raise lighter dust particles, especially those derived from unconsol-idated volcanic ash. During the monsoon season, dusty gust fronts caused by the outflow of rain-cooled air often precede thunder-storms. In the winter, cold fronts wedging southward into a dry air mass can produce similar gusts. But the most severe usually occur on dry spring days, when gusting winds can carry dust high into the atmosphere and cause gray-white powder to fill the skies, thicken-ing the air and dimming the sunlight.[23]

Most of the time, however, the wind is calm, skies are clear, and the sun beats down with unrepentant fury. The region is hailed as the sunniest place in Texas and one of the sunniest in the United States. On clear days, with the region registering on the "very high" to "extreme" end of the ultraviolet index, bare skin can burn in a matter of minutes. Because of this, no one traverses the desert for long without protection. In these borderlands, sombreros are much more than a cultural icon.[24]

The Living Earth

Other than climate, bedrock holds the greatest sway over soil formation. The length of time soil is exposed, the topographic character of the landscape, and organisms that live within it are important, but the composition, grain size, and hardness of the "parent" rock are the qualities most central to arid land soils. Of the five main soil types in Pinto Canyon, gravelly residuum (weathered bedrock) and Holocene alluvium (water-deposited sediment) derived from sedimentary or igneous rocks are most common. But on higher landforms—escarpments, hills, and mountains—soil tends to be shallow if it exists at all. In many places, bedrock is all that shows. These three categories (gravelly residuum, Holocene alluvium, and bedrock) constitute about 85 percent of the total area, with arroyo channels and floodplains taking up most of what remains.[25]

Desert soils worldwide share many similarities, being light in color and having low organic content and limited moisture available for plant growth. Due to evaporation rates that far outpace rainfall, desert soils accumulate salts and other insoluble minerals. In the northern Chihuahuan Desert, soils retain calcium carbonate. Because rainfall is not sufficient to leach out such carbonates, they tend to build up, often at shallow depths. As a result, rocks in contact with soil tend to develop a characteristic white limestone rind.[26]

A Diverse Flora

Although the low country is poorer rangeland for domestic livestock, it has greater overall biodiversity than the uplands. There are many reasons for this disparity, notably the substantial topographic relief, a wide range of soil types, and the overlapping of distinct habitats that create what are known as ecotones. Taken together, they create an abundance of microniches that host plants and animals outside of their normal ranges. This pattern is especially apparent in Pinto Canyon, where plants of the Chinatis and the Sierra Vieja overlap. The soils of both the Chinatis and most of the Sierra Vieja are derived from a variety of volcanic rocks, but just north of Pinto are the most extensive limestone deposits in the entire range. The interlacing of distinct soil types, in turn, supports different vegetative communities.[27]

The few trees that grow to any reasonable size are generally found only along watercourses and within the recesses of shaded canyons. The tallest—the Rio Grande and Arizona cottonwoods—grow along Pinto Creek and associated tributaries, often among Goodding's and coyote willows. Higher-elevation areas of the Chinatis and Sierra Vieja, as well as mesic canyons, contain stands of gray oak, Gambel oak, and chinquapin oak along with Arizona walnut, papershell pinyon pine, velvet ash, alligator juniper, and even the vibrantly colored bigtooth maple. Smaller trees such as netleaf hackberry, Mexican buckeye, and mountain laurel are common along stretches of Pinto Creek.[28]

By and large, however, Pinto Canyon and the surrounding areas are dominated by midsized shrubs. Robust sacahuiste and sotol, common on north-facing slopes and higher-elevation areas, give way to the smaller, and more savage, lechuguilla farther downslope. Lower yet are dense stands of mariola along with the spiky-armed ocotillo and the stately Spanish dagger. Creosotebush, mesquite, and whitethorn acacia are ubiquitous across the lowlands, along with sometimes-impenetrable thickets of the dreaded catclaw. In the foothills and extending out to the lowland breaks beyond are candelilla, leatherstem, skeleton-leaf goldeneye, prickly pear, and the pencil-thin and thorny tasajillo, along with a broad range of smaller cacti.[29]

Grasses were much more abundant historically, but fire suppression, drought, and grazing over the last 120 years have caused a substantial decline in density and diversity. Shrubs, which can easily outcompete grasses given the chance, have largely replaced many former grasslands. Even so, a large number of grasses occur and, in wet years, respond with vigorous growth. Higher-elevation areas and moister environments support blue and sideoats gramas, tanglehead, green sprangletop, and a number of species of bluestem. At lower elevations slim tridens, Arizona cottontop, Hall's panicum, bush muhly, fluffgrass, and burrograss predominate. Along Pinto Creek grow dense stands of the hearty deer muhly, especially in shaded sites. Finally, the rolling hills west of the mountain front—extending almost to the banks of the Rio Grande—are often loosely furred by tufts of chino grama and threeawn.[30]

This summary description grossly oversimplifies the fact that a great many species occur either higher or lower in elevation than their normal range due to the great differences in effective moisture that result from slope, aspect, and the protection afforded by variable relief. North-facing slopes and steep-walled canyons, for example, host far more mesic species than would be there otherwise. Ocotillo and lechuguilla can be found close to the rim as well as at the bottom of canyons, and drainages such as Pinto Creek support vegetation like mountain mahogany and oaks normally found at higher elevations.

A Refuge for Wildlife

Vegetative mixing serves to create microhabitats that, in turn, help support a broad range of wildlife species—some well known and others more cryptic. The area has been so little studied that there are almost certainly animals still unknown to science or that occur outside of their known range. The same factors that make the region a refuge for rare and relict species of plants allow it to serve a similar function for wildlife. A biological survey conducted in 1949 examined a cross section of the northern Sierra Vieja, dividing the area into three life zones that roughly correspond to the Marfa Plain, the Vieja Mountains, and the Rio Grande Plain. Out of a total of forty-six mammals that were identified, half of those were restricted to the Viejas—the largest number in any of the three zones. Significantly, the researchers concluded the lowlands and the plains were best treated as two distinct districts within the larger "Chihuahuan biotic province."[31]

With the results of the earlier Sierra Vieja research and a more recent study in the Chinati Mountains, we know the region hosts at least fifty-two native mammals. In addition to ten species of bats, there are at least twelve species of mice, six species of rats, three

Raccoon kit in deer muhly (grass) on Pinto Creek. Photo by D. Keller.

species of squirrels, and two species of gophers. Midsized mammals include desert cottontails, black-tailed jackrabbits, striped skunks, Western hog-nosed skunks, ring-tailed cats, raccoons, porcupines, badgers, and the pig-like collared peccary, or javelina.[32]

Larger herbivores often pursued by hunters include three species of deer. Mule deer are the most common, followed by white-tailed deer and Coues deer, also known as Carmen whitetail deer or "fantails"—a subspecies of whitetail deer unknown to science until 1940. Because of its rarity—being restricted to juniper woodlands in the mountains of West Texas and northern Mexico—it is a prized quarry. Elk, presumably from a native population, are commonly seen lounging in the upper parts of the canyon locally known as the "elk hangout." Although native bighorn sheep largely died out during the early twentieth century, mostly as a result of disease and competition with domestic sheep and goats, they have since been reintroduced by Texas Parks and Wildlife to the Sierra Vieja and may soon expand as far south as the Chinatis.[33]

Several exotic species have become naturalized. Among those exotics are burros, whose wild herds may date as far back as the Spanish *entradas*. Aoudads—an African sheep imported in 1957 to

the Texas Panhandle—occupy some of the roughest and least acces-
sible portions of the canyon. Feral pigs are sometimes seen in the low
country, and for a couple of years there were numerous sightings of
an East African sable antelope that had escaped from the Crown X
Ranch, a nearby exotic game preserve that boasts more than thirty
wild species, mostly from Africa. The animal briefly enjoyed a free-
range existence until someone decided it would make a nice trophy,
leaving its headless body beside the Pinto Canyon road.[34]

Large predators have always been an integral part of the local
ecology, and all but a few of the region's most iconic predators
remain. Bobcats, gray foxes, coyotes, and mountain lions still roam
the area canyons and mountains. Confirmed sightings of black bears
in the Chinatis suggest they are naturally recolonizing, or perhaps
they were never completely extirpated. Wolves, once wide-ranging,
are now notably absent, although the area was one of the animal's
last strongholds. In 1942, a gray wolf killed on the Cleveland Ranch
above Pinto Canyon was believed to be among the last in Texas.
Mexican wolves persisted a while longer, until the last two were
killed in the nearby Davis Mountains in 1970. Even so, in December 1972 a "wolf-like" animal was shot on the Capote Ranch about
fourteen miles north of Pinto Canyon, suggesting that some individuals might remain undetected. As a whole, however, predator
populations have probably varied more in the last century than ever
before due to control efforts. Sheep and goat ranching caused predator populations to spike, leading to targeted eradication efforts.
The result, over the years, has been populations of coyotes, bobcats,
wolves, mountain lions, and eagles that have been in constant flux.[35]

Reptiles and amphibians are both abundant and among the most
commonly seen of vertebrate fauna. One study conducted in the
Sierra Vieja logged fifty-three species of reptiles and amphibians,
including one salamander, two turtle, ten frog and toad, sixteen liz-
ard, and twenty-four snake species. Among the latter are five species of rattlesnakes that keep pedestrians ever wary. In addition
to the prairie, black-tailed, and banded rock rattlesnakes there are
the aggressive Western diamondback and the extremely poisonous Mojave rattlesnake. Among a number of snakes infrequently
encountered are the rare lyre snake and the beautiful gray-banded
kingsnake, both favored by collectors.[36]

A 1948 survey of birds in the northern Sierra Vieja documented sixty-three species, twenty-four of which were restricted to the Viejas. In addition to the commonly seen (and heard) mourning dove, scaled quail, and Chihuahua raven are rarer species, including the electric blue lazuli bunting, the bright yellow Scott's oriole, the mimicking phainopepla, and the threatened yellow-billed cuckoo. Commonly sighted raptors include red-tailed hawks, Swainson's hawks, and golden eagles. On summer nights, the plaintive peents of common nighthawks fill the skies with sound. Although there have been no formal entomology studies, insect fauna is known to be correspondingly diverse. One inventory conducted in Big Bend National Park—about ninety miles to the southeast—identified some four thousand species and estimated that at least another thousand remain to be recorded.[37]

Pinto Canyon Prehistory

Archeological evidence suggests humans were occupying Pinto Canyon and the southern Sierra Vieja by at least thirteen thousand years ago. In an environment significantly different than today, one still under the sway of the Pleistocene ice age, Paleoindian hunters

Scaled quail in Pinto Canyon. Photo by D. Keller.

in the region appear to have sought a more diverse fauna than traditional interpretations suggest. Although mammoth remains and other vestiges of Pleistocene megafauna have been documented in the region, none have been found in good association with human artifacts. Projectile points dating to this period in the region are often diminutive forms, perhaps suggesting they were targeting smaller prey than those hunted on the southern plains, where the points were first discovered. What seems certain is that the size of the human population here was small enough that some have suggested Paleoindian presence in the region was incidental or at best seasonal. More likely there was some combination of seasonal and permanent residents, even if those residents were so mobile that their remains imply a more ephemeral occupation.[38]

By at least the Early Archaic period, beginning about eighty-five hundred years ago, a more diversified hunting and gathering "archaic lifeway" developed, which would persist for the remainder of the prehistoric period. It was the beginning of more intensive place-based adaptations—ones focused on hunting medium- to small-sized game and a wide pantry of vegetative foods gathered from the desert. Among the most important were succulents, berries, roots, and seeds. Pinto Canyon dwellers roasted prickly pear tunas, various agaves, and sotol in subterranean "earth ovens," which converted their complex chain of carbohydrates into more easily digestible sugars. They also cooked various wild game, mesquite beans, grass seeds, and a host of other food items on hot rock roasting platforms, the ubiquitous remains of which suggest their use throughout time.[39]

The dramatic physiography undoubtedly undergirded the attractiveness of the area to prehistoric visitors. Site density in Pinto Canyon and the southern Sierra Vieja is extremely high, and the time periods represented and archeological signatures are broad. For most of prehistory, the area was likely part of a seasonal round—a favored hunting and gathering ground returned to year after year. For others, it might have been more central to their adaptation, perhaps even a place to overwinter. Specialized resources such as siliceous stone for tools, rare plant-derived medicines, or species of animals more common here than elsewhere probably drew people throughout time. The abundance of surface water and rock shelters quenched thirst and provided cover while the many boulders and cliff faces served as perfect canvases for prehistoric art, the remains of which abound.[40]

If this broad suite of resources became favored in prehistory, then we might expect they also generated competition. Indeed, a number of sites contain stacked rock features suggestive of a defensive posture. Conflict between human groups occurred throughout time, although during earlier phases of prehistory, when human populations were quite small, there would have been less reason to defend a territory against another group. That changed as population increased during the Late Prehistoric period and especially after the Euro-American invasion caused a huge reshuffling of native tribes across the continent, leading to a storm of conflict.[41]

The Late Prehistoric period, beginning around thirteen hundred years ago, marked the arrival of new technological innovations, notably the bow and arrow and ceramics. The latter attended the spread of agriculture—the corn-bean-squash cultivar that allowed the first sedentary adaptations in the region. By at least eight hundred years ago, La Junta de los Rios—the region surrounding the junction of the Rio Grande and the Río Conchos—became the easternmost extension of Puebloan culture in the Americas. As part of the larger Casas Grandes interaction sphere, La Junta villagers were a peripheral but regionally significant hub of trade. Early Spanish chronicles suggest the villagers may have specialized in making tanned deer and bison hides as well as sinew-backed bows and may also have served as trade intermediaries between the bison-hunting groups to the northeast, the Puebloan agriculturalists to the north, and the major trade center at Casas Grandes to the northwest.[42]

The La Junta cultural district extended from near modern-day Redford, northwestward past the river junction to Candelaria—the northern extent of the Presidio Bolson. Over time, villages appear to have expanded and contracted within this district. Although most of the major pueblos were sited at, or below, the confluence of the two rivers, where the flow was more dependable, there was at least one settlement upstream as well. San Bernardino, located about twelve miles northwest of the river junction, was first mentioned in Spanish records in 1582. Although distinct and separate from the other pueblos, the people of the so-called Tecolote (owl) nation were linguistically and ethnically related to other La Junta people. The chronicler of the Espejo expedition into present-day New Mexico in 1582–83 noted the Tecolote also lived in small groups for some distance up the Rio Grande, above San Bernardino, moving their temporary fields with the shifting river channel.[43]

Archeological evidence supports this historical record. A dozen or more sites bearing prehistoric ceramics have been documented along the river as far north as Candelaria. If these sites, typically containing mounds of burned rock, chipped stone, and ceramics—but with no evidence of houses—represent *rancherías* (outposts) of the Tecolote, then it may indicate something of a hybrid adaptation. Indeed, most evidence suggests that even the downstream villagers were at best semisedentary—that they grew crops and traded but also made hunting and gathering forays at least as far as the bison plains to the north. Employing a broader-spectrum adaptation helped reduce the risk associated with the many conditions they faced: drought, floods, and raids from enemy tribes being chief among them.[44]

Despite the agricultural adaptation at La Junta, across the rest of the greater Big Bend only the bow and arrow were universally adopted, as Archaic-style mobile hunting and gathering remained the primary lifeway. Ceramics and agricultural products are infrequently found in rock shelters but are otherwise usually absent from the archeological record. We would normally expect a similar pattern in Pinto Canyon and the southern Sierra Vieja, but that has not been the case. Instead, there has been a consistent presence of pottery in rock shelters, along with a mysterious near-absence of arrow points in open campsites. The latter trend is especially striking when compared to similar collections in other parts of the region that produce arrow points in far greater numbers. What the pattern may suggest is that the lowlands of Pinto Canyon and the southern Sierra Vieja served as specialized procurement areas for semisedentary groups from La Junta such as the Tecolote, who may have chosen to inhabit rock shelters while defending their gathering grounds from more nomadic Indians.[45]

If Pinto Canyon and the surrounding lowlands were indeed the chosen gathering grounds of the Tecolote, their claim on the area ended abruptly around 1720. When the Spanish visited La Junta again in 1747, they found the pueblo of San Bernardino and their upstream *rancherías* abandoned. The reason for their disappearance appeared to be increasing conflict with their greatest enemy—the Apache. Believed to have migrated southward from their Athapaskan origins in the northern Great Plains starting around 1450, the Apache had reached the plains of New Mexico and the mountains of Far West Texas by the early 1600s. As they went, they waged war on existing sedentary and nomadic tribes alike. The Tecolote may have fled, but others weren't so lucky. The entire Chisos tribe was probably annihilated. The Jumano, having few options, appear to have been assimilated; by 1750, the Jumano name had disappeared entirely from Spanish documents.[46]

Thereafter, most of the Big Bend beyond the confines of a more centralized La Junta became the domain of various Apache groups, most notably the Mescalero. Despite being relative newcomers, the Apache left an indelible mark upon history. Although many have long believed the place-name Chinati to be an anglicized version of the Nahuatl-based word *chanate*, meaning "blackbird" or "grackle," recent evidence suggests the word may mean something else entirely. Based on historical, linguistic, and ethnographic data, some scholars argue that the word is more likely derived from the Western Apache word *ch'íná'itíh*, which means "mountain pass." If so, then the mountain range may have inherited a name originally applied to Pinto Canyon—one of only two major passes between the desert lowlands and the Marfa Plain along an eighty-mile stretch from Chispa summit to the southern end of the Chinati Mountains.[47]

Troubled Spanish and Mexican Frontiers

For a brief time, La Junta served as the Spanish gateway to lands beyond the northern frontier. Following Álvar Núñez Cabeza de Vaca's chance passage through the area in 1535, several Spanish *entradas* would pass by the mouth of Pinto Creek along a riverside trail that connected La Junta and the Jornada Mogollon pueblos near present El Paso, including the Rodriguez-Chamuscado expedition in 1581, the Espejo expedition in 1581–82, the Mendoza-Lopez expedition in 1683, and the Ydoiaga expedition of 1747. Remaining close to the Rio Grande and with eyes set on the far horizon, however, none ventured up Pinto Canyon.

The first Europeans to set foot in the region were not part of a formal Spanish entrada at all but wayward travelers from a failed expedition to Florida. Yet Álvar Núñez Cabeza de Vaca's account of his epic journey across what would become Texas served as the first written description of the pueblos at La Junta. Significantly, his passage marked the beginning of the historic period in the Big Bend and inspired several expeditions that followed. Shipwrecked off the coast of Galveston, Cabeza de Vaca and his three companions were attempting to return to the Spanish colonies. After resting two days in the villages of La Junta in the late summer of 1535, the men continued their journey up the Rio Grande and thence to the southwest, where in several months' time they finally reunited with their fellow Spaniards. Nearly half a century later the Rodríguez-Chamuscado expedition traveled down the Río Conchos to La Junta before turning upstream to follow the Rio Grande northward to a pueblo south of Socorro. The next year, Antonio de Espejo spent nine days at La Junta before marching upstream, likely passing the Ruidosa area on 17 December.[48]

In response to pleas from a Jumano named Juan Sabeata for the siting of a Spanish mission that would provide protection from the dreaded Apache warriors, Juan Dominguez de Mendoza launched an expedition from El Paso in December 1684. Following the Rio Grande downstream, the expedition likely passed the mouth of Pinto Creek—probably the day after Christmas—before continuing to La Junta, where Mendoza's men erected a cross, left a priest to minister to the Indians, and continued northeastward to the Edwards Plateau. Capt. Joseph de Ydoiaga, who commanded one of three expeditions to La Junta in 1747, provided much greater detail about the region and the people living there. Taking the traditional route down the Río Conchos, the party spent about three months in La Junta before reconnoitering up the Rio Grande to El Cajon, as far as the Eagle Mountains, where they discovered *rancherías* that—like those of the Tecolote—had been abandoned due to Apache raids.[49]

In 1715 Don Juan Antonio de Trasviña Retis established the first formal Spanish missions at La Junta and provided substantial ethnographic data on the various pueblos. The native populations did not universally welcome the Spaniards, and between 1717 and 1746 the missions at La Junta were intermittently abandoned as a result of Indian uprisings. While missionaries sought to convert Indians to Christianity and compelled their subjects to work in the fields, soldiers garrisoned at the associated *presidio* (Spanish fort), being less concerned about the Indians' souls, were much more prone to abusing the villagers. Over the natives' vocal objections, the Spaniards constructed a presidio at La Junta in 1760. But several hundred Apache warriors, joined by some of the more daring villagers, soon attacked the new fort, Presidio de Belén. With the protection of the thick adobe walls of the fort and having much greater firepower, however, the Spaniards were able to repulse the Apache advance. Losing seven of their warriors, the Indians fled northward to near present-day Ruidosa. A month later, a hundred soldiers from Chihuahua City were dispatched for a retaliatory expedition. Taking the camp by surprise, the soldiers killed nine and took prisoner forty-seven warriors along with more than a hundred women and children. Those who managed to escape fled into the nearby Chinati Mountains.[50]

Facing recurrent problems, by 1767 the presidio had been abandoned, and the soldiers moved farther up the Río Conchos in defiance of the viceroy's instructions. The outpost was reestablished at La Junta in 1773, the same year the Spanish Crown commissioned a cordon of presidios along the northern frontier as a line of defense against Apache and Comanche depredations. In addition to the presidio at La Junta, others were built at San Vicente (near present-day Boquillas), at San Carlos (near Lajitas), and at El Principe (near Pilares). Largely ineffective and difficult to supply and maintain, all were abandoned in 1781 except for the one at La Junta. By that time, the presidio was almost all that was left of the once thriving villages. The unwelcome military presence, along with European-introduced disease epidemics and raids from enemy tribes, left the valley largely deserted.[51]

In 1821 Mexico won independence from Spain. But in gaining control of the lands that had belonged to Spain, it also inherited the rash of problems plaguing the northern frontier, with which it was ill equipped to deal. Sensing a vulnerability, the Apache and Comanche stepped up their raids. In hopes of colonizing the frontier, the nascent Mexican government opened lands along the Rio

Grande to settlement and decreed that prisoners sentenced to hard labor could be reassigned to frontier military colonies. After working off their sentence, they would be given land on which to settle.[52]

One such penal colony was established at Vado de Piedra (meaning "rocky ford"), about twenty-four miles northwest of Presidio del Norte, as La Junta was now known. Here, several hundred convicts were sent to do agricultural work in fields along the Rio Grande and to guard cattle and horses against Indian raids. But the colony was beset with problems. To supplement the larder, detachments of prisoners went to the Marfa Plain to hunt antelope and bison. On one occasion, the convicts killed a small herd of bison near San Esteban Spring. As the men fell to the task of skinning and butchering their kill, they were completely unaware of being watched by a Comanche band. Once the convicts had finished their work, the opportunistic band descended, killing them all and making off with the meat.[53]

Indian attacks became increasingly brazen; even the colony proper was no safe haven. In October 1831 more than a hundred Comanche warriors attacked the remaining convicts and soldiers, killing two men before stealing their precious horse herd. Captain Ronquillo from Presidio del Norte took off with a column of soldiers and pursued the thieves, to no avail. Four years later eight hundred Apache fighters swept through Chihuahua, stealing livestock and captives and leaving death in their wake. The situation had grown so dire that the state government forbade citizens to leave their homes without military-approved firearms. Then, in 1837, the governor began employing scalp hunters, paying murderous roving gangs for the scalps of Indian men, women, and children alike.[54]

The US-Mexican War and the Border Surveys

In 1848, the United States wrested from Mexico the newly formed state of Texas, in addition to 525,000 square miles of land in yet another expansionist move. To map and explore the new southern boundary, the US government launched a number of expeditions and surveys, including the Whiting and Smith Expedition and the

United States and Mexican Boundary Survey, both of which provided some of the earliest detailed scientific information about the Texas-Mexico borderlands.[55]

William Henry Chase Whiting left Fredericksburg in early 1849 with a retinue of sixteen men bound for Fort Leaton at Presidio del Norte. On 25 March, after a five-week march, they arrived tired and bedraggled, having barely eluded an Apache band under Chief Gomez. Four days later, the party continued their march up the cottonwood-lined banks of the Rio Grande opposite the Sierra Grande in Mexico, the sides of which looked as if they "had been combed down by the teeth of some titanic harrow."[56]

The next day the party passed Chinati Peak. Not realizing it already had a name, Whiting—in the spirit of conquest—proposed to call it Mount Barnard. (Fortunately for posterity, the name didn't stick.) By noon the party found itself opposite the military fort at Vado de Piedra. "Built for defense against the Indians," Whiting wrote, "it is, in its crumbling desolation, a mournful monument of Mexican weakness." Near present-day Candelaria the party pitched camp across from a prominent landform of sandstone columns, which Whiting claimed presented "a resemblance to old feudal castles with the round towers of those days."[57]

In 1851, William H. Emory arrived at La Junta to oversee the formal expedition charged with mapping the sinuous course of the Rio Grande as the new US boundary with Mexico. It was also the first expedition to make significant scientific observations in the region. Moritz von Hippel, a civilian surveyor who mapped the area between El Paso and Presidio del Norte, held a favorable opinion of Presidio Valley, finding it to be good grazing country and the soil to be "of easy cultivation." Also less critical of the dilapidated condition of Vado de Piedra than Whiting, he noted a substantial community of an estimated three hundred people. "Here are large cultivated fields, which are watered by acequias, and yield abundant crops of wheat and corn," he wrote. Later that day the party passed the mouth of Pinto Creek but found little of note until they approached the old presidio at Pilares some forty miles upriver. It lay abandoned and bore "abundant signs" that the "smelting of silver ore was carried on extensively."[58]

Early Settlement

Following the end of the war with Mexico, settlers began to test the land beyond the confines of Presidio del Norte. In 1848, entrepreneurs established the Chihuahua Trail through the dusty town that briefly became a major port for commerce between San Antonio and Chihuahua City. It didn't take long before the business potential became evident. The merchant John W. Spencer and a freighter named Ben Leaton met John D. Burgess in Chihuahua City, and the three agreed to move to the north bank—now within the United States—to profit from the trail traffic. If Leaton and Spencer had been involved in the scalp trade, as some believe, they decided to turn their attention to more humane pursuits. Burgess took up freighting while Leaton settled in a fortified trading post just downstream from Presidio del Norte. Spencer established a horse ranch at Indio, several miles upstream, but was soon beset with problems. Horses were favored plunder for Apache and Comanche raiders, and after suffering recurrent thefts, he finally switched to cattle in 1854 to market to soldiers at the newly established military post of Fort Davis, about eighty miles to the northeast.[59]

In the wake of Leaton, Spencer, and Burgess's efforts, William Russell arrived in Presidio del Norte to become the first to farm and ranch in the lower Sierra Vieja. Born in 1839 in Kentucky, Russell left home around 1854 and supposedly established a trading post and freighting business at Horsehead Crossing on the Pecos River before joining the Confederate army during the Civil War. After the war, Russell moved to Presidio, where he opened another trading post and married Tomasita Rodriguez in 1865.[60]

Russell's ambitions soon took him beyond the growing settlement around Presidio del Norte. In 1868, he established a farm at Candelaria to raise grain for the troops at Fort Davis and Fort Stockton. It was a short-lived experiment. The moody Rio Grande jumped its banks one day, washing his fields away in a raging flood. Shifting strategy, Russell sought higher ground and began a sheep ranch above the Rimrock at Capote Mountain and installed his father-in-law, Dario Rodriguez, as manager. Yet, miles from the relative safety of civilization around Presidio, they were easy targets. In July 1873, just downstream from Russell's sheep ranch, seventeen-year-old Regino Nuñez was in camp with a small party of men at the mouth of Capote Canyon when a band of Mescalero Apache raiders suddenly attacked. Eight hours later, when the Indians finally retreated, five of Nuñez's party were dead. In a letter to the *El Paso Herald*, Russell complained, "The Indians made their appearance a few miles below this place yesterday, carrying off a woman and her two children . . . and for the last six months during every moon, they depredated upon the ranches above and below here, killing and driving off stock."[61]

Russell's concerns grew increasingly personal. In 1876, Apaches raided his sheep ranch at Capote Mountain, killing four of his herders. His brother-in-law, Matildo Rodriguez, was the only one to escape. Undeterred, in 1872 Russell decided to give farming a second chance farther downstream at a place long known as Ruidosa. Meaning "noisy" in Spanish, the name referred to the sound of the Rio Grande rushing through a rocky channel just above what would become the town. After clearing and leveling his fields, Russell constructed the first irrigation ditches in the area. Optimistic that others would follow suit, he also built a toll mill that would provide flour to the area for the next thirty years. More an entrepreneur than a laborer, however, Russell spent most of his time at his home in Presidio, relying on a man named George H. Brooks to faithfully manage both the farm and the mill for the next twenty-odd years. Delegating the risk to others proved a smart move.[62]

In 1879, Mescalero raiders attacked Russell's farm at Ruidosa, killing four of his laborers, wounding three more, and making off with a large number of horses, cattle, and sheep. The next year they struck again. Capt. Louis H. Carpenter of the 10th US Cavalry stationed in Presidio wrote his superiors that "on the 13th of March last, a party of six Indians on foot, made their appearance near Russell's [farm]. Killed a Mexican boy of nineteen and carried off a younger boy. The party of Indians divided, three of them driving a horse and with the boy crossed over the Chenati [*sic*] Mts, taking a northeasterly direction. The other three Indians with two horses went up the Rio Grande or concealed themselves in the Mountains of the Capote or the Vieja."[63]

The Runway of the Indians

Even as families settled beyond the safety of Presidio, Pinto Canyon and the surrounding mountains remained the domain of the Apache, just as had been the case since the days of Spanish control. As early as 1777, the Spanish military commander Hugo O'Conor wrote of the Chinatis as one of several ranges inhabited by those he called the *indios bárbaros*. More than a century later, Robert Smither of the 10th US Cavalry would disparage the same range. "I did not visit the Chinati Mountains in the northward of this point," he wrote, "although I am informed . . . that the 'run-way' of the Indians has always been in that vicinity and to the north of it." The pattern made sense. Following a raid, the Indians would flee in the direction of the nearest range offering protection and a means of escape. Around Presidio, that was usually the Chinatis and Sierra Vieja.[64]

Historical accounts of raiders fleeing into these wastelands abound. There is also tantalizing physical evidence that either adds to the picture or hopelessly complicates it. Within steeply incised canyons eroded out of ancient bolson deposits are areas strewn with small boulders of black vesicular basalt. More than two hundred of these boulders have been discovered displaying images carved or pecked into the rock. Some of these "boulder glyphs" appear to be horses with riders, some appear to be figures dancing, while others are more abstract. But a great many bear a striking resemblance to livestock brands—symbols with curlicue flourishes, some that look like anchors, many that resemble letters. Despite their abundance, the glyphs remain enigmatic. We don't know what they are, why they are there, or even when they were made. However, some evidence suggests that horse nomads may be the artists. In the protohistoric and early historic periods, such nomads were usually Apache.[65]

If the boulder glyphs are somehow related to these latter-day raids, what is certain is that the practice had a short life. Following years of recurrent conflict with the Apache, the US military stepped up counterefforts. After the master tactician Col. Benjamin H. Grierson took charge of the District of the Pecos with his famed "Buffalo Soldiers" of the 10th Cavalry, the days of Apache raids were numbered.

Identifying the weakest link in their hit-and-run, fleet-footed methods, he began to fortify and dispatch sentries to guard springs and waterholes across the Trans-Pecos, essentially driving the Apache out of the region for want of water. Whenever attacks were reported, he pursued the raiders relentlessly, whittling them away until they were too few and too weak to fight.[66]

Victorio, perhaps the last great Apache war chief, led his people on their final raids across Far West Texas. His flight from the wretched conditions at the San Carlos reservation in southeastern Arizona sparked the last rebellion and emboldened others to join his ranks. In 1880, at Viejo Pass in the Sierra Vieja, a group of Victorio's men ambushed Lt. Frank H. Mills and a detachment of Tigua scouts, killing scout Simon Olgin. However, the final battle against Victorio, at Tres Castillas in Mexico the following October, left him and most of his men dead. Two weeks later, a party of thirty to forty survivors attacked a small cavalry patrol at Indian Hot Springs about sixty miles northeast of Presidio, killing two soldiers. The warriors resumed raiding in the desert basins farther east, but in the Diablo Mountains the Texas Rangers defeated them in the last Indian battle in the Trans-Pecos. Although a few small groups occasionally escaped their New Mexico reservation to conduct small-scale raids, the war against the Apache had ended.[67]

Even so, the country remained wild and dangerous. Outlaws and "road agents" filled the void the Apache had left. In 1880, a band of outlaws under Jesse Evans briefly made the Chinati Mountains a staging area for raids on Fort Davis. As leader of the "Jesse Evans gang" in New Mexico, he had been one of the hired killers who sparked the Lincoln County War. Eventually fleeing New Mexico, he and his gang turned southward, where within a span of two months they were credited with some twenty incidents of assault and robbery in the area between Fort Davis, Fort Stockton, and Presidio.[68]

After a month of intense searching, the Texas Rangers tracked the outlaws to their Chinati Mountains stronghold. Spotting four men on horseback above them on the mountainside, the Rangers spurred their mounts. The outlaws turned and fled, opening fire on the Rangers, who followed in hot pursuit. The running battle stretched on for two miles before the outlaws took cover behind a rock ledge. Following a fierce gun battle, the Rangers charged the outlaws' position, forcing their surrender. By the time the dust and gun smoke had cleared, one Ranger and one outlaw lay dead. But Evans was only one of many who would seek refuge in these trackless mountains to avoid detection. Over the years, there would be many more, among them runaway Indians, Mexican bandits, and *sotoleros*.[69]

Founding of Ruidosa

With the Apache driven out of the state and outlaws safely behind bars, the land along the banks of the Rio Grande began to be settled in earnest. Most who came were of Mexican descent and arrived with the skills required to start new lives along the untamed river valley. In 1876, in the wake of Russell's irrigation project and toll mill, Presidio County created the "Rio Doso" precinct and funded the construction of a road up from Presidio. A population census taken four years later tallied more than 260 people living in the precinct. Up- and downriver on *ancones* (flats between the wide meanders of the river), new settlers cleared and planted fields, built adobes and jacales, constructed pens of upright ocotillo and sotol gathered in the hills, and began scratching out a living dependent on the fertility of the floodplain and the bounty of the river.[70]

Although the first ditches in the area were likely rudimentary affairs intended to divert water from tributaries—the kind of structures that had to be rebuilt following every flood—by the mid-1880s larger ditches were being constructed to take water directly from the Rio Grande. In 1885, a state school land agent, William Baines, described Russell's Ruidosa farm as "an irrigated farm in fine state of cultivation" and added that "the leasee [*sic*] of the farm claimed for Russell three leagues [nine miles] of river front, but I am sure his farm extends at least eighteen miles along the river." Indeed, Baines had been sent in part to report such abuses. But he hadn't seen the worst of it.[71]

The March of the Iron Rails

Early in 1883, the Southern Pacific and the Galveston, Harrisburg and San Antonio Railway completed construction of a railroad through the region, giving birth to a host of towns along the length of it, including Marfa, which became the seat of Presidio County. It also ushered in the first major wave of settlers beyond La Junta and rendered the Chihuahua Trail obsolete. Presidio no longer had pre-eminence as a port of entry. After freight traffic dwindled, the loss sent Presidio into freefall—an isolated backwater destined to remain little more than a dusty suburb of a dusty Mexican border town.[72]

Meanwhile, with access to the region thrown open by the railroad, the land rush was on. Around 90 percent of the pioneer stockraisers in Presidio County arrived between 1880 and 1890, with the greatest increase between 1884 and 1885. By the end of 1885, more than sixty thousand head of cattle were said to be grazing across the county, and all of the springs and surface water had been claimed. A lot of reshuffling remained to be done, but by and large in the span of just a few years what had been a howling wilderness was now a settled frontier.[73]

Free-Grass Days in Pinto Canyon

After settlers claimed the better farmlands along the river and the grazing lands on the Marfa Plain, a handful of others tried their luck in Pinto Canyon. Around 1884, John E. Gardner built a rock house that was likely the first permanent structure in the canyon. When Baines, the state land agent, rode through a couple of months later, he noted that while Gardner's application denied that his section fronted any stream, Baines had "stayed all night at [Gardner's] house and it is not fifty feet from a lasting stream of water." If abuses were rampant, however, perhaps here it didn't really matter. The agent found Pinto Canyon to be poor range and too remote to recommend for settlement. Despite the abundance of water there, Baines was struck by the wild character of the canyon, and he saw little agricul-tural promise. "Being surrounded by terrible mountains," he wrote of the area, "the State might do well to let them have it."[74]

But Gardner, for one, decided he didn't want it. He stayed only long enough to entice a friend, W. H. Cleveland, to the area before moving on. In 1885, Cleveland drove about three hundred head of rangy longhorns from Dimmit County down into Pinto Canyon, although the trail was so narrow in places he could barely get through with a packhorse. Having turned his cattle out, he set up camp in the rock house Gardner had built. "There wasn't much living to it in those days," Cleveland later recalled, adding, "we just did the best we could." Although he found the canyon "wild and beautiful," he ultimately established his permanent homestead a few miles to the northeast and about a thousand feet higher in elevation, where his better-grassed ranch overlooked the scabrous lands below.[75]

∧∧∧

By the close of the nineteenth century, Pinto Canyon remained a hinterland, used only for the free grass it had to offer. Without landowners, use of the land was largely governed by prior appropriation, although such squatters' rights were not to last. In 1899, a new law began enforcing payment for the use of state-owned lands, essentially putting an end to the fleeting free-grass days. Even so, Pinto Canyon would remain unsettled, awaiting homestead laws that offered a more reasonable chance at success. Although that would not happen for more than two decades after the initial land rush, once such laws were passed, Pinto Canyon would be settled with an earnestness many would later regret.

3

Cañon del Pinto

The Settling of Pinto Canyon

We were lucky that Father had made friends with a man named José Prieto. With papa in bed with a broken leg, we would have been helpless without José to help us.

—Millie Wilson

IT WAS AN early evening in January 1907 as Dora Wilson turned the two-horse team down into Pinto Canyon. There was no road, and the wagon swayed ponderously as it rolled down the rock-strewn slope. Fearing the wagon would overrun the horses, Dora's husband, James, ran a pole through the spokes of the wagon wheels, locking the rear ones in place. But with everything they owned strapped down inside, the weight was still too great. The metal rims of the wheels skidded across rocks and cut furrows into the soil. James's friend, Robert Greenwood, quickly tied his rope to the rear of the wagon and dallied it around his saddle horn. When he reined his horse back, the rope pulled taut and the wagon stopped. He gave slack and the wagon lurched downslope, bucking clumsily over rocks and crashing through sotol plants. The descent was agonizingly slow, and they had not gone far before darkness was upon them. With nowhere else to go, the family pitched camp on a small bench halfway down the slope.[1]

The next morning, the Wilsons rehitched the team and resumed their slow descent. Finally reaching the canyon bottom, they sighed with relief. With their small cattle herd already grazing peacefully beside Pinto Creek, Greenwood and his son—having fulfilled their obligation—bid their farewells and headed back to Marfa. The Wilson family continued on. A short distance farther, Dora, her sister Josie, and the three young Wilson girls gasped at the sight of their new home—the squat one-room rock house built by John E. Gardner some thirteen years earlier. Since then, the structure had been used as W. H. Cleveland's goat camp and had fallen into disrepair. It wasn't much to look at, but it was theirs.

At forty-six, James Wilson was older than most of his homesteading peers. Born in Coldwater, Michigan, in 1861, James was struck with polio at the age of two. One leg became so weak and underdeveloped that the doctor had to fit him with a brace so he could

James E. Wilson in Pinto Canyon around 1915. Photo courtesy of the Marfa Public Library, Marfa, Texas.

walk. Even then, his leg remained feeble and prone to injuries. It was a physical deformity that made work—and life itself—more difficult. He moved to New Mexico with his family before shifting again in 1884, to Texas, where his father filed on land at Paisano Pass between the nascent railroad towns of Marfa and Alpine. There he and James built a house where the family lived, and they ran sheep for about a year. Either restless or uninspired, the family soon pulled up stakes again and turned southeast, eventually landing in Mississippi. While living in Pass Christian, James married Dora McCollister, who would give birth to four children over the next decade, although only two of those first four—Millie and Mamie—survived. Around 1906, James moved his new family to Texas, where they reoccupied the property near Paisano Pass that his father had left him. Within a year, Dora gave birth to their third daughter, Ora. Then, for reasons obscured by time, James decided to sell his place at Paisano Pass and try his hand elsewhere. Had it been greener pastures he sought, he might as well have stayed put, for there was no better land to be had. Instead, what probably lured James—and others who would follow—had less to do with quality than quantity.[2]

In 1905, the state legislature passed the Eight-Section Act, which doubled the acreage one could homestead—allowing a full eight sections (5,120 acres) to the highest bidder. In exchange for occupying the land for three years, making three hundred dollars' worth of improvements, and paying the principal and interest (typically a dollar an acre, payable over forty years at 5 percent interest), the state granted full title. The only problem was that much of the unsold land was already under lease by ranchers, leaving it unavailable for homesteading. The state land office addressed the problem in 1907 by suspending all leases of state land and offering the parcels to the highest bidder, with no preference given to lease holders. The new act drew the ire of ranchers, some of whom had leased sections of state land for more than a decade. It also ignited a second land rush, with newcomers scrambling for the unsold land that remained.[3]

By that time, the heyday for homesteading in the Big Bend had already passed. The top-quality grazing land—the Alpine valley, the Davis Mountains, the Marfa Plain—had long been claimed. To find eight contiguous or near-contiguous sections, homesteaders were forced farther into the hinterlands, since most of the unsold land was lower in elevation and more desert than grassland. The greater land allowance helped offset the fact that country like Pinto Canyon had far less agricultural potential than other areas. However, with the meager but reliable flow of Pinto Creek, water was plentiful. Over the course of only five years, Pinto Canyon would be transformed from free range to homesteads and ranches.

Being one of the first to arrive, James Wilson gained the advantage of having first choice of land sections, including one that already contained a structure, meager though it was. "It wasn't a very pretty home," recalled James's daughter Millie, "but it was shelter for us." Still, to accommodate a family of six—including Dora's sister Josie—there was much to do. While James made rounds to check on his cattle, Dora and Josie set to driving out bats, tossing out packrat nests, and sweeping out loose dust and rodent droppings that lay on every surface. They sprinkled water on the dirt floor to keep the dust down. They righted the door to swing properly and with a few buckets of mud patched holes in the roof and gaps in the mortar joints that held together the angular stones. Having brought scarcely more than trunks, they fashioned crooked but functional beds from cottonwood branches, with fence wire for bed slats. Using chino grass gathered from the canyon slopes, they stuffed gunnysacks for makeshift mattresses. With their drafty rock house and handmade furniture in order, the Wilsons settled in to start their new life in the canyon.[4]

Their second day found James loose herding his cattle, trying to keep them contained within the canyon and off the neighboring ranches. One particularly stubborn cow refused to leave a thicket. Dismounting, James picked up a rock and hurled it toward her. It had the desired effect, but as she charged out, the commotion spooked James's horse, which reflexively kicked out a hind foot, hitting James's weak leg and breaking it. Somehow James remounted and rode back to the house, where Dora put him to bed. Two agonizing days later Dr. Yates arrived from Marfa, but aside from splinting James's broken leg to the brace he had worn since childhood, there was little the doctor could do. Before leaving, he showed Millie how to adjust the straps to accommodate swelling. Now only time and rest would allow his frail bones to heal. Yet there were endless chores to be done, and with only Dora and Josie and Millie able to work—the others being too young—James was in a tough spot. Being in such a remote location, he could not have counted on help from others. But if despair overtook him, it didn't last long, for not far down the canyon their only neighbor was riding toward their tiny house.[5]

A Friendship Is Born

James first set eyes on Pinto Canyon during a scouting trip, likely following an inspection of the records at the Presidio County Court-house. There, a map tacked to a wallboard would have shown the blocks and sections within the county; nearby a sheaf of papers would list the parcels that remained unsold. Riding out of Marfa, he would have headed west, a few days' supplies in his bulging saddlebags beneath his one-blanket bedroll. Coming finally to the edge of the Rimrock, he would navigate the narrow horse trail down into the canyon, leaving the blanket of grassland behind. Descending, James may have felt as if he were entering the very bowels of the earth, trading a world mostly horizontal for one defined by verticality. As the canyon walls closed around him, the hulking mass of Chinati Peak rose to the south while to the north rose the coral-colored seabed cliffs for which the canyon is named. As James followed the creek downstream, he passed a whole separate range of hills and mountains within the broad shoulder of the canyon where the mountain flanks met the marl and limestone of an earlier age. A few miles farther, he came upon a herd of Angora goats and, before long, spotted the shepherd—a dark, squat Hispanic man with a broad smile and easy manner—quietly watching over his herd.[6]

As James approached, the man greeted him in Spanish, and James responded, having learned the language in his childhood. The man's name was José Prieto, and he'd been grazing his animals in the canyon for years. James explained that he was scouting for land, trying to find a good place to settle with his family. José would have told him there were many good places, that in spite of the land being rough and brushy, there were many springs and the water was cool and sweet. He may have mentioned the names of mountains as the two men rode over narrow goat trails, James leaning closer to hear the location of springs and good pasture, places where natural shelter might be had, about suitable ground for building a house or a set of pens. For having free-ranged his goats there for most of the last fifteen years, José was one of very few people who knew the canyon intimately.

José Prieto

José Prieto was born in 1866 in a sleepy village along the banks of the Río San Pedro, about forty-five miles southeast of Chihuahua City—a place known as Santa Cruz de Rosales. His parents, Francisco Prieto and Basilia Acosta, like many from that village, may have descended from a common lineage—one of the many tribes that had been enslaved to work in the silver mines of northern Mexico. Although Mexico outlawed slavery following independence from

José Prieto in his Sunday finest around 1920. Photo courtesy of the Prieto family collection.

Spain, the low social standing of the native people had kept them
in servitude, with virtually no way to rise above it. The same tribal
bonds that held the families together also propelled them to leave
together. José was only eight years old when his family, along with
several other families, immigrated to the United States. Like many
who came before and after, the Prietos and their kin sought to escape
a life of drudgery for a chance at the American dream.[7]

Haggard from the 175-mile-long journey, the families crossed
the border at Presidio. Francisco and his wife led a packhorse with
their belongings, while José trailed behind on another, his bare feet
hanging loose along the horse's flanks. After paying a small fee at
the customs house, they turned northward, plodding upriver on
the winding trail beside the Rio Grande. The next day the Prietos
reached the budding community of Ruidosa—scarcely two years
old—where Russell's fields appeared neat and orderly while others
up and down the river were in the process of being cleared and
houses were being built even as they passed. Finding a place to
their liking that had not been claimed by some prior squatter, the
family halted. In the coming months, Francisco and his young son
set to work building their house, clearing fields, and planting corn,
beans, squash, chiles, tomatoes, various melons, and anything else
that might grow. Along with a herd of Spanish goats, their small
farm was just enough to feed the family and—in good years—provide a modest income. But the work was hard, and money was
almost nonexistent. José was thirteen years old before Basilia gave
birth to their next surviving child, Francisco Jr., followed closely by
two daughters, María and Clara. In his late teens, José began working on ranches above the Rimrock—the only ones whose owners
could afford to pay wages. Rounding up cattle, branding calves,
and repairing fences much of the time, he also developed particular
skill in hand-shearing sheep and goats, a job that took him away
for months at a time as he traveled from ranch to ranch, often as far
away as Sanderson—130 miles to the east.[8]

In 1891, at twenty-five years of age, José married seventeen-
year-old Juanita Barrera from Barrancas, Chihuahua, just across the
Rio Grande from Ruidosa. They settled on a piece of land just west of
Cerro de la Cruz—a prominent mountain on the north side of Pinto
Canyon along the upper edge of the Presidio Bolson. Here, José built

a dugout along a tributary to Arroyo Escondido, a place known as
El Tanque de la Ese, or the S-tank, named for the sinuous shape of
the drainage. Probably meant only as a temporary shelter, the dugout would serve as the family home for nearly two decades. Juanita
gave birth to their first child, Victoriano, only ten months after their
wedding. Despite several pregnancies, she would not have another
surviving child for six years. After Sotera was born in 1898, however,
Juanita had a child nearly every other year thereafter: Petra in 1900,
José Jr. in 1902, Gregorio in 1904, Pablo in 1907, Frances in 1909, and
Guadalupe in 1911. Juanita would be in her forties when Jane, their
last, was born in 1915. But by then, their living situation had long
since changed.[9]

There had been a few men, including Gardner and Cleveland, who
drove livestock into Pinto Canyon before José arrived, but they had
only been passing through, grazing their herds for a season or two.
They never intended to stay. José Prieto would be the first to settle
permanently in Pinto Canyon, but, unlike others who followed, his
was not a decision based on land laws. He had come from a different
direction and with more lasting intentions. Legally, he was no more
than a squatter. Close to his childhood roots in Ruidosa, he sought
little more than to build a simple life for himself and his family, just
as his own father had done. Although the others had come to raid
and run, José had come to stay. Yet of all those who soon arrived to
try their hand at ranching below the Rimrock, José would outlast
them all.

The Rescue

James Wilson, bedridden in the cramped *jacal* with his family, faced
an impossible dilemma. He'd risked everything to make the move to
Pinto Canyon—what seemed now no more than a fool's errand. His
injury had come at the worst possible time, with everything on the
line and no one to turn to for help. They could not have anticipated
José's kindness. Whether Dora called on him or José happened by,
it wasn't long before he stepped through the low doorway to find
James on his back, helpless and wincing in pain. For weeks, while
James slowly healed, José rode dutifully the six miles up Pinto Creek
to check on the Wilson family, to help Dora and Josie and the girls

(who now bore the brunt of James's mishap), to gather his cattle when they began to stray. Once every week or two José led a pack-horse down to the store in Ruidosa and returned with groceries for the hungry family. Quiet, but always smiling, José must have seemed like some guardian angel. Whatever may have set them apart—ethnicity, language, customs—shrank into irrelevance, bound as they were by isolation even more profound.[10]

Although the whole family shared the burden, of them all, eleven-year-old Millie suffered the most and now was being crushed under the weight of her family's troubles. The stress of the move, her father's accident, and the uncertainty of the future bore down on her relentlessly. In addition to her regular duties of helping to care for her younger sisters, she now assumed more of the ranch work while also caring for her father. If she learned to tolerate his screaming when she adjusted his brace, to fend off self-pity as she emptied his bedpan in the morning, at some point the stress became too much. In protest, her muscles and joints swelled until any movement at all became painful. But Millie was lucky, for a known cure lay only eight miles to the west. Over the next three weeks Millie stayed at the Ruidosa Hot Springs taking the cure: applying hot mud to her aching joints, drinking the mineralized water, and soaking daily until she finally felt well enough to return home.[11]

Weeks passed. The days grew warmer. When the fracture in James's leg had sufficiently fused to the point where his leg could bear weight, he limped outside and slowly set to work. He cut down a small Mexican walnut tree and nailed and wired poles across the open forked end of the timber. Removing the singletree from his wagon, he fastened it to his handmade sled and harnessed a gentle mare before setting off to gather rocks. Piling on as many as the sled could hold, James clicked to the mare as she struggled against the weight until the sled slowly began to move, her hooves clattering over river cobbles back toward the house. Once he'd collected enough, he mixed a stiff mud mortar and began laying up his first course. Over days, the walls slowly rose to ceiling height. He cut cottonwood trees along the creek and slid them into place for roof beams. Over these he placed saplings, followed by mats of grass, before slathering on several layers of mud to form a mostly water-proof roof. Once complete, the new room was to serve as their kitchen. After months of sweating over an open fire, Dora was finally able to cook on their cast-iron woodstove—one of the treasured possessions that had made the trip.[12]

The Gift

Although José's ranch at the S-tank marked the beginning of his long tenure in Pinto Canyon, it was, like many other ranches, his only by virtue of possession, not ownership. Squatting carried risks. But the power to change his condition was elusive. As with so many Hispanic residents in the borderlands, the odds were stacked against him—an illiterate Mexican peasant in a country ruled by literate, English-speaking Anglos. It was a disadvantage that prevented many Hispanics in the borderlands from capitalizing on the generous land laws of the state. The greatest barrier of all, however, was not knowing how to file on land. The procedure, like the arcane language in which it was written, seemed the privilege of the fair-skinned gentry, an exclusive club that did not readily accept newcomers. But if there was some unspoken rule against sharing such secrets, James Wilson ignored them completely. He explained the new Texas land law to his new friend—the expanded acreage allowed and the terms for filing. If all worked out as planned, José could finally own ground he had occupied for nearly two decades. He was a quick study. And in gaining fluency in the language of deeds—the dialect of the landed—it also unlocked his hidden potential, ultimately allowing him to assemble the largest and most successful ranch in Pinto Canyon.[13]

However, filing was seldom straightforward. It was one thing to fill out the application and pay the fees, but it was entirely another to have that application represent the right piece of land. Even if a section were properly filed on, there was no guarantee it could be located accurately on the ground. It was a problematic legacy born of haste. In the race to build more railways, Texas offered railroad companies alternate sections of unsold state land—sixteen sections for every mile of track laid—almost all of which was in the Trans-Pecos. As a condition of ownership, the companies were required to survey the land. As might be expected, the frenzied surveys were so haphazard that many claimed they had been done off the back

of a running horse or in the backrooms of offices without any surveyor ever putting a boot on the ground. The upshot was havoc:
homesteads built miles from their purported location, fences erected through neighbor's pastures, springs falsely claimed. The whole
mess was a constant source of irritation to county surveyors, title
companies, and the state land office, which was unable to monitor
its vast landholdings, much less enforce policies. These were only
some of the problems facing state land agent William Baines, who
found much to gripe about after riding through the Ruidosa area.
"Nothing in land matters is secure here," he complained. "There are
many irregularities and but few of the surveys were actually made
in the field; at least there are no corners established to most of the
blocks, much less the sections; and until such is done and the land
classed as to range, soil, and timber, by sections, and a local agent
established who is honest and true, the State's interest here cannot
be protected or enhanced."[14]

If the poorly done surveys frustrated state agents like Baines, they
were far more worrisome to landowners. Nevertheless, such problems didn't stop the feverish effort to file and settle. In January 1908,
José Prieto filed his first land application at the Presidio County
Courthouse, choosing three sections in Pinto Canyon in addition
to his home section at El Tanque. Although the four sections were
only contiguous at their corners, they lapped against the flanks of
the Chinatis, embracing some of the better grasslands. Haphazard
though the surveys may have been, filing was still the greatest security José had ever known. Putting fresh eyes on what he now could
claim as his own, he decided to relocate his homestead to one of his
sections just south of Pinto Creek that included a spring he called
Ojo Acebuche (Wild Olive Spring). Less than a mile to the southeast,
on a small bench, he built a simple mud-mortared rock house less
than three hundred square feet in size and an attached open shed.
It left much to be desired, but for the first time he had built on land
that was his and his alone.[15]

Wilson's Road

Ten days after José Prieto left the Presidio County Courthouse, James
limped through the doorway to file his own application for seven

mostly contiguous sections at the head of Pinto Canyon. That land encompassed a sizable section of Pinto Creek and one of the strongest flowing springs in the entire canyon. With the clock ticking on his three-year occupancy requirement, James turned his focus to improving access. Pinto Canyon had long been more connected to Ruidosa and the primitive road that linked the river hamlets to Presidio. With Marfa now serving as county seat and being slightly closer than Presidio, a road east made sense. In November 1908, the Presidio County Commissioners Court honored James's request and appropriated $1,000 for the construction of a road through Pinto Canyon. In addition to allowing access, the project provided a modest but much needed source of income for James and other local men, like Regino Nuñez, who helped build it. But constructing a road was no simple task, the work being performed manually with picks and shovels, sweat and blood, and a good deal of cussing. After seven months of back-breaking labor, the *Marfa New Era* reported, the Pinto Canyon road was in good condition for a visit to the Ruidosa Hot Springs—the key destination for most travelers venturing below the Rimrock.[16]

Dora and James Wilson assist an automobile up the steep road out of Pinto Canyon. Photo courtesy of the Marfa Public Library, Marfa, Texas.

Although the road out of the canyon was complete, the descent was precipitous and the ascent too challenging for most of the 20-horsepower engines of the day. Living close by, the Wilsons frequently found themselves aiding travelers. Tying their ropes to the front bumper of a Model T or other early motorized contraption, they used horses to pull the vehicles up the hill—true horsepower making up for what was lacking under the hood. Despite better access to Marfa, Ruidosa remained the primary supply depot. Every couple of weeks, James saddled a packhorse and rode to Ruidosa for supplies. By horse the round-trip took all day. When larger loads of supplies were needed, he had to use a wagon—turning the supply venture into an overnight excursion.[17]

James soon modified his ranch operation to suit the more desert-like conditions of the canyon by trading his cattle for goats. Perhaps at José's suggestion—or by his example—he purchased from W. H. Cleveland one hundred Angora nannies, animals adapted to the terrain and forage. Being smaller in size and nimbler of foot, goats are better suited to broken and brushy country than cattle. More importantly, because they are primarily browsers rather than grazers —eating shrubs more often than grasses—they rely on plants little used by other classes of livestock. Able to survive on little more than twigs and thorns, they make good use of mesquite, oak, yucca, dalea, wild olive, and other shrubs and forbs generally ignored by cattle. Also, being agile and sure-footed allowed them to better navigate uneven terrain and across precipitous rock ledges affording only the tiniest footholds. And with their ability to stand on their hind feet they had the added advantage of reaching higher vegetation. While sheep shared many of the same traits and often occupied the same range as goats, they lacked the intelligence of goats and were less adept at avoiding poisonous plants. Of the predominant classes of livestock, Angoras were easily the best choice.[18]

Lacking the funds to hire help, James pressed his daughters, Millie and Mamie, into service as herders. They kept watch over the animals to make sure they did not stray too far, always with an eye out for coyotes and wolves. The girls found the goats easy to manage and thrilled at their ability to turn on command. This trait allowed the girls to drive the goat herd to places they themselves wanted to explore—hillsides where they found petrified wood and fossils or hidden caves with smoke-blackened ceilings and stone tools thousands of years old.[19]

The mohair from the Wilsons' first shearing fetched a good price, so they celebrated their first real income since their arrival in the canyon. Having cash in hand, James determined to build a more permanent house for his family. He decided on a spot about a mile and half east of their rock hut. Just up a tributary canyon known as the "old hole" was a beautiful location right beside a strong spring that fed a natural pool in volcanic rock below a grove of oak, wild cherry, and walnut trees. Wild grape vines entwined the branches that spread into a wide canopy, providing deep shade. After clearing a terrace just south of the drainage, James hired a crew of men from Mexico to build a three-room adobe house, the earthen bricks bearing a grayish cast from the ashes of ancient campfires. Having lived in a dusty jacal for the past two years, the Wilsons considered their new house, appointed with little more than a handmade dining table and beds, luxurious.[20]

In April 1911, James filed proof of three years' residence, paving the way toward gaining full title from the state. An added benefit

Wilson ranch house at the "old hole" around 1920. Photo courtesy of the Watts family collection.

The Wilson family around 1912 at their Pinto Canyon home. From left:
Josie McCollister (Dora's sister), Millie, James, Ora, Dora, and Mamie.
Photo courtesy of the Marfa Public Library, Marfa, Texas.

of his proof of residence was that James could now offer the land as
collateral. Having depleted his coffers on the house, he did not wait
long before securing a loan from two of Dora's uncles for $2,000,
guaranteed by five sections of land and fifty head of cattle—nearly
everything except the home section and his herd of goats, a small
insurance policy against total loss. It was to be paid back in four
years, at 10 percent interest, or else he stood to lose the land. The
loan certainly reflected need. With ongoing payments on the land
and expenses from the new house, James likely needed the cash. It
may have been James's first loan, but it would not be his last.[21]

Seeking Greener Pastures

**They were land people, always looking for more land. They
migrated to land deals. It was in their blood.**

—Jackie Sutherlin

In 1910, two brothers arrived with their families to become the most
recent homesteaders in Pinto Canyon, and their presence expanded

the population of the canyon to more than thirty-three people—more than would ever live there again. Their journey had been a long one. The Sutherlins had been slowly migrating southwestward for decades. Hailing from the large and illustrious Sutherland clan that wielded considerable power in northern Scotland, their ancestors probably immigrated in the late 1700s to Virginia or the Carolinas. Among their descendants was Francis Marion Sutherlin, born in Orleans, Indiana, in 1836. His family had moved to Texas by the time he was twenty, and by 1861 they were living on a farm in Young County, where Francis married Nancy Ann Bragg—a farmer's daughter from Alabama—shortly before joining the Confederate army. Assigned to the 2nd Texas Cavalry, he fought Union forces in defense of Louisiana, where he lost his brother to combat. It was also there he suffered a bullet wound to his ankle that left him crippled for life.[22]

Despite his injury, Francis eked out a living, and over the years Nancy would give birth to three sons—Mart, Charles, and George. Although Charles died at age thirteen, Mart and George grew to adulthood and soon had families of their own. By 1900, Francis and Nancy were living with George and his wife, Mollie, on a farm in Sterling County. Mart and his wife, Eliza, along with their growing family of ten children, lived a short distance down the road. But being in poor health, Francis had little to show for his life. In January 1906, he applied for a well-deserved Confederate pension, claiming indigent status, as he was "in actual want, and destitute of property and means of subsistence." With only seventeen head of cattle and two horses valued at $120 to their name, they bordered on poverty. Even so, Francis's pension was denied, with no reason given. He'd reached the end of his rope. Something had to give. As they had done so many times before, the family packed up and moved once again.[23]

The Sutherlin migration across Texas had been mostly linear and always to the southwest until they arrived finally at the far corner of the state, where they could go no farther. By now an ailing man of seventy-two, Francis, along with his wife, their two sons, and their families, made the bold move to Presidio County—one of few counties in the state where land could be had under the Eight-Section Act. By the time they arrived, Mart's wife had died, leaving him to

Nancy, Charles,
and Francis Marion
Sutherlin around
1870. Photo courtesy
of the heirs of Francis
Marion and Nancy
Ann Sutherlin.

care for their twelve children, the oldest twenty-two years of age
and the youngest an infant. The Sutherlins began to file on land in
earnest, with every adult and about half the children eventually
staking claims. Between 1907 and 1915, they would file on nearly
fifty Presidio County sections, much of which was dry land in scattered parcels across the Marfa Plain—the leftover sections along
the edges of blocks that had already been claimed. Even so, their
filing frenzy was primarily speculative, the intent being to sell to
neighboring ranchers or latecomers at a higher price. It was a chance
for the Sutherlins to finally get ahead, an opportunity to escape the
hardships of their past.[24]

Francis filed on five sections along Alamito Creek adjacent to a
new reservoir under construction by the St. Stephen Land and Irrigation Company—a commercial land development venture. San

Esteban Lake, which was completed in 1911 at a cost of $62,000, was intended to irrigate small farms of ten to forty acres. The company planned to settle some six hundred families in the area in what would be the first large-scale row-crop effort in the county away from the waters of the Rio Grande. Like so many other lofty schemes, it was destined to fail. Within seventeen years the lake would be filled with silt, the farmers would abandon their fields, and the company would fold. But by that time, Francis and his family had long since moved on.[25]

Francis was fortunate to have two healthy sons who could help provide for the family. Even so, their financial position may not have been strengthened much by George, who reportedly preferred gambling and carousing to holding down a steady job. Years before, Francis had given him money for seed to sow his fields. However, the money disappeared—invested elsewhere or gambled away. According to another story, George once hired a man to break some colts for

George Sutherlin playing cards. Photo courtesy of the heirs of George Barton and Molly Geneva Sutherlin.

George Sutherlin homestead on Pinto Creek. Photo courtesy of the Marfa
Public Library, Marfa, Texas.

Mart Sutherlin homestead on Horse Creek. Photo courtesy of the Harry
Ransom Center, the University of Texas at Austin.

him, but when it came time to pay, George was nowhere to be found. Since Mollie had no money, the hired man helped himself to one of the horses instead.[26]

Three years after arriving in Presidio County—and having bought and sold other sections in the county—Mart and George filed on several sections below the Rimrock. It may have been the live, flowing water of the river or the lofty vistas that drew them. More likely it was the availability of contiguous sections. Mart settled on six sections along upper Horse Creek, a northwestward-trending tributary to Pinto Creek. George and his family, however, arrived with little to their name. Fortunately, they befriended José Prieto, who took the family under his wing and even sold George a section on which he could build his home, where Horse Creek joins Pinto Creek.[27]

Considering their timing, the Sutherlins needed all the help they could get because the weather was not cooperating. The drought of 1910–11 resulted in less than half the average yearly rainfall, making it a particularly bad time to get into the ranching business. As grass cover and browse diminished, livestock began to range more widely. And in the free-range days before fencing, keeping the animals close by was not always an easy task. Mart, in particular, was having problems. In May 1911, W. H. Cleveland asked the district court to issue a restraining order against Mart declaring him to be "perpetually enjoined from turning sheep loose upon, or driving or herding sheep on or upon the lands described in plaintiff's petition." Since Cleveland's holdings literally surrounded Sutherlin to the south and east, it was a tall order. Fortunately, by mid-July 1911, the drought had broken and rain was falling across the area. In November, the drought—by general consensus—was declared over.[28]

The Pinto Canyon School

With the arrival of the Sutherlins, the canyon now had twenty-six children spread between the Prietos, the Wilsons, and the two Sutherlin families. Five of those children were eighteen or older and six were under six years old. The remaining fifteen, however, were school age. Because the nearest school was sixteen miles away in Ruidosa, in 1910 Presidio County signed off on construction of a one-room adobe schoolhouse and adjacent teacher's residence in

The Horse Creek schoolhouse around 1915. Photo courtesy of the Marfa Public Library, Marfa, Texas.

Pinto Canyon. It was as centrally located as it could be: three miles west of the Wilsons, a little more than three miles northwest of Mart Sutherlin, and two miles east of the Prietos. George Sutherlin's three boys had the shortest commute of all; their house was within horse-shoe pitching distance of the schoolhouse.[29]

When Millie Wilson rode beside her father in their wagon to pick up the new schoolteacher in Marfa, it was her first trip out of the canyon in three years. Climbing the steep road to the edge of the rim and seeing the grasslands open before them likely induced feelings of agoraphobia, the plain lacking the containment of geological struc-tures to which her mind had grown accustomed. Whatever Millie might have felt, it likely paled in comparison with what teacher Sue Woodward was met with when they made their return descent into Pinto Canyon—likely an audible gasp to see the land drop away into chaotic depths unlike anything she had ever seen. The remote and ragged nature of the landscape might have appealed to some, but no single teacher remained in the canyon for more than a couple of years. After Sue Woodward left, Etta Brown took charge before being replaced by Anna DeVolin, who was making the rounds of the Presi-dio country schools, having taught at a ranch school near Candelaria and, later, in Presidio.[30]

Class photo at the Horse Creek schoolhouse around 1912. From top left: Annie Sutherlin, John Sutherlin, Daisy Sutherlin, Mamie Wilson, Bill Sutherlin, Millie Wilson, Eddie Dee Sutherlin, Sotera Prieto, Petra Prieto. Bottom row: Charlie Sutherlin, José Prieto Jr., Bill Sutherlin, Gregorio Prieto, Barton Sutherlin, M. D. Sutherlin Jr., Dolan Sutherlin. Photo courtesy of the Marfa Public Library, Marfa, Texas.

Having a schoolhouse, Pinto Canyon suddenly felt something like a community. Just across the road, the towering jointed mountain known as Cerro Hueco—in whose shadow the schoolhouse resided—seemed to visually anchor the canyon, just as the school anchored the neighborhood of scattered ranches. For a handful of years, the voices of children echoed off canyon walls above the constant clamor of bleating sheep and goats. With the first rays of sunlight striking the high summit of Chinati Peak, the canyon still shrouded in shadow, mothers packed lunches as the children caught their burros. Sometimes with two or three piled on a single animal, they rode down well-worn trails toward the schoolhouse. Staking their burros a short distance away so they could graze, the kids scurried inside. Released for the noon lunch break, they led their burros down to Pinto Creek for a drink. After school let out, they mounted up once again for the long and often hot afternoon ride home.[31]

Although school was not yet mandatory in Texas, the state constitution of 1845 established a public school system, and most residents embraced the opportunity for an education. In the days before good roads and school buses, the school district accommodated the children of far-flung ranch families by taking the school to them. Schools were built in both Ruidosa and Candelaria in 1902. Although larger in size than the Pinto Canyon school, they were otherwise of similar specifications: one structure divided into two rooms to separate the children into two classes according to age. Mary Kilpatrick, who would marry customs inspector Jack Howard, served as the teacher at Candelaria for the first eight years it operated, while her counterpart Alice May Campbell held class at the Ruidosa school. By 1911, the Ruidosa precinct, encompassing the Pinto Canyon school as well as several other small schools, boasted a student population numbering 287.[32]

The Horse Creek school, unlike those in Marfa and Alpine, was not segregated. With so few students, ranch schools by necessity had to serve students of any age or ethnicity. Although classes were held in English, recesses were bilingual, allowing Anglo kids greater

Cerro Hueco, a volcanic intrusion with pronounced vertical jointing stands prominently across from the old Horse Creek schoolhouse in the heart of Pinto Canyon. Photo by D. Keller.

exposure to the language of the borderlands. On weekends, after finishing their morning chores, the children sometimes gathered at the upper end of Pinto Canyon to flush wild burros out of the bottoms. Descended from stock that had escaped captivity and gone feral, the burros were a nuisance, forming herds that competed with domestic livestock. If the children could catch one, they did, but most of the time they chased them for the sheer joy of it. Splitting into two groups, the children spread out along the upper slopes on opposite sides of the canyon. At a signal, the two lines of children would start down the slope yelling and waving their arms, some laughing hysterically as the startled burros—their haunches flexing, heads cocked sideways—ran wild-eyed and braying before them.[33]

José's New Home

In 1911, José filed proof of occupancy on four sections, his certificate corroborated by the affidavits of W. E. Love, D. L. James, and J. E. Wilson. It was cause for celebration. Having fulfilled one of the critical requirements for homesteading under the Eight-Section Act, he had reached another milestone. But the small rock house José had built on his land was no longer adequate. By the time their last child, Jane, was born in 1915, the tiny house could not contain them all. Gregorio Prieto later recalled how the house was so crowded the family had to sleep in rows, "like pigs."[34]

José set to work. Enlisting the help of his brother Francisco, they began making adobe bricks. After digging a pit about two feet deep and six feet in diameter, they filled it with water and added shovelfuls of clay-rich earth and chopped grass. With their pants rolled up to their knees, they stomped through the pit in their bare feet like high-stepping dancers, mashing the three ingredients together to make a thick but soft adobe mix. They shoveled the wet adobe mixture into a wooden form twelve by eighteen inches in size set flat upon the ground. Pressing the mud into the corners, they struck the top smooth, then pulled the form. The process was repeated, brick by brick, until more mud was needed or they ran out of flat ground to work on. After a couple of days, the bricks were turned on end for further drying. Once firm, the bricks were carefully stacked. Considering that building a house required approximately twenty-five

Prieto homestead in Pinto Canyon below Chinati Peak (*upper right*). Photo courtesy of the Prieto family collection.

hundred bricks, each weighing between fifty and sixty pounds, it was backbreaking work. To six-year-old Frances, the stacked adobe bricks looked like a corrugated pathway. She hummed to herself as she walked down a row of them until reaching the end, where the last brick fell and broke in two. She looked up to meet the eyes of her frowning uncle, who asked José what he was going to do about it. But José only smiled. There was no harm done. They could always make more bricks.[35]

Once the bricks were all stacked, José and Francisco set batter boards and mason's lines in a rectangle fourteen by forty-eight feet catty-corner to the rock house. Then they mixed more adobe for mortar, gathered river stones, and began laying them in a thick mud bed to make a cobblestone foundation. On top of this low rock wall, they began laying the adobe bricks, one course at a time, leaving openings for four doors and ten windows, until the walls had reached a height of about eight feet. Meanwhile, José laid in hardware and milled lumber, doors and windows—all of which had to be transported from Marfa by wagon. Cutting the lumber to size, they built the door and window bucks, then installed a top plate on the uppermost course upon which they set *vigas* as the base for a flat, earthen-topped roof. With the addition of a porch on the west

Prieto family around 1915. From top left: unknown, Gregorio, Sotera, Petra, Victoriano, Tomasa Barrera (Juanita's sister), Pablo, and José Jr. Seated: José and Juanita with Francis (*left*) and Lupe. Juanita was pregnant with her last child, Jane. Photo courtesy of the Prieto family collection.

side of the house and the doors and windows installed, the brothers stepped back to admire their work. After decades of living in dug-outs and jacals, the Prieto family would finally be living in style.[36]

It had been a costly endeavor, in terms of both time and money. The house was part of José's honest fulfillment of the requirements for homesteading. But it also represented the greatest investment he'd yet made toward improving his ranch—one that signified another rung up the socioeconomic ladder. Although the new house was relatively luxurious, José was not one to spend frivolously. By the sweat of his brow and the skin of his own hands, he had slowly and steadily increased his net worth. In 1892, the first year José filed property taxes with Presidio County, he reported only two horses and thirty head of goats for a total worth of $73. Twenty years later he filed proof of occupancy on his four sections and rendered for taxation two burros, three horses, twenty cows, and six hundred goats, for a total worth of $4,065. His first two decades had not been a meteoric rise in wealth, but his holdings would soon multiply beyond anything he could have anticipated. José's best years were still to come.[37]

Homestead Refugees Albin and Belle O'Dell

In early 1913, a young family arrived in Pinto Canyon eager to start a new life. Although one of the last families to arrive, as a result of the revolution spreading across northern Mexico, they would also be the first to leave. A farmer from Balmorhea, a small cotton and truck farming community eighty miles to the northeast, Albin O'Dell was born in 1879 in Collins County, Texas. He was the last of nine children born to Philip and Emily Hancock O'Dell and one of only three O'Dell children who would survive into adulthood. In his early years, death was a constant companion; his mother died when he was a mere four years old, his older sister a couple of years later.[38]

Albin was sent to live with friends of the family until his fourteenth year, when he joined his father and older brother in Roscoe, Texas. He took employment where he could find it. He cowboyed in northern Texas and southeastern New Mexico before moving to Balmorhea in 1906 to work as a farm manager. Three years later, at age thirty, Albin married fifteen-year-old Belle Jones, and they had their first child, Jennie, in August 1910. For the next couple of years

Albin O'Dell and Belle Jones O'Dell on their wedding day in 1909. Photo courtesy of Frank O'Dell.

the newlyweds lived in Hagerman, New Mexico, on a small alfalfa farm left to him by his father, and it was there that Belle gave birth to their second child, Pete. Perhaps tiring of the flatlands, or inspired by the prospect of owning his own ranch, Albin traded the farm in late 1912 for eight hundred sheep and moved his family south to take up land in the rocky confines of Pinto Canyon.[39]

In February, Albin filed on a full eight sections below the Rimrock. One of these, along an upper stretch of Pinto Creek bordering the Wilson Ranch, became his home section. Here, Albin built a small house, pens, a dipping vat, and an outdoor shed. It was a considerable amount of work, far too much considering how little use he would get from them. As attractive as the land prices and the stunning location of their new homestead were, the lowlands proved a greater challenge than he expected. The rough terrain and the abundance of predators were difficult enough to deal with. More troubling yet was spillover from the Mexican Revolution, whether starving peasants or opportunistic bandits.[40]

Perhaps not anticipating their vulnerability, Albin had built their homestead in a conspicuous and largely indefensible location. Sited only a few hundred feet from the Pinto Canyon road, they were exposed to more visitors and more traffic than anyone save George Sutherlin. Their accessibility would prove to be their downfall. Albin began to lose sheep. But with no sign of predation by the normal suspects, he attributed the loss to bandits who had started raiding along the border. The loss of sheep was one thing, but fearing for their personal safety was entirely another. Before long, their resolve to settle in the canyon was put to the test.[41]

One day when Albin was in El Paso for jury duty, Belle watched through the window as a group of Mexican revolutionaries rode up to the house. They wore oversized sombreros and *bandoleros*. Pistols rested on their hips, rifles in their scabbards. The men sat on their horses as the leader called out toward the house. Belle's heart pounded like a piston inside her chest. But determined not to betray her fear, she composed herself and stepped out to face the men. The leader stood up in the saddle. Wiping the sweat from his brow, he greeted Belle in Spanish and said he and his men meant no harm; they sought only water for their horses, then would be on their way. Belle pointed to the concrete trough at the edge of the corral and told them they could get their water there. The man tipped his hat as he clicked to his horse, his men following closely behind.

Belle slowly exhaled. She went inside the house and wrapped several tortillas and slices of *asadero* cheese inside a piece of cloth. When she walked back outside, the men had dismounted to loosen the horses' girths so they could drink. At one end of the trough, a man splashed water on his face. The leader walked around his horse toward Belle. He extended a dark hand, creased and scarred from a life of toil, and took the bundle from her, slightly bowing his thanks. She nodded to him and turned to walk back to the house—a lone woman resolute and upright, poised in her long prairie dress. Minutes later, her heart still pounding furiously in her chest, she watched as the men continued down the road.

When Albin returned a few days later, he could scarcely believe the story Belle was telling him. Although the men had been respectful, there were no guarantees the O'Dells would be as fortunate the next time. Their vulnerability was too great. Albin cursed his decision to build so close to the road. With increasing reports of stolen livestock and bandits along the river, he felt the risk was too great. He'd lost enough loved ones for one lifetime. In May 1914, Albin sold most of his land to Mart Sutherlin's oldest son, Grover, and returned to Balmorhea only fifteen months after having arrived. If their time in Pinto Canyon was fleeting, they were only the first of many who would follow.[42]

Sutherlin Fortune

As Albin struggled with the looming threat of the Mexican Revolution, the Sutherlin brothers seemed to be holding their own. In 1912, Mart purchased two additional sections, bringing his holdings to a full eight sections. A year later, he and George individually filed their three-year proof of occupancy on their homestead sections. It helped that the weather had been agreeable following the drought of 1910–11; for the three years thereafter, rainfall was abundant. In 1914, the nearest weather station reported more than twenty-three inches—the wettest year since the turn of the century. The men's substantial wool and lamb crop reflected the increased moisture. George's wool harvest was high enough from his flock of twenty-five hundred sheep that he was inspired to report it to the *El Paso Herald*.[43]

By all appearances, George seemed to be enjoying a relative spell of success. Since moving to Presidio County, he had filed on and sold eight sections of land on the Marfa Plain in addition to the section purchased from José Prieto. With those, and the income from the wool harvest and lamb crop, he was probably better off financially than ever. Maybe the money in his pocket gave him confidence to try his luck elsewhere. More likely—as was the case for Albin O'Dell—the danger of the Mexican Revolution, coupled with his vulnerable location, weighed on him. In 1914, only a year after proving up on his land, he sold his homestead to customs agent O. C. Dowe and moved his family to Albuquerque. There, for a time, he found work in a wagon yard. But with him being itinerant, and periodically estranged from his family, permanence wasn't in the cards. After a stint as a laborer at the International Brick Company in El Paso, he progressed through a series of jobs in La Mesa, Anthony, and Gallup, New Mexico. Much like Albin O'Dell and scores of others over the years, George's time in Pinto Canyon proved evanescent.[44]

Following Francis's death in 1915, Nancy moved in with her son Mart and his family on Horse Creek. As his homestead grew more crowded, it undoubtedly added to stress brought on by increasingly drier weather. By 1916 the region was gripped by yet another drought, forcing Mart to sell nearly half his livestock. In September, the *El Paso Herald* reported he'd sold twelve hundred head of old ewes and yearling muttons to a "Weatherby from San Angelo," most likely Sim Weatherby, who would later figure prominently—and tragically—in the history of Pinto Canyon. In spite of the drought, Mart's optimism was unflagging, and he claimed the range to be "in better condition than ever before." If so, it didn't last for long. The following year, rainfall would amount to only about a third of the average.[45]

∧∧∧

For most of the early history of Pinto Canyon it had been a place hidden and peripheral to human endeavors. The Eight-Section Act changed that. But of all those who homesteaded in the canyon, José Prieto suffered fewer setbacks than the others. Starting with nothing but a crude dugout and a small herd of goats, by 1915 he held title to four sections along the flanks of the Chinatis, a well-appointed ranch house, and hundreds of goats. Meanwhile, the Horse Creek schoolhouse provided an anchor that, for a brief while, allowed the canyon to become an insular community where the children of four families—Anglo and Hispanic alike—would learn and laugh and play together, would drive braying burros out of the canyon or meet by moonlight at the creek bed swimming hole, and would share stories and dreams in tongues not always their own. Only recently settled and still largely untamed, the canyon offered a sense that with enough work and perseverance anyone could access the American dream. For some, like José Prieto, that dream would be attainable. For others, like the Wilsons and the Sutherlins, it would prove forever elusive.

4

Signal Fires in the Chinatis

Bandit Raids and the Mexican Revolution

Even more bandits made their headquarters in the village
of San Antonio del Bravo, directly across from Candelaria,
Texas. One of its most noted desperadoes was Chico Cano—
feared, hated, and hunted for six bloody years.

—**W. D. Smithers,** *Chronicles of the Big Bend*

ON AN EARLY spring morning in 1919, Hilario Nuñez—a rancher
in lower Pinto Canyon—woke to find he was short about twenty-
five head of cattle. The tracks on the ground told the story: they had
been stolen. Because the trail of prints headed toward Mexico, he
suspected bandits. He hurried over to the Wilson Ranch, where he
knew James had been entrusted with a military telephone. Hearing
Hilario's story, James looped the leather strap of his telephone over
his saddle horn and rode the four hundred yards to the twisted telephone wire the Signal Corps had installed between Marfa and Camp
Ruidosa. From a pole just beside the road dangled two wires bearing U-shaped terminals at their ends. He attached the connectors
to the posts on the telephone and tightened the wingnuts to secure
them. Then, spinning the crank of the magneto, he rang the camp.
Capt. Hans E. Kloepfer at Camp Ruidosa answered and promised
they would investigate immediately. But Wilson and Nuñez weren't
about to wait. They set out down the Pinto Canyon road, gathering

reinforcements as they went, including Mart's oldest son Grover
and Texas Ranger Nate Fuller, who was on patrol. The Pinto Canyon
contingent crossed the Rio Grande at Ruidosa and headed toward
the plume of dust rising from the column of US Cavalry soldiers
already following the bandit's trail. The men fell in line behind the
column, and together the ranchers and soldiers headed southwest
across the dry desert bajadas toward the rocky and forbidding Sierra
Quemada, where they knew the bandits would be hiding.[1]

The Nuñez Ranch Raid, as it came to be called, was the only bandit raid of consequence that took place in Pinto Canyon during the
Mexican Revolution. Because most thefts of horses, sheep, and goats
were opportunistic in nature, such crimes couldn't compare to the
orchestrated raids on the nearby Brite and Neville Ranches. Considering the chaos and violence that raged only thirteen miles away
across the border, it is curious that Pinto Canyon remained as quiet
as it did. After 1916, when US troops became a continuous fixture
at Camp Ruidosa, the military presence certainly served as a protective barrier against incursions. Even so, the absence of conflict in
the canyon prior to the establishment of the camp was mostly good
fortune. After all, the road through the canyon—as one of precious
few passages to the Marfa Plain along the seventy-mile-long barrier
created by the Sierra Vieja and Chinati Mountains—was as useful to
bandits as it was to ranchers below the Rimrock.

By the time the Mexican Revolution erupted in 1910, the stage
for conflict had been set. Throughout the thirty-five-year dictator-
ship of Pres. Porfirio Díaz, landless peasants suffered from the ever-
widening gap between themselves and the wealthy few. The rigged
re-election of Díaz in 1910 served as the tipping point that sparked
the revolution and provided Francisco Madero the rebellion he
needed to finally force Díaz out. Born into a wealthy landowning
family and lacking both leadership and military experience, Madero
was an unlikely revolutionary. Even so, the democratic reforms he
advocated appealed widely to the peasant class hopelessly shack-
led in servitude. From Ojinaga southward to Chihuahua City, the
wealthy Terrazas and Creel families had driven hundreds of Mexicans from land they had lived on for generations to sell it to wealthy
investors for the Kansas City, Mexico and Orient Railway. The dis-
placed were ready sources for rebel troops. When violence broke out

near Ojinaga in December 1910, four contingents of the 3rd US Cavalry were deployed from Fort Clark to positions along the border, including subposts at Presidio, Ruidosa, and Candelaria. In early February 1911, US troops reported rebel signal fires burning in the Chinati Mountains. Fires near El Mulato, about sixteen miles southeast of Ojinaga, appeared to answer them. Then, on 7 February, the rebels routed a much larger force of *federales* at El Mulato, the first of a string of local victories that served to fan the revolutionary flames.[2]

In June 1911, revolutionaries under the command of José Inés Salazar—who had helped Pancho Villa capture Juárez a month earlier—raided the ranch of Lamar Davis just across the river from Candelaria, confiscating his entire store of food, guns, saddles, horses, and mules. As rebel troops massed, terrorized families fled across the river for safety—the first of many such cross-border flights during the blood-soaked decade of the 1910s. In October, following a string of revolutionary victories, Mexico elected Madero president. Expecting a return of order, US troops withdrew from the border in November. But the move had been premature. Instead, the revolution became increasingly fractured and regionalized. Four months after pulling out, US troops were redeployed to the Big Bend, where they would maintain a presence for the remainder of the Mexican Revolution. With Marfa serving as headquarters for the Big Bend Military District, much of the US troop activity was focused along the border from Presidio northward as the swath of country across from the Chinatis and Sierra Vieja became the local stage for the revolution.[3]

Madero's tenure was short-lived. In 1913, Gen. Victoriano Huerta, a Díaz loyalist, seized the presidency and eliminated the highly anticipated Madero reform plan. Madero's imprisonment and assassination a few days later reignited the revolution. Pancho Villa and other revolutionaries joined forces with Venustiano Carranza—a wealthy landowner serving as governor of the state of Coahuila—as open warfare spread across the state of Chihuahua. In November, Villa took Juárez with hardly a fight and claimed Chihuahua City on 8 December. From there, the victors marched northeast to Ojinaga, where in January 1914 Villa led a decisive victory. Villistas by the thousands descended upon Ojinaga, leaving an ominous dust cloud rising in their wake. The spectacle of their approach terrified the *federales*, many of whom abandoned their posts, threw their rifles into the Rio Grande, and waded across into the custody of US troops.[4]

Villa's attack was swift and brutal, the rebels victorious within an hour of the charge. Mangled bodies littered the hillsides and lay beached along the banks of the Rio Grande. Offering no quarter, Villa's men shot the wounded where they lay. Meanwhile, firing squads systematically dispatched captured *federales* against the few adobe walls that still stood among the rubble. For two days, volleys of rifle fire announced each execution. The bodies were stacked like cordwood and set ablaze, smoke from searing hair and flesh wafting through the air with a nauseating stench. Meanwhile, more than four thousand *federales* and civilians—men, women, and children— who fled across the border were marched northward under military escort. To some, the twelve-mile-long column looked more like a cattle drive than a march. Arriving in Marfa, the refugees found themselves herded into huge makeshift wire corrals at the edge of town. Soldiers then loaded the refugees onto trains and shipped them to an internment camp at Fort Bliss.[5]

The Mule Trains of Pinto Canyon

During the first five years of the revolution, Pinto Canyon remained largely isolated from the growing strife. That soon changed. Between 1916 and 1921, when the US Cavalry maintained a permanent camp on the outskirts of Ruidosa, the canyon served as the key supply route from headquarters at Camp Marfa. The passage of mule trains, wagon trains, and—later—steam-powered tractor trains became near daily reminders of the revolution raging just across the Rio Grande. Supply trains most often consisted of sixty-four pack mules and a bell mare. The one-way trip took two days. The packmaster headed the procession, followed by the cook leading the bell mare, which the remaining mules followed instinctively. Ten pack mules trailed the bell mare, followed by two packers—the men in charge of saddling and loading the mules. With five or six sets of such ten-mule processions with their two packers, plus the blacksmith and *cargador*, or saddler, bringing up the rear, the mule train lined out of Camp Marfa. Following the Pinto Canyon road thirty-two miles

Military wagon train descending Pinto Canyon in the best-known historic photograph of the canyon, taken by W. D. Smithers around 1918. The Horse Creek schoolhouse is visible in the middle center of photograph. Photo courtesy of the Harry Ransom Center, the University of Texas at Austin.

across the Marfa Plain, they halted at their one overnight camp at Cleveland's tanks just above Pinto Canyon. Early the next morning found the pack train winding through Pinto Canyon before crossing the rolling chino grass hills the last nineteen miles.[6]

Being completely dependent on rations and supplies allocated by the quartermaster at Camp Marfa, US troops along the border considered the supply trains critical lifelines. For a cavalry camp, with hundreds of horses and mules, feed was the most vital commodity. Of the approximately thirty-two hundred pounds of cargo in each train, the majority consisted of hay and oats, the remainder being rations for the troops. If provisions arrived late, both men and animals suffered. As a result, pack trains generally operated with the efficiency of a well-oiled machine. Even so, occasional wrecks were inevitable. One morning as the mules clopped through Pinto Canyon, they passed a herd of Mexican goats grazing quietly by the roadside. Unbeknown to the packers, mule number seven was a goat hater. Nearing the goats, the mule suddenly broke out of the line and charged, its two packers and the packmaster in pursuit. As the goats scattered in terror, the mule seized one in its teeth and, raising it up, tossed it to its death before going after another. The packers grabbed the mule's halter just before the animal clamped down on a third victim. All the while the rest of the pack train plod-ded on, the remaining packers laughing hysterically.[7]

The Rise of Chico Cano

The mass exodus of Mexican refugees triggered by Villa's capture of Ojinaga brought the immediacy of the revolution to the doorstep of the Big Bend. After August 1912, following the first wave of the revolution, US troops were stationed intermittently at Candelaria. Although Ruidosa would not be reoccupied until July 1916, after that date the camp was manned almost continuously for the remain-ing years of the revolution. Meanwhile, the scale of conflict escalated and stories of the most feared bandit in the region rolled in as fast as the telegraph operators of the Signal Corps could relay them. Claim-ing the area across from Ruidosa as his home turf, he would become Public Enemy No. 1 for the Texas Rangers as well as the US Cavalry, both of whom focused obsessively on one man whose position on the local stage was second only to that of Pancho Villa himself. His name was Chico Cano.[8]

Born in 1887 to Catarino Cano and Juliana Venegas near Pilares, Chihuahua, Chico Cano grew up helping his father farm a plot of land and raise a meager herd of livestock. As a young man, Chico earned a reputation for breaking horses, and he began working ranches on both sides of the border and as far east as the Pecos River. Tall, handsome, and big chested with piercing blue-green eyes, Cano stood out from his countrymen, both in appearance and in action. By 1911, as the revolution gained strength, bandit groups began form-ing along the border. Not one to pass up such opportunity, Chico formed a family-based gang consisting of his four brothers—Manuel, José, Antonio, and Robelardo—along with a number of cousins. A natural leader who inspired men to follow, Chico began leading his men on raids on both sides of the border. At first stealing only horses and cattle, they soon extended their quarry to food, clothing, rifles, ammunition, and anything else needed to keep the band supplied

Maj. Roy J. Considine (*left*) and Captain Matlack of the 8th Cavalry bracketing infamous bandit Chico Cano before the latter two men became sworn enemies. Photo courtesy of the Harry Ransom Center, the University of Texas at Austin.

and fed. Like many bandits, Chico held greater loyalty to his land and his people than to any particular faction. Still, he would play a significant role in northern Chihuahua during the revolution, even if his frequently shifting alliances kept the US military always guessing about his true intentions.[9]

Cano was already notorious along the border after his daring escape from US authorities left one man dead. Joe Sitter, a customs inspector assigned to the Sierra Vieja borderlands, had developed a particularly bitter hatred for Cano, believing him to be behind nearly every theft in the region. In January 1913, tipped off that Cano would be attending a wake near Pilares, Sitter finally got his chance. With fellow customs inspector Jack Howard, as well as J. A. Harvick, an inspector for the Texas Cattle Raisers Association, Sitter rode up on the wake being held on the Texas side of the Rio Grande. Standing his horse, Sitter shouted toward the house for them to send out Cano. After a brief wait, Chico's father, Catarino, stepped outside and told the men his son was not there. But they weren't having it. If Chico didn't come out at once, Sitter threatened to burn the house down and everyone inside with it. The tension grew. Shortly, another man emerged and pleaded with them to at least allow the women and children to escape. When they agreed, Chico's youngest brother, Robelardo, slipped out dressed as a woman. With the men distracted, he broke out of the group and ran for help.[10]

Inside, Cano knew he was trapped. Buying time for his brother to bring reinforcements, he addressed Sitter directly, asking what assurance there was he wouldn't be killed. Finally, having stalled as long as he could, Cano agreed to surrender. As Sitter and his men covered the doorway with their pistols, the men slowly filed out, their hands held high. After the last man had passed, Cano emerged, his head held high. After handcuffing him, Sitter's men put him on a mule and bound his feet beneath the animal's belly. With their prisoner secured, the men set out for Valentine. But they had gone only a short distance when, rounding a curve, Cano's rescuers opened fire. One bullet grazed Sitter's skull. Another hit Harvick in the leg. A third passed through Howard's chest as his horse tottered and fell from another bullet, pinning Howard to the ground. Amid the gunfire, Cano's mule, goggle-eyed, bolted for the *bosque* (thicket) along the river. Bouncing like a rag doll, Cano held on for dear life as the mule bounded through the dense mesquite thicket and into the river. Finally reaching the opposite bank, scratched and bleeding, he regrouped with his men. Once free of his bindings, Cano swatted the mule, sending it back across to the US side. Meanwhile, at the sound of gunfire, people in Pilares had rushed to help the injured men. Thanks to their aid, Sitter and Harvick would survive with minor injuries. But it was too late for Howard. His internal injuries were so severe he died early the next morning. Outraged at his friend's death, Sitter swore revenge.[11]

Revenge came, but not through Sitter. And, rather than effecting a resolution, it only served to intensify the conflict. Hearing the news of Jack Howard's death, his father-in-law, J. J. Kilpatrick, a farmer and storeowner in Candelaria, offered a machine gun to anyone who could deliver the body of Manuel Cano, Chico's brother, believed to have shot Howard. Two Mexican men accepted the offer and promptly kidnapped and murdered Manuel in 1914. When Chico learned who was responsible, he killed them both. But he wasn't done yet. The very next year Sitter, along with several other men, were lured into a narrow canyon by following suspicious horse tracks and the drag marks of lead ropes. As the walls closed in around them, the bandits opened fire. All except for Sitter and Texas Ranger Eugene Hulen were able to escape. Unable to leave their position, the two men held off their attackers until the ammunition ran out. The next day R. M. Wadsworth, another US customs inspector, saw Sitter's battered corpse. "He looked like he died in great agony," he claimed. "His knees drawed up, cramped up, his hands and fingers like that, drawed up over his face; you could see where his flesh had been knocked off his knuckles with rocks; his left eye in his head had been caved in."[12]

More than ever, Cano was a wanted man. The attack had been premeditated, brazen, and on US soil. This time he had gone too far. The US military and local law enforcement were placed on alert. "In regard to band of Chico Cano," wrote Chief of Staff W. H. Hay from Fort Sam Houston, "you are directed to use every effort to locate and capture the band of outlaws . . . without waiting for request to do so from state or customs authorities."[13] Such open-ended directives were rare; the US military was not supposed to intervene with civilian affairs. But Cano was a special case. Even so, he would prove notoriously difficult to capture and frustrated nearly every attempt to pursue him. His elusiveness became legendary and, to the US Cavalry, maddening. The War Department began gathering intelligence on his movements at least as early as 1916, when he was reported to be a Carranza captain as well as an inveterate thief. "[Cano] has long been known in that district as a bad character, is somewhere up river from San Antonio with sixty to seventy-five men, probably stealing TO or Terrazas cattle," read one intelligence report.[14]

Troop Buildup on the Border

The Big Bend Military District remained a shoestring operation through 1915. Although troops from some seven different cavalry regiments had cycled through, the number of soldiers stationed in the region at any one time had been small. The buildup of border troops awaited more alarming events, and they weren't long in coming. In 1915, with the rebels gaining ground, Victoriano Huerta fled into exile and Venustiano Carranza became president of Mexico. However, Carranza's presidency did little to unify the various factions of the revolution. Instead, the other major leaders of the revolution—Pancho Villa and Emiliano Zapata—turned against Carranza, whom they had never really trusted, and stepped up their separate revolts. But major defeats at the Battles of Celaya and Agua Prieta left Villa's Division of the North in shambles, and US support of Carranza made things worse for the rebel leader. Infuriated, Villa raided Columbus, New Mexico, in March 1916, triggering the famous Pershing Punitive Expedition into Mexico and raising alarm across the borderlands. After bandits raided Glenn Springs in the lower Big Bend on 5 May, leaving three soldiers and one civilian boy dead, President Wilson mobilized Texas, New Mexico, and Arizona National Guard troops, which, along with the 6th and 14th Cavalry, established camps at regular intervals along the Big Bend border with Mexico. Just five miles north of the mouth of Pinto Creek, Camp Ruidosa would be garrisoned almost continuously

Troop B, 1st Texas Cavalry, National Guard, descending into Pinto Canyon, 1916. Photo courtesy of the Harry Ransom Center, the University of Texas at Austin.

through 1920 and bear witness to much of the revolutionary activity in the region—and a good deal of the banditry—over a violent five-year period.[15]

In July 1916, the entire 1st Squadron of the 1st Texas Cavalry pitched camp on a patch of ground just north of the Ruidosa school-house—a broad stream terrace contained within towering walls of sandstone. For the nine months of the National Guard occupation, soldiers outnumbered townspeople. One of the four troops periodically rotated out to Candelaria for a month or two at a time, but the remaining three troops, consisting of three to four officers and some 280 soldiers, kept station at Camp Ruidosa. Living in a small city of conical Sibley tents situated in neat rows—sandwiched between a horse picket line and a mess hall—the men's accommodations were meager. Even the constructed buildings were a motley assortment of makeshift wood-framed structures. Two pack trains, which maintained separate quarters adjacent to the soldiers' tents, completed the camp.[16]

In spite of the picturesque setting, most of the men found the assignment tedious and dreary. G. W. Fuller, seventeen years old when he arrived with the 1st Texas Cavalry, recalled, "We thought

First Squadron, 1st Texas Cavalry at Camp Ruidosa, 1916. Courtesy of the Texas Military Forces Museum, Camp Mabry, Austin, Texas.

we were going right in to join Pershing, but instead they gave us the job of guarding the border, with headquarters at Ruidosa, where our principal job was to try to halt the depredations of another border bandit, one Chico Cano." Some guardsmen found the conditions too much to bear. "Some of the city boys couldn't take it there," Fuller noted, "and there were some desertions. Most of the deserters tried to get out through Mexico, however, and were returned to us by [Mexican] Federal Troops."[17]

Cano's Elusive Loyalties

Meanwhile, intelligence reports indicated Chico Cano had defected to the Villistas after a break with Colonel Riojas at Ojinaga. With two hundred rebels, he helped recapture Ojinaga in May, securing the port with an eye toward exporting thousands of stolen cattle through it. Officers at Ruidosa and Candelaria focused on Cano's movements with an almost lyrical obsession, as if he alone embodied every threat the revolution represented. But his proximity exaggerated his infamy and his shifting loyalties primarily reflected which faction was brutalizing his friends and neighbors the least. Constantly struggling to maintain his supply of arms and ammunition, Cano sometimes found himself unable to mount an effective force. In January 1917, he was sighted near Santa Barbara, across from Ruidosa. The next day, when Carrancistas invaded and confiscated everything in the town, however, Cano was nowhere to be found. Terrified citizens fled across the river for safety, and some of them continued marching inland. "All roads leading from the river to the Southern Pacific Railroad in this district, have refugees, who have recently fled from Mexico," wrote Col. J. A. Gaston, of the 6th Cavalry in Ruidosa. But Gaston exaggerated the possibilities. There was only one direct route from Ruidosa to the railroad, and that was through Pinto Canyon.[18]

In February 1917, as the country prepared to enter World War I, President Wilson recalled the National Guard from the border. After Germany resumed unrestricted submarine warfare and made an overture to forge an alliance with Mexico—promising to help it reclaim Texas, New Mexico, and Arizona—US neutrality came to an end. On 6 April, the United States declared war on Germany. As troops deployed overseas, the 6th Cavalry arrived in the Big Bend to replace the National Guard. Only seven months later, however, the 6th Cavalry was replaced by the 8th Cavalry, which would hold the border single-handedly for the next two years—longer than any other cavalry unit during the entire Mexican Revolution. While the American Expeditionary Forces were engaged overseas, the 8th Cavalry, under command of the legendary Col. George T. Langhorne, effectively confronted the spillover from the revolution even as it witnessed the greatest period of banditry the Big Bend had ever seen.[19]

By the time the 8th Cavalry arrived, Mexico had descended into a quagmire of ambiguous alliances. Defending Carranza's new regime, Carrancistas had become the new federal army of Mexico, and both the Villistas of the north and the Zapatistas of the south vowed to defeat them. Sometimes collectively known as *legalistas*, the revolutionaries had in common only their wish to see Carranza deposed. Bandit gangs, not bound by ideology or politics, joined whichever faction was in power. Notably along the northern border, loyalties shifted like the wind. Carranza's rise to power had served only to fuel a multifaceted civil war—one increasingly regional, idealistic, and brutal. For the upper Big Bend, that meant an uptick in raids on US soil, culminating in a triad of violent events.

Battles and bandit raids had taken place along the border since the beginning of the revolution, but between 1917 and 1920 the frequency and scale grew to unprecedented proportions. In the Big Bend, the primary stage was the forty-mile stretch between Vado de Piedra and Pilares, with much of the activity centered on Barrancas and San Antonio del Bravo, across the Rio Grande from Ruidosa and Candelaria, respectively. By mid-1917, Gen. José Inés Salazar, commanding two hundred Villistas, and Chico Cano, commanding perhaps a hundred more, dominated the area. Just south of Barrancas on 21 February, Cano and a band of forty Villistas attacked a Carrancista patrol that had been sent north from their camp at Vado de Piedra. Taken by surprise, the Carrancistas fled, with Cano's men in hot pursuit. It was a decisive victory. By the battle's end, fifteen Carrancistas lay dead and two thousand rounds of ammunition had been captured. Cano lost only one man. In the melee, several Carrancistas, five of whom were boys under sixteen years of age, fled across the river to seek amnesty. On 8 May a second battle broke

out above Barrancas, with Salazar's forces defeating a troop of Carrancistas, killing twenty-four and causing another ten to desert their
posts.[20]

Despite sharing a common enemy, Cano and Salazar could not
maintain their alliance. For one thing, Salazar did not share Cano's
loyalties. A native of Casas Grandes, Salazar initially was a "Red
Flagger" under Pascual Orozco and fought against Villa in the first
battle of Ojinaga, in 1914. After being imprisoned by Carranza and
freed by Villa, however, he had a change of heart. By April 1917,
Salazar was one of Villa's most trusted lieutenants. But he had
never won Cano's trust, and after Salazar's men reportedly raped
several local women, Cano was livid. Outside Barrancas, the two
men dismounted one hundred yards apart and advanced with their
rifles drawn. Spewing streams of profanity, they halted with only a
hundred feet between them, their rifles aimed to kill, fingers tight
against the triggers. Their men stood in silence, anxiously watching and awaiting the outcome. After several tense minutes, the men
finally lowered their guns. The showdown was over, but there was
to be no reconciliation. Chico Cano was done with Salazar. He sent
word across the river that Salazar was planning to attack Candelaria
but promised he would strike him from the rear if he did. Because
the battle never happened, some believed Cano invented the threat
simply to gain the Americans' trust. Instead, Salazar retreated to
Juárez in disgrace with a small handful of men who remained loyal
to him. Having run out of supplies and seeing their leader humili-
ated, most of his men deserted and joined Cano's side. For Salazar,
it was the end of the road. Only three months later he would die in
battle outside Nogales Hacienda, Chihuahua.[21]

Following his break with Salazar, Cano made it known he would
stay on his side of the border and remain on good terms with the
United States. To demonstrate his goodwill, he reportedly pursued
and killed bandits who had stolen sixteen horses form the Brite
Ranch in early February, returning eleven of them to Den Knight, the
ranch foreman. Cano's motivation to do so may well have been sin-
cere, but he also had practical reasons. For one thing, he needed supplies. His men were so desperately short on ammunition that Cano
offered one cow for every ten cartridges. The arms-for-cattle trade
had been a brisk and fluid, if illegal, exchange ever since the revo-

lution began and one that many ranchers along the border found too good to pass up. Although such smuggling was technically the domain of US customs inspectors because it was a violation of American neutrality, the US military had also taken an interest. "The temptation to ship ammunition to the other side is very strong," wrote Col. J. A. Gaston. "Several seizures of ammunition have been made during the past week."[22]

Meanwhile, Cano's loyalty to the Villistas began to waver. Although he had served directly under Pancho Villa for a while, Cano had an arrogance that grated on the "Lion of Chihuahua." At one point, Villa threatened to have him shot. Cano now seemed focused on reuniting with the Carrancistas and repairing relations with the Americans, but neither side seemed to be buying it. After Carrancistas attacked Cano's gang above Candelaria on 19 May, they scattered to the US side, where the US Cavalry almost intercepted them. Had it not been for the signal of his *avisadores*—a specialized class of bandits who sent messages by reflecting sunlight with handheld mirrors—they would have been killed or captured. Both sides were now squeezing Cano. With everyone after him and nowhere left to run, Cano knew the border had become too volatile, even for him. He retreated to his ranch in the mountains for a few months to let things cool off. He knew that by the time he returned, everything would be different.[23]

Meanwhile, battles continued to rage across the Rio Grande, constantly redefining the contours of the revolution. When not attempting to stop bandits, the US Cavalry provided temporary refuge for deserters and ravaged civilians. In October 1917, a group of Villistas attacked and routed a small Carrancista garrison across from Ruidosa, causing the men to flee across the river. Soldiers detained them and, when the coast was clear, ordered them to return to Mexico. About the same time, seventeen women—some of whom had been raped—fled from Barrancas to seek protection and medical treatment at Camp Ruidosa. Then, on 14 November, Villa and some sixteen hundred men, desperate to secure arms and ammunition, recaptured Ojinaga. But because the river port at Presidio remained closed to him, Villa departed after just a few days, leaving the town under the command of Gen. Alfonso Sánchez. Moving north from Ojinaga, Sanchez attacked Carrancistas under

Lt. Col. Jorge Meranga at San Antonio del Bravo, causing Meranga's entire garrison to desert to the US side until Sánchez's departure made it safe for them to return.[24]

Bandit Raids in the Sierra Vieja

I have worked from Eagle Pass to El Paso as a mounted inspector along the river, and the conditions have been bad and gradually getting worse all along. The bandits have crossed and stolen horses and cattle all along and raided ranches. . . . I have lost horses and mules from my ranch . . . only about 12 miles from the border.

—O. C. Dowe

Raids soon became more frequent. Whether due to the reduction of troops along the border or because Mexico was growing increasingly destitute, from the final months of 1917 through August 1919 there was a dramatic uptick in cross-border thefts and retaliations. On 30 November bandits stole a number of cattle from the J. F. Tigner Ranch near Indio, about twenty miles southeast of Ruidosa. Tigner, along with ranch foreman Justo Gonzales and twenty-one soldiers from Camp Ruidosa, pursued the bandits, following a clear trail that led to the hamlet of Buena Vista, only a half mile west of the Rio Grande. As they drew close, some 250 bandits concealed in the brush opened fire. The troopers dropped from their mounts and returned fire with far greater effectiveness—killing at least eighteen bandits before the others fled. During the battle, a soldier was wounded and another was killed, as were five cavalry horses. Assessing the casualties, soldiers found the body of Justo, whose hands had been tied and his head crushed with a rock. As for Tigner's cattle, they had already been slaughtered. One more problem remained—the soldiers could not locate Tigner. Second Lt. Leonard Matlack returned with his men three different times the next day before they finally found him concealed in brush, where he had been hiding in terror for almost two days.[25]

A few days later, a group believed to be Villistas fired on Troop L of the 8th Cavalry as they patrolled across from the riverside villa of Los Mimbres, about a mile and a half south of Buena Vista. Most shots fell hopelessly out of range, but one managed to hit a private in the leg. Cavalry units from Indio and Ruidosa, joined by a machine gun unit from Presidio, crossed the river to exact revenge. At Buena Vista, the troops fired some four hundred rounds, killing twelve men before burning both Buena Vista and Los Mimbres to the ground—a scorched-earth strategy that became standard procedure for future military strikes across the border. In explaining his actions, Matlack claimed he had been alerted by Texas Rangers, customs officers, and soldiers that both towns were known bandit hangouts. If he was looking for an excuse to strike, he'd finally found a good one.[26]

But the worst of the bloodshed lay ahead. In perhaps the most notorious raid in the region's history, bandits struck the Brite Ranch on Christmas Day in 1917. That morning, the ranch foreman's father, Sam Neill, sat drinking coffee at the headquarters when he saw six men galloping their horses toward the compound. Drawing close, they pulled out their guns and began yelling, "Matan los gringos!" With that signal, several dozen bandits suddenly appeared from hiding and began firing. Neill and his son grabbed their rifles and,

Captain Matlack (*reclining*) with Troop K, 8th Cavalry, and their canine mascot. Photo courtesy of the Harry Ransom Center, the University of Texas at Austin.

amid the intense gunfire, were able to kill one of the bandits. After some thirty minutes, they called an uneasy truce, and the raiders turned their attention to Brite's store, where they began loading sacks of food and clothing on horses stolen from Brite's remuda. Meanwhile, having no knowledge of the raid in progress, postman Mickey Welch drove up in his mail hack. Probably because his two Mexican passengers recognized the bandits, both were shot and killed. For his part, Welch was given the opportunity to escape, but being a hot-headed and obstinate Irishman, he refused to relinquish his mules to the thieves. It was a poor decision. After hanging Welch from the rafters, the bandits cut his throat. Finally, after more than five hours at their task, the raiders departed with some twenty stolen horses heavily burdened with loot from the store. By then, news had reached Marfa, which mobilized a motorcade of soldiers, ranchers, and Texas Rangers. Arriving late, however, they watched helplessly as the bandits escaped over the Rimrock and dashed across the Rio Grande to safety. The next day the cavalry crossed in pursuit, reportedly killing ten bandits and recovering some of the stolen goods.[27]

Citizens were outraged. The raids had grown more frequent, were taking place farther inland, and were becoming more brazen. Something had to be done. Posses began armed patrols of Marfa, Valentine, and nearby roads. Some two hundred ranchers, law officers, and townspeople met at the Stockmen's Club in Marfa, where they resolved to disarm the entire Hispanic population in Presidio and surrounding counties. Then, on a cold night in late January 1918, a number of ranchers and Texas Rangers, escorted by a detachment of soldiers from nearby Camp Evetts, descended on the tiny town of Porvenir, about thirty miles north-northwest of Candelaria. There they exacted a cold-blooded revenge in an event that would live in infamy.[28]

With some of the ranchers wearing masks to conceal their identity, they woke the residents out of their slumber and searched the houses for weapons or items stolen in the Brite Ranch raid. Despite finding nothing, the party separated out fifteen men and, marching them about a mile out into the desert, lined them up against a low rock bluff and shot them to pieces. The Rangers outrageously claimed they had been fired upon first from the brush and that the fifteen men had been killed in the crossfire. Troop G of the 8th Cavalry, who had been ordered to accompany the Rangers on their mission, knew better. Having patrolled the area for months and befriended the townspeople, they knew they were innocent. Even so, the soldiers were clearly complicit. Following the massacre, the terrified villagers fled to Mexico. A few days later the cavalry burned the village to the ground. The US town of Porvenir—whose name means "future" in Spanish—ceased to exist.[29]

In the back-and-forth cycle of strikes and counterstrikes, of attacks by one party followed by revenge from the other, on 25 March bandits retaliated by raiding the Neville Ranch. Located in the northern reaches of the Sierra Vieja along the Rio Grande, the ranch made for an easy target, though not a very appropriate one considering that Ed Neville had been a far better friend to border Mexicans than most Anglo ranchers. Arriving at the ranch late in the afternoon, Neville was greeted by his son, Glen, and his Mexican cook, Rosa Castillo, and her three children. After supper, Neville heard footsteps outside and looked in time to see several bandits take cover behind his chicken house. When they emerged again, they began shooting into the house, the bullets easily penetrating the walls. Neville ran outside to take cover amid a volley of gunfire that knocked his rifle out of his hands. Finally reaching a ditch, he hid until the cavalry arrived several hours later. He found his house a scene of utter ruin—the bandits having emptied it of everything that could be carried off. Worse yet, he found his son dying from a gunshot to his head, his face terribly disfigured from being beaten. Rosa lay dead on the kitchen floor from multiple gunshot wounds. Although one son bore a head injury, all three of her children survived.[30]

Gathering reinforcements, the US Cavalry crossed the Rio Grande in pursuit, following the trail of the bandits deep into the Sierra Pilares. After many miles, they camped for the night at Alto Puerto, just west of the mountain range. Setting out early the next morning, the company followed the treacherous trail the bandits had taken. But, burdened with their plunder, they could not flee for long. The trail eventually circled back to the town of Pilares—once the site of a Spanish presidio but now little more than a dozen squalid *jacales*. As the troopers approached the village, the bandits opened fire. Pvt. Theodore K. Alberts fell dead. According to official military reports, in the battle that followed, the soldiers killed some thirty-three Mexicans. Upon searching the dwellings, they found Mauser rifles, explosives, and much of the loot taken from Neville's ranch in addition to US mail from Mickey Welch's mail hack, thus tying them to the Brite Ranch raid as well. After retrieving everything of value, the troopers torched the town and returned to the US side after a two-day march of some seventy miles.[31]

The Neville Ranch raid was the last major bandit attack that caused loss of American lives. Still, smaller raids were far from over. F. A. Spence, who purchased Neville's ranch following the raid, lost more than one hundred head of cattle over the course of only four months. For the next couple of years the action trended southward, and for a time the raiding was centered just across from Ruidosa. By August 1918, Chico Cano was back and sharing command of fifty Villistas with Idelfonso Sánchez at Barrancas. At daybreak on 3 August, his troops attacked a garrison of thirty-five Carrancistas at Vado de Piedra. Catching the troop completely by surprise, Cano's men killed seven and took many prisoners. They also recovered several horses stolen from the US side. Sánchez, either in a show of goodwill or as an act of strategic diplomacy, returned them to the commanding officer at Ruidosa, along with one of O. C. Dowe's horses that had disappeared months earlier.[32]

The Villistas' return of stolen horses underscored the fact that much, if not most, of the horse theft along the border was due to the Carrancistas. It also gave the Villistas, in perpetual need of smuggled ammunition, a public relations card they could play. Reports indicated that Cecilio Estrella—a Carrancista captain stationed at San Antonio del Bravo—was carrying out official orders to capture horses from the American side for use by Carrancista troops. On 25 September, he and his men stole fifteen horses from the Cleveland Ranch, just above Pinto Canyon. Despite Estrella admitting to having taken the horses, he obstinately refused to return them.[33]

The Pinto Canyon Contingent

The revolution had edged into Pinto Canyon in the form of supply trains and minor thefts—O'Dell's sheep, Dowe's horses—and at least one refugee flight. But for the most part the bandit raids and the revolution had remained outside the canyon. That changed in early 1919, with the raid on the Nuñez Ranch. After James Wilson telephoned Camp Ruidosa with news of the theft, the bugler sounded the Call to Horse, and within twenty minutes Captain Kloepfer and forty men with three days' worth of provisions rode out of camp. They easily found a clear trail showing where the cattle had been driven into Mexico. Civilian scout Jim Watts led the company across the Rio Grande. About a half mile in, the unit was joined by the Pinto Canyon contingent, Nuñez and Wilson having been joined by Grover Sutherlin and Texas Ranger Nate Fuller. The combined forces followed the trail as it led southwestward through the river thicket, finally crossing a plain for some sixteen miles before turning due west into the mountains, where the country grew increasingly rough and rocky. As the sun set and light began to dim, the group entered a canyon, where they found the tracks of about ten mounted men who had joined the bandits.[34]

As darkness descended, the soldiers spotted the bandits, already beating a hasty retreat out of the canyon. About two-thirds of the way up the mountain, the bandits halted and opened fire. The soldiers scattered, one contingent galloping forward to secure a better

Scout Jim Watts with Troop M, 8th Cavalry, at Ruidosa. Photo courtesy of the Harry Ransom Center, the University of Texas at Austin.

position, another maneuvering into a flanking position. Meanwhile, one of the machine guns sprayed the front of the bandits' firing line, sighting only by the flashes of their rifles. Two of the bandits' horses reared and fell backward from a cliff, crushing their riders, while another bandit hurled obscenities at the Americans. After about fifteen minutes, the shooting stopped. Six bandits and as many horses were dead. Several more bandits had been wounded but escaped. Better armed and better trained, the soldiers emerged unscathed, without suffering a single casualty. Meanwhile, the stolen cattle had been discovered in a side canyon. With darkness almost absolute, the company retreated the way they had come, the herd of cattle driven by the Pinto Canyon contingent girded fore and aft by a platoon of soldiers.[35]

The horses plodded slowly across the trackless country as an approaching storm dimmed the starlight, the company proceeding only with the aid of occasional flashes of lightning. After about nine miles, the darkness was so complete that the company was forced to halt. The soldiers formed a large circle around the cattle and the pack train. And then they sat on their horses and waited. For two hours rain and hail pelted the company, water streaming off their hats onto their saddles and cascading down their horses' legs in rivulets to the waterlogged ground beneath them. When the rain finally stopped, they resumed their march—the horses' hooves alternately slapping and sucking at the wet earth. With the first rays of light, the soggy company crossed back onto US soil. In his subsequent report, Kloepfer praised the civilians. "Great credit is due Scout Watts and the Rangers and ranchers who acted as trailer and also took an active part in the engagement," he wrote. "Much of the time they followed the trail through rocky, hard ground at a trot and through the canyon beds at a gallop."[36]

Waning Days of the Revolution

Only a few days after the Nuñez raid, three of Chico Cano's men crossed into the United States about two miles south of Candelaria and proceeded cross-country to Pinto Canyon, where they stole several horses and saddles from O. C. Dowe and Grover Sutherlin.

Believing they had found the bandits' trail, troops from Candelaria and Ruidosa set out in pursuit. Crossing into Mexico, they followed the horse tracks northward past San Antonio del Bravo until the trail split, the tracks of six horses turning west, the remaining three heading north. The troops split and followed each branch separately. The western trail continued westward for about ten miles before turning northwest and finally eastward, heading in the direction of the riverside village of El Comedor. Nearing the village, the soldiers had entered a dense thicket when they were suddenly fired upon. They had stumbled upon Chico Cano and his men taking their afternoon siesta. In the firefight that followed, the soldiers killed five bandits and wounded two more who escaped, one of whom was the elusive Cano. As the soldiers recovered six horses and two mules, it was clear Cano had given up trying to win over the Americans. But given his new commission, he may have just been following orders. He was now a captain in the Carrancista army, in command of the *sociales*—the Mexican border patrol.[37]

Things quieted for a while, the days resuming a routine manner until a climactic event in the summer of 1919. Following the end of World War I, the fledgling air corps was used as the 1st Surveillance Group of the US Army Air Corps. From Marfa, the pilots began conducting regular flights eastward as far as Boquillas and westward to near El Paso, patrolling the border. Although they were a potent deterrent to banditry, the planes were notoriously undependable. Derided as "flying coffins," the planes had DH-4 engines that tended to overheat, and the landing gear often got snagged in forced landings in the rough terrain of the Big Bend. Mechanical and design flaws left little room for error.[38]

On Sunday, 10 August, during a routine flight, Lt. H. G. Peterson mistook the Río Conchos for the Rio Grande and followed it more than a hundred miles deep into Mexico. Then mechanical failure set in—the rigging on one of the wings collapsed, causing the plane to go down not far from Falomir, Chihuahua. The pilot and his gunner survived the emergency landing, but their luck ended there. After an exhausting march across the desert wastelands and nearly drowning in the waters of the Río Conchos, the men were captured by a bandit gang under Jesús "Pegleg" Rentería. Secreting his prisoners outside San Antonio del Bravo a week later, he sent word to the cavalry across the river: bring $15,000 by the end of the day following or the aviators would be killed.[39]

The military had no way to come up with the money in time. Instead, ranchers attending the Bloys Cowboy Campmeeting near Marfa raised the money within minutes, whereupon it was rushed to the border. After tense negotiations, Rentería finally agreed that Captain Matlack could make two trips across, taking half the money each trip in trade for one aviator. The first exchange went without a hitch, but during the second, Matlack—overhearing the bandits' plot to kill them both—pulled the aviator up behind him in the saddle, told the bandits to go to hell, and fled across the river by a different route. Early the next morning, Matlack led five cavalry troops and one machine gun troop across the Rio Grande in pursuit. The bandits' trail passed by Ojo del Carrizo, where the soldiers captured four men wanted for various crimes in the United States. But in a breach of protocol, Major Yancey from Camp Ruidosa turned the men over to Texas Rangers who were accompanying the troopers. Once the soldiers were out of sight, the Rangers summarily shot them—an incident that resulted in Yancey's court-martial.[40]

The trail continued deeper into Mexico. On their third day of pursuit, as the troops neared Coyame, a DH-4 dropped a message from the commanding officer, Colonel Langhorne, ordering the troops back. Defying orders, Major Yancey pressed on to the town. Once there, however, he learned that twelve hundred Carrancistas had been dispatched to engage the American invaders. Narrowly averting what could have been a slaughter, they wisely turned back. After six days of fruitless pursuit, the troops crossed to the US side with little to show—none of the money having been recovered and none of the bandits having been killed. By some accounts, Rentería would live another fifty years—reportedly in style.[41]

Following what amounted to the last punitive expedition into Mexico, things began to quiet down. Zapata had been assassinated at Chinameca in April, and Pancho Villa's Army of the North suffered defeat at the last Battle of Juárez, in June, before being trounced in Durango, forcing him into retirement. The revolution had lost its most influential leaders, causing momentum to wane. In late September 1919, ten months after the end of the Great War, the 5th Cavalry arrived to relieve the 8th. With their departure, the number of shooting incidents and thefts reported along the border declined markedly. Greater stability in Mexico certainly helped, but the removal of the 8th Cavalry may have been the most significant factor.[42]

Meanwhile, the War Department began a major push to improve military roads and living conditions in border camps. "For the last five years [soldiers] had been in tents and shacks, resulting in a lowering of morale, as well as being detrimental to enlistments," reported the captain of the Quartermaster Corps. In 1919 the secretary of war authorized nearly $4 million for construction projects along the Mexican border, including the camps at Ruidosa and Candelaria. By the time improvements were complete a year later, Camp Ruidosa boasted four officers' quarters, two enlisted men's barracks, a lavatory, a mess hall, three hay and grain sheds, four stables, two blacksmith shops and storerooms, one power house, one rolling kitchen shed, and one tower and water tank with a ten-thousand gallon capacity.[43]

But the left hand didn't seem to know what the right hand was doing. The War Department had been too quick to act, for the improvements came just before the army began to close the border outposts. In 1921 the adjutant general wrote, "Very little trouble along the Mexican border, necessitating the use of troops, occurred during the last fiscal year. . . . As the results of the comparative quiet that prevailed along the border during the past fiscal year, many outposts and smaller posts were abandoned." Over the next eighteen months, troops on field duty gradually consolidated at Camp Marfa. In October the *Cavalry Journal* reported, "All outposts in the Big Bend District have now been withdrawn to Marfa, so that now all troops of the regiment are together once more. Frequent patrols are sent out from Marfa, covering the river from Glenn Springs to Candelaria."[44]

Even as troops withdrew, Chico Cano remained the most wanted outlaw in the region, free and at large. At a Christmas dance in Candelaria, Cano appeared out of the darkness and sat his horse, observing the festivities. After being spotted, he rode away on his galloping mount, firing his pistol into the air as he went. The *Marfa New Era* version of the story reflected his mystique: "[Chico Cano] suddenly appeared on his fiery grey horse. His huge sombrero and great bandoleers were plainly seen as he rode several times in front of the store of the Mexican who last April stamped on the faces of four of his men after they were dead." Apparently Cano was still gunning to avenge the deaths of his comrades killed during the ambush by US troops. A month later, just downriver, Carrancistas under Cano's command robbed the T. D. Baldwin Ranch store in Ruidosa of eighty dollars' worth of provisions and later threatened to kill Baldwin if he didn't send over some tobacco. When Baldwin failed to comply, a bandit returned a few days later armed to the teeth and attempted to coax Baldwin out. Two Texas Rangers happened upon the scene and promptly arrested the outlaw. But they had ridden only a few miles before Cano's men suddenly opened fire. The Rangers shot back, but in the melee their prisoner escaped.[45]

Incidents of theft and violence continued to dwindle as Mexico regained political stability and US troops pulled out of the border region. But after a decade of bloodshed and instability, Mexico had been left in shambles. More than a million people had lost their lives. If the revolution caused distress among the American citizens and soldiers along the border, it paled in comparison with the suffering endured by the civilian population of Mexico. The chaos that had allowed the strong to prey on the weak made life across the border a near constant horror. The repeated thefts, destruction of property, murders, and rapes exacted an awful toll on landed gentry and peasants alike, enough that by the time US troops finally pulled out of the region, most of the homes and farms across from Ruidosa and Candelaria had been abandoned. A proposal to limit the opening of the port at Ruidosa to one day a month instead of weekly caused the town's residents to protest that it would cause even greater hardship to the few who remained: "The people on the Mexican side of the river are now reduced to the direst poverty and have no means of procuring the necessities of life except by the sale of chino grass . . . nor have they any other place to buy their supplies except at this place." If denied this port, they claimed, the Mexicans would be left to either steal or starve to death.[46]

The seclusion of Pinto Canyon was probably its saving grace during the decade of the Mexican Revolution. In the last years of the revolution Camp Ruidosa helped by serving as a bulwark against bandit incursions. Aside from the Nuñez Ranch Raid and a few minor thefts of horses and livestock, the role of Pinto Canyon was primarily as a conduit for supplying the US Cavalry. For the larger Sierra Vieja region, however, the story was far different. Other than the major raid at Boquillas and Glenn Springs, the vast majority of banditry in the Big Bend centered on the remote hinterlands of the Sierra Vieja. The resulting violence triggered a protective cordon of US military camps far more active than anywhere else in the region. But in spite of the depredations and bloodshed that struck fear into ranchers and their families, by far the greatest suffering was endured by the Mexican people themselves. By the end of the revolution, the tiny towns strung along the right bank of the Rio Grande had been laid to waste.

5

El Camino Afuera

Adjustments and Departures

I held him in awe—utmost respect. He was quiet but always smiling.

—José Manuel "Joe" Prieto, grandson of José Prieto

JOSÉ PRIETO stepped out of his house into the predawn light, closing the door gently behind him. Down at the corrals, he opened the gate and stepped inside, where two mules stood waiting. They nuzzled him as he poured a coffee can full of oats into the half-barrel feed trough wired to the railing. As they ate, he brushed each in turn, then saddled the large bay and put a packsaddle on the other. He loaded four salt blocks—two per side—in the panniers hanging off the wooden forks of the packsaddle and secured the straps. Taking the lead rope in one hand, he swung into the saddle and headed south on a trail that led back into the mountains as the first rays of light struck the rocky summit of Chinati Peak, casting a broad, long shadow across the canyon below.

An hour later José topped out on a bench on the northside slope of the mountain and stepped down into a herd of a hundred Angora goats. Unloading one of the salt blocks, he started for a clear spot at the edge of the bench, where the action of tiny hooves had left the ground completely bare. As he walked, the goats closed in around him. Bending over to set the block down, a kid goat lipped the edges of his sombrero, staring through horizontal pupils. José chuckled as he placed his hand gently upon the goat's knobby head. A broad smile spread across his face as he walked slowly through the herd, his hand grazing the mohair on their backs as he went.[1]

The year was 1928, and José Prieto was at the top of his game. It had taken him most of his sixty-two years, but he had slowly and steadily assembled a twelve-thousand-acre ranch in the shadow of the Chinati Mountains and become an honored member of the community below the Rimrock. The son of a hardscrabble desert sharecropper, José had been born into a life of labor. And though his lineage was not far removed from peonage, he had risen far above the constraints of his past, enjoying social and financial success beyond his wildest imagination.

During the first decade and a half of the twentieth century, four families arrived in Pinto Canyon determined to make it their home. Aside from the brief drought of 1909–10, rainfall had been ample and the market strong—a period sometimes referred to as the Golden Age of Agriculture. The next fifteen years would prove much more variable and volatile, with droughts in both 1916–17 and 1921–22, as well as wild market fluctuations driven by World War I. The postwar recession of 1921 was the most precipitous decline of agricultural commodities ever seen in a single year. But prices would soon rise to new heights as the Roaring Twenties witnessed unprecedented economic prosperity across the country. For ranchers in Pinto Canyon, it was a period of adjustment and—for many—departure. Although there would be new arrivals—such as O. C. Dowe—they were more than offset by the departures of the Wilson and Sutherlin families. Of them all, only José Prieto would remain. By the end of the 1920s, on the cusp of the Great Depression, José would find himself at the pinnacle of his economic success.[2]

A Customs Inspector for Pinto Canyon

In the summer of 1914, thirty-one-year-old O. C. Dowe, a US mounted customs inspector, rode slowly across the McGee Ranch just north of Capote Creek. He and his men had been sent on assignment out of Valentine to apprehend several Mexicans suspected of smuggling. Topping a small rise, Dowe's right hand dropped instinctively to

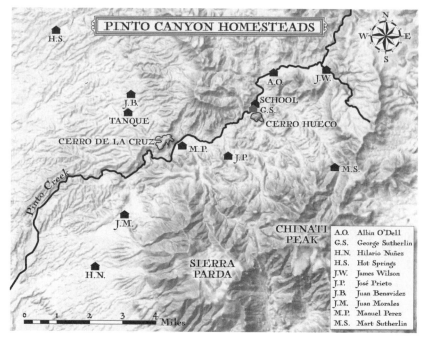

Map showing locations of Pinto Canyon homesteads. Map by Tish Wetterauer and David Keller.

rest on the grip of his revolver after seeing something move in the distance. Through his field glasses he watched tall, slender poles moving among the brush, but could see no men. Finally, he saw a head rise above the brushline. But the head was covered by a bonnet rather than a sombrero. The poles were stalks of lechuguilla, and the figures were only children playing. He brought the glasses down and exhaled. Ten minutes later, Dowe and his men pulled up short of the yard of Bill and Dixie McGee and shouted a greeting. When Bill came out, Dowe negotiated the sale of a calf to feed them for a few days while they rested their horses. It was late in the day and clouds had gathered, darkening the sky. Rain spattered the parched earth as they drove the calf back to their camp for the night.[3]

The next day, the men butchered the calf and jerked the meat—hanging thin slices to dry in the desert sun. Dowe took a quarter of the carcass back to the ranch house to give to the family. It was only then he noticed Millie Wilson—one of the children he had seen playing the day before. At nineteen, she was older than the others but still looked young for her age—young, except that her dark eyes had a deeper knowing about them, a certainty of her place in the world. Dowe looked at the girl, sensing in her a fortitude born of years on the ranch, of the long days riding horseback through the whitebrush, of gathering goats and building corrals and mending pasture fences. Nor could he help but notice near the waistline of her dress a discolored band where she sometimes wore a holster.[4]

A few weeks later, Dowe rode up to the Wilson house to ask about a section of land James had for sale. Located just northwest of Cerro Hueco, it was a dry and rocky piece of land, but it was only part of a bigger plan he had in mind. Within a month of buying that section, Dowe added three more to his holdings before purchasing the George Sutherlin homestead on Pinto Creek. With a total of five contiguous sections bordering Wilson on the west, he now had a place of his own—one conveniently close to Millie Wilson, in whom he'd taken an interest. Theirs was not a particularly drawn-out affair. Within a short six months, Millie and Dowe were married at the

O. C. Dowe and Millie Wilson Dowe on horseback in Pinto Canyon around 1915. Photo courtesy of the Marfa Public Library, Marfa, Texas.

Wilson Ranch. Dowe later recalled, "It was a pretty big wedding for that part of the country." There had been a total of four attendants.[5]

Orin Curtis Dowe was born in Cuero, Texas, southeast of San Antonio, in 1883. He completed the eighth grade before deciding he'd had enough schooling. He caught a train west, got hired on the Cox Ranch near Organ, New Mexico, and spent the next three years working as a cowboy. On his first trip home, in 1907, he stopped in Del Rio to see his uncle, Luke Dowe, a US deputy collector of customs at the crossing to Villa Acuña, and the elder Dowe convinced his young nephew to fill out an application to be a mounted customs inspector. Before leaving Cuero, young Dowe received his acceptance letter. His first assignment was in the rough river outpost of Lajitas, forty-five miles southeast of Presidio, where he would replace the previous customs inspector, who had been ambushed and killed by Mexican smugglers.[6]

Dowe's primary job was to enforce the payment of import duties, which meant patrolling the river and apprehending anyone illegally bringing horses, cattle, or liquor across from Mexico. It was sometimes dangerous but most often monotonous work. With his partner, Dowe rode across the *mesillas* and *bosques* (terraces and river

Dora Dowe with baby ducks at the old Wilson Ranch in Pinto Canyon around 1921. Photo courtesy of the Watts family collection

thickets) cutting for sign. After arresting a smuggler, it was a long two-day horseback trip to Alpine to have him processed, making for exceedingly uneasy travel and a long night of fitful sleep.[7]

In 1909, Dowe married Delia Shoemake of Marathon, and she gave birth to their daughter, Eva. But in 1911, Delia fell sick and died, forcing Dowe to place Eva with relatives. When he finally returned to work, he received a new assignment in Marfa and Valentine and ultimately in the mountains and canyons below the Sierra Vieja Rimrock. Now married to Millie, he was finally able to bring Eva home to live with them after a five-year absence. Over the course of the next three years, Millie would give birth to two daughters—Nell in 1916 and Dora in 1918. When not on assignment, Dowe tended his ranch and a small herd of cattle. But the days were long ones for Millie, with a five-year-old and two infants and the many chores that fell upon her in his absence. That, coupled with the growing threat of bandits, encouraged her to stay at their rent house in Marfa for company as well as safety.[8]

Growth of the Prieto Ranch

Even as the revolution raged across the border and banditry loomed closer, José Prieto was enjoying greater financial security than ever before. With his new home and the state land parcels being steadily paid down, José slowly started to expand his ranch. In 1917, he purchased a section on the western slope of Sierra Parda originally filed on by his eldest son, Victoriano, in 1915. Over the next three years José purchased additional land from James Wilson, O. C. Dowe, and downstream neighbor Hilario Nuñez so that by 1920 he had acquired five additional sections, bringing his total holdings to ten. Now with a solid block of six contiguous sections around his homestead and two sets of outlying contiguous sections, his ownership looked more like three ranches than one and may have made for difficult livestock logistics. But it was an inconvenience José would rectify in the years to come.[9]

José also had another long-standing issue that required his attention. Despite having lived in the United States for fifty years, owning land, paying taxes, and raising a family, José had never gained US citizenship. Whether a matter of practicality or pride, in 1922 he

José Prieto's petition for naturalization, filed in 1922. Courtesy of the Presidio County District Court, Marfa, Texas.

finally filed a petition for naturalization in the district clerk's office at the Presidio County Courthouse. In his petition he avowed he was neither an anarchist nor a polygamist and that he had resided in the United States for the previous five years. Either the bureaucratic wheels were slow to turn or he was unable to schedule the required interview, for it took two more years before his citizenship was finally granted. But in 1924, at age fifty-eight, more than half a century after immigrating to the United States, José became a citizen. If he voted in the presidential race between Calvin Coolidge and John Davis that year, it would have been his first ballot cast.[10]

By this time, José's children were beginning to start families of their own. In 1924, his oldest, Victoriano, married Martina Jaquez at El Corazón Sagrado de la Iglesia de Jesús in Ruidosa, about the same time their first child, Francisco, was born in Barrancas. Two years later, at a ceremony in Marfa, José Jr. married Francisca Vasquez—

daughter of Miguel Vasquez, whose family may have immigrated to the United States with the Prietos in 1874. Then, in 1928, Gregorio married Pilar Nuñez—the eldest daughter of Hilario Nuñez—at the church in Ruidosa. Victoriano and Martina soon moved to Ojinaga, and José Jr. had no interest in ranching. Gregorio did, however, and that may have been José's incentive to purchase nine additional sections—including a house built by Juan Nuñez about a mile west of the Prieto house. Wanting all his sons to have ranches of their own, José offered the house to Gregorio and Pilar and set them up—at a reasonable rate—with their own starter herd of sheep and goats.[11]

It was probably no coincidence that the year José doubled the size of his ranch was also a pinnacle for mohair. The market had been uncertain since 1910. Starting at a low of around twenty-six cents per pound, it rose steadily to fifty-eight cents a pound before the crash of 1921 drove prices to a low of seventeen cents. The market rose out of the slump in fits and starts throughout the 1920s until reaching an all-time high of seventy cents per pound in 1928. Although wool also took a nosedive in 1921, it did not soar to the heights that mohair did during the 1920s. With José's business sense and a lifetime of thrift, he had been able to save enough between 1922 and 1928, when prices were good and rainfall was ample, to purchase the additional sections of land in one fell swoop for $2.39 an acre from S. V. Field, a speculative buyer from San Patricio County.[12]

With the new purchase, José had increased his holdings to nineteen sections, or 12,160 acres of land. By fate or by design, they would also be his last. Still, the added sections went a long way toward consolidating his ranch—in effect combining the three separate holdings into one large, contiguous block. There remained three sections amid the south-central portion of the ranch that did not belong to him, but, being mostly on the northwestern flanks and slopes of Sierra Parda, they would not have offered as much as his lower acreage. In effect, José had slowly acquired most of the land along the northern and northwestern flanks of the Chinati Mountains, which ranged from thirty-six hundred feet above sea level at Pinto Creek to seven thousand near Parda's summit. His ranch also encompassed a diverse array of landforms and had enough space to allow José to split the operation into seasonal ranges, with the lower elevation areas grazed mostly during the winter months and the higher ranges during the summer months. As a localized form of small-scale transhumance—annual cyclical migrations from lower to higher elevations—José was engaging in an age-old practice common to herders across the globe.[13]

The Wood and Sutherlin Ranch

By 1918, Mart Sutherlin was enjoying the greatest financial success of his life. Although the drought of 1916–17 forced him to sell nearly half his livestock, the aridity was also a likely reason the neighboring Pool Ranch came up for sale. With the combined income from the sale of his livestock and probable inheritance following Francis Sutherlin's death, Mart was able to expand his ranching interests far beyond the confines of Horse Creek. Forming a partnership with his son Grover and a neighboring rancher, S. T. Wood, a recent arrival from Crockett County, allowed Mart to ranch on a scale grander than ever before. Their new spread consisted of thirty-six sections of land spanning the eastern half of the Chinati Mountains and lapping downslope along the length of Oso Creek, a tributary to Cibolo Creek. "The ranch is one of the largest in the Big Bend country and will be improved by its new owners," claimed the *El Paso Herald*. But one problem remained: "On the ranch is a large orchard and this year hundreds of bushels of fruit are being eaten by hogs. Arrangements have been made by Mr. Sutherlin to gather what fruit has not been used up."[14]

Of Mart's six sons, Grover had been the most active in ranching and would remain in Pinto Canyon the longest. The first of Mart and Eliza's children, Grover was born in Del Rio in 1885. Moving with his itinerant family, he helped his father on the ranch until the age of twenty-nine, when he married twenty-year-old Elsie Bell Petty during the same year he purchased six sections from Albin O'Dell before the latter fled the canyon. Grover turned around and sold all but one section at a nice profit to J. F. Parker and filed on four state sections adjoining his father's ranch. Three years later he filed proof of occupancy on his land and put them up as his part of the new partnership.[15]

The founding of the Wood and Sutherlin Ranch came at a good time, for the years following the drought were exceptionally wet.

The nearest weather station reported eighteen inches of precipitation in 1918, almost 30 percent above average. The *El Paso Herald* reported, "Good rains have fallen on the Pinto Canyon section, and crops and grass are doing well. Cattle are in good shape and farmers throughout the county are expecting record crops. The Mark [*sic*] Sutherlin orchards in Pinto Canyon are reported especially fine." Indeed, Mart was doing well enough financially to invest beyond rangelands surrounding the Chinatis. Along with his son-in-law, L. L. Hay, he also purchased seven irrigated farming tracts near Socorro, New Mexico, for $26,000.[16]

Grover's Burden

Only two weeks after Mart secured his farms in Socorro and was poised to reap greater profits than ever, tragedy struck. On 27 October, Mart fell deathly ill from influenza. Ten days later, he would be dead. His illness was part of a much larger pandemic ravaging the global population, one that ultimately left more than fifty million dead worldwide. The first wave of "Spanish influenza" appeared in Kansas in the spring of 1918, followed by much more deadly second and third waves in the fall and winter of 1918. It was the second wave that struck Mart. The contagion had followed the railroads across the country, arriving in the southwestern United States in October. The first death from flu in Presidio County was registered on 14 October. By December, at least thirty people had died in the county. But those were only the deaths that had been registered. Dozens more went unreported, especially those of the poor and illiterate living in cramped, primitive conditions along the border and among whose ranks the disease exacted the heaviest toll. According to church records, between the bloodshed of the Mexican Revolution and the Spanish flu, an estimated 141 people died in Candelaria alone—so many, in fact, that they had to be buried in a mass grave. All were believed to be of Mexican descent.[17]

Flu epidemics typically cause death in the very young and very old. This time it was different, disproportionately targeting young, healthy adults who were around twenty times more likely to die from this strain compared to previous epidemics. The reason remained a mystery for decades, although recent research indicates it may have been caused by an overreaction of the body's immune system—that the body's own frantic effort to fight the virus ultimately killed the host. Severe infections of the respiratory tract that accompanied the viral infection suggest opportunistic bacteria were ravaging the lungs, resulting in an accumulation of bloody fluid that suffocated the victims.[18]

The progression of the illness was rapid. As Mart's death grew imminent, he summoned an attorney to draft his final will and testament. In it, he elected Grover to be executor, granting him "title to all of my estate, both real and personal, until my youngest daughter, Vera Sutherlin, becomes twenty-one years of age."[19] Thereafter, the estate was to be divided equally among Mart's eleven surviving children. It seemed a reasonable request. The problem was that Vera was only twelve years old; there was still nearly a decade to wait. In the meantime, Mart directed Grover to "keep my ranch well stocked with livestock according to his best judgment and to buy out of sales of such livestock only such animals, according to his best judgment, as are needed in keeping up and improving such livestock." Two days after signing his will, Mart died.[20]

Grover's duty in fulfilling his father's wishes was an incredibly tall order. At thirty-three years old, he had a wife and family of his own to care for. Along with their two toddlers, he and Elsie also had his seventy-six-year-old grandmother, Nancy, and his youngest two siblings—Mart Jr., age fifteen, and Vera. And in gaining his father's assets and operating capital, he also gained his debt. Mart was estimated to be worth about $42,750, but nearly all of it was tied up in land and livestock. He still owed on almost everything. In the final calculation, Grover was left with less than $5,000 in total assets. It was precious little to work with. And without his father, Grover was now totally on his own.[21]

In July 1919, the second shoe dropped when the court ordered the sale of the Wood and Sutherlin Ranch. In addition to the thirty-five-section ranch, the partnership owned 956 head of cattle, 23 horses, and 5 mules and had $656 in the bank. It also had indebtedness of $43,230 for livestock and land plus $12,192 owed Wood and $2,564 owed the M. D. Sutherlin estate. Mart and Grover had owned one-quarter interest each, and Wood owned the balance. It was a web of ownership and debt that was difficult to untangle, and, in any case, Grover couldn't afford the interest. After Wood brought suit in district court to recover debt, the judge ordered all assets of their partnership sold and divided. It was yet another blow for Grover. In addition to losing the Wood and Sutherlin Ranch, he had to give up the four homestead sections that had been his contribution. The value of the family ranch—the six sections originally homesteaded by Mart and four more subsequently acquired—was barely enough to support his family. The partnership with Wood had promised Mart and his family greater financial security. Now that chance was gone and, with it, Grover's ability to prosper.[22]

Even so, Grover persisted. Running the thirteen-hundred-odd sheep remaining on his father's ranch under his own Spear 4 brand, he dug in for the long haul. Though relieved of the obligations of the partnership, he had long days ahead. Spanning rolling grasslands and broad stream channels, the Wood and Sutherlin Ranch stood in stark contrast to the broken country below Chinati Peak. Aside from the valley of Horse Creek, there was little level ground—and what there was of it was rocky and brushy. Predators remained a constant problem. With every nook and cranny providing a place to hide, they were almost impossible to control. As little more than fluffy treats to eagles, coyotes, wolves, and mountain lions, sheep required constant protection. For the most part, that protection amounted to a war against predators. In 1916 alone, Mart, along with a couple dozen others, had been paid bounties for 123 wolves and 39 mountain lions in addition to a slew of lesser varmints.[23]

In 1920, the weight on Grover's shoulders began to lighten when Vera moved in with her sister Annie in El Paso to attend school. The following year, his grandmother Nancy died and was buried in the family cemetery along Horse Creek, in a fenced plot that also contained the grave of May Sutherlin, Grover's younger sister, who had died in 1914 at the age of twenty-two. He was left now with only Elsie and their two children—George, age six, and Alice, age five. A ranch house that had once been filled with his parents and siblings was quieter, and perhaps lonelier. But there were also fewer mouths to feed and fewer personal concerns to attend to, which likely eased his burden.[24]

It certainly helped, too, that 1920 had been an exceptionally wet year across much of the Big Bend, with the nearest weather station reporting close to twice the average rainfall. Even so, the moisture seemed self-limiting. The next two years were increasingly dry, with 1922 bringing only about half the average rainfall. Still, Grover sheared six thousand pounds of wool that year, selling the better part for forty-eight cents per pound—far above the average price for shorn wool, which stood at only twenty-seven cents per pound. After the United States entered World War I and launched the "raise more sheep" campaign, prices doubled. But, following the signing of the Armistice, they quickly returned to prewar levels. Without the heightened demand for uniforms and blankets, wool was again at the mercy of common market fluctuations.[25]

The Pinto Canyon Cattle Company

Meanwhile, just down the creek from Grover, at the homestead built by his uncle George years before, O. C. Dowe had determined to shift his base of operations. Only a month after filing proof of occupancy, he sold the five-section ranch to Lela O. McKinley, a widowed absentee owner from Pearsall, Texas. With the proceeds, he purchased the six-section Wilson Ranch from his father-in-law for $3.80 an acre and 118 head of cattle. James Wilson was probably eager to sell the ranch to Dowe, keeping the fruits of his thirteen years of hard labor on his Pinto Canyon ranch within the family. It was also an excellent time for James to get out of the business, just as it was a bad time for Dowe to reenter. As a result of the drought of 1921–22, ranchers were dumping livestock on the market, driving beef prices down by 44 percent and wool and mohair prices by 70 percent in the span of only three years. "I lost my shirt," recalled Dowe. "I had given $85 for the cattle and they went down to $10 [per head]." In 1921 he sent 1,302 pounds of wool to the San Angelo Wool Growers' Central Storage Company along with a note: "Please sell this wool for me as soon as possible, or can you let me have some money on same?" With the market collapsing and the weather growing drier, Dowe was increasingly in need of cash.[26]

The drought and economic crash were probably the motivation for Dowe to pursue a partnership with three businessmen from Detroit. Their introduction likely came during a trip to Michigan in 1919, when Dowe received a summons to appear in Henry Ford's libel suit against the *Chicago Tribune*. During the Mexican Revolution, newspapers criticized the automobile magnate for his pacifist stance. When the editors charged him with being an anarchist, Ford sued, and Dowe was one of many witnesses called to testify about conditions along the border. Courting his new friends, Dowe guided hunting trips in the Big Bend and Mexico and occasionally held holiday blowouts that left him penning apologies for his drinking. Apparently undeterred, in 1922 Joseph E. Boyer, Harry Andrews, and James S. Smithwick formed the Pinto Canyon Cattle Company, with Dowe as the organizer and operator. For his part, Dowe entered the partnership with his six sections of land and improvements thereon, along with 115 cows, 15 yearlings, 90 calves, 25 saddle horses, 10 mules, 100 goats, 140 head of sheep, and one Dodge automobile— for a total value of $21,000 minus the debt owed to James. The remaining $7,500 equity became Dowe's share, which was matched by each of his three partners. With their pooled $30,000 of operating capital, the Pinto Canyon Cattle Company was born.[27]

Dowe immediately began scouting for more land in which to invest. He entertained an offer on the Chinati Ranch from T. D. Wood. But at $6.00 an acre for patented land and $8.50 an acre for school land—about double the going rate—the price was far too steep. "It is one of the best ranches in the country, but he has the land up awful high," Dowe wrote his partners.[28] He believed they could do better in Mexico, where land prices remained depressed following years of revolution. "From what I can see, we will have to go into Mexico in the near future to ever be able to make any money out of the ranch business," he wrote.[29] But with conditions dry and getting drier, there was little to offer hope in the short term. By March, the situation had grown desperate. "If it don't rain here in two months 'tis sure going to be hard, everybody is feeding," he complained.[30] It was an untenable situation, and he'd been left few options. In a letter to Lela McKinley, the widow who'd bought George Sutherlin's ranch, he wrote, "I am going to turn my place back to Mr. Wilson and 114 head of cattle in about ten days. I can't see how I can come out, as I owe so much, and can't even pay my interest. If you should want my ranch here too, I will deed it and 114 head of cattle to you free, as I would rather let you have it than to turn it back to Mr. Wilson. He and [I] never could get along, and I want to get away."[31]

But Lela McKinley didn't accept Dowe's offer, so on 15 April he deeded the ranch back to James Wilson and set his sights on bigger and better things. After resettling briefly on Capote Creek some twenty-eight miles to the northwest, he moved to Coyame, Mexico, where he would ranch for most of the next twenty years. The transfer likely didn't endear James to his son-in-law, for regaining his ranch during a drought and depression was the last thing James needed. As late as September, the *Marfa New Era* wrote that James was in town to spend a few days with his family, and he reported that his range was "badly in need of a good rain." At sixty-one years of age and having struggled with a painful disability his entire life, he was ready for a break. He had enjoyed one, briefly, for the twenty-two months Dowe had run the ranch. But now he was faced with resuming the drudgery he thought he had left behind forever.[32]

A Dose of Levity in Ruidosa

Meanwhile, a celebratory air was building down by the river. Election Day pulled people from the far-flung ranches between Candelaria and Indio and the rim to the river—in some cases, as much as a day's travel away. It was a time when people set aside their daily routines and became part of a larger borderland community. The day before, families had gathered for a barbecue at the old cavalry barracks above the town, after which some napped in the shade of cottonwoods, while others fished in the river to await cooler temperatures and the rodeo that would follow.

Livestock for the rodeo was supplied solely by donation. The Jacquez and Vasquez families provided wild cattle and goats, while José Prieto and Wert Love supplied bucking horses. Among the events was a match roping between "old-timers"—Grover Sutherlin and Gallie Bogel—and the younger Wert and Alonzo Love. The "elders" (Grover was now forty-one years old) handily beat their juniors. But the competition was soon eclipsed, the crowd roaring with laughter, as one of the boys jumped on a wild cow and, facing backward, emphatically fanned its flanks with his hat as it ran. Later, revelers danced at the old barracks and passed bottles of sotol—the moonshine of the borderlands. As the hour grew late, some staggered off in the night when their feet could no longer keep the beat, others dancing until the dawn. In the morning, they stood bleary-eyed before the schoolhouse awaiting their turn to vote before heading home to nurse their hangovers.[33]

Wilson's Departure

There had probably always been tension between James Wilson and O. C. Dowe. It might simply have been a personality clash, but James was known to be difficult. His disability had long been a festering frustration, the pain a constant, unwelcome companion. His advancing years would not have helped and his financial situation had grown dire. James had done well enough to support his family, well enough to improve his ranch, well enough to purchase a few lots in town, but he had never been able to set money aside or expand the ranch beyond the original six sections. To resume ranching, he needed operating capital. The ink on the deed had scarcely dried before he secured a loan against the ranch at the Marfa State Bank for $7,300, which allowed him to squeeze by for the next two years. But it was not enough. In 1924, he upped the ante to $10,000 by adding several town lots as additional collateral, which got him by for another two years. By 1926, however, he was even further in debt. This time he put it all on the block: the ranch, the town lots, and "all my stock," amounting to eleven hundred Angora goats. For that, he received $13,000. He was able to secure a loan extension in January of the following year, as well as another in March. But by then James had run out of options. If he didn't find a way out soon, the bank would be in the ranching business. Fortunately, by September 1927 he'd found a buyer, one J. H. Young, who purchased the ranch for $4.70 an acre—a decent price, all things considered. With that, James was able to leave Pinto Canyon behind for good. Still, for his eighteen years in the canyon, everything earned by the sweat of his own brow and the pain of his disability, after paying off his loan, he finally retired to his Marfa house with no more than $5,000 to show for his years of effort.[34]

If James had been grouchy toward his wife and daughters during their years in Pinto Canyon, the added financial strain toward the end probably didn't help. But now Dora was left alone with him, as all three of their daughters had moved on. Millie had been the first, but Mamie followed two years later when she married James Watts (son of army scout Jim Watts) in 1917. Ora, the youngest, was the last to leave, marrying William McDaniels in 1926. Whether a result of his temperament or not, a year and a half following the sale of the ranch, Dora filed for divorce. On 6 February 1929, the district judge agreed "that the allegations in plaintiff's petition are true and . . . that the plaintiff, Dora E. Wilson be and she is hereby divorced from the defendant, J. E. Wilson."[35]

The divorce may have been the final nail in the coffin. James remained in Marfa for about a year before renting an apartment in Sanderson. Perhaps James sought a fresh start. More likely he went there to die. Now sixty-nine years old, divorced from his wife, likely estranged from his children, nearly broke, and with only the company of his endless pain, life had become too much to bear. At nine thirty in the morning on 6 May 1930—scarcely four months after his arrival—James Wilson shot himself in the head with a .38 caliber pistol. "The barrel of the pistol had been placed in the mouth and the bullet came out at the top of the skull," reported the *Sanderson Times*. A service was held in Marfa two days later. "He was well known all over this section and a large number of friends attended the funeral," reported the *Big Bend Sentinel*. It was a tragic end to a difficult life. Somewhere it had all gone wrong, perhaps decades before, when he left the well-grassed Paisano Pass for the scrub desert of Pinto Canyon, setting in motion the screenplay of his own demise. And yet, one thread of his legacy lived on through José Prieto, who never forgot the gift James had bestowed upon him— all of it now a distant memory as José paid his last respects to his old friend.[36]

Sutherlin's Final Years

Grover kept saying how he needed to unload those sheep. "The price is good, we need to do it now," but his neighbors would tell him, "Just hang in there, the price is going up." And then the bottom fell out of the market and he lost his shirt on it. Basically, they failed.

—Martin Sutherlin, grandson of Grover Sutherlin

Grover Sutherlin might have enjoyed a respite during the market boom of the war years, as well as some relief during the abundant rainfall of 1920. But that did not change the oppressive nature of his task. Operating a ranch that would have to be sold in only seven years left him with little incentive to expand the operation or invest in long-term improvements, little reason to improve the bloodlines of his flock, and almost no freedom to plan for the future. As the drought doubled down in 1922, Grover petitioned the county court for relief. "The small amount of land now owned," Grover claimed, "owing to its rough and barren nature and the small quantity there-of, will not permit a sufficient number of livestock to be kept to make it profitable." To continue running the ranch he would be "heavily encumbered and a great and unnecessary hardship will be worked on this executor." County judge K. C. Miller, in acknowledging Grover's plight, directed him "to sell all the remaining lands belonging to this estate at private sale as soon as convenient."[37]

It was a victory for Grover, but not one that offered much promise. Still, he had been freed from the burden of constantly struggling to hold the ranch together. No longer imprisoned by the terms of his father's will, he could pull up stakes and leave the whole mess behind as soon as he found a buyer for the ranch and the livestock. Even so, Grover would continue to run the ranch for another five years. Perhaps no one made an offer. Or perhaps the wool market, climbing out of the postwar slump, along with a few wet years, encouraged him to hold out a while longer. It was not quite convenient enough to sell just yet.

When wool prices bottomed out in the 1921 crash, the average US price for shorn wool sat at a low of seventeen cents a pound. But there was a fast recovery. By 1923, it had already returned to the prewar price of nearly forty cents a pound. Mounting good fortune, rainfall followed. From a low of around eight inches for 1922 it rose to more than sixteen for 1923. As early as March of that year—normally one of the driest months—the *Marfa New Era* reported that Grover, in town for supplies, claimed "good rain down his way." The good fortune seemed to temper Grover's drive to leave.[38]

Nevertheless, Grover's departure was eminent. Despite the rain and stabilized market, by the end of 1926 Grover had begun putting his affairs in order. The ranch simply wasn't enough—especially with his growing family of four children now between the ages of three and nine. In December, all the remaining heirs to Mart's estate deeded their portion of the ranch to Grover for only $2,500. The transactions were pragmatic, giving Grover cleaner title to the ranch and allowing him to transfer it without being encumbered by existing claims. It is also possible Grover's siblings—some gone a decade or more now—felt Grover deserved whatever he could get for the ranch, especially considering his good faith execution of his father's dying wishes.[39]

If the ranch was advertised for sale, it was not advertised widely. Another eight months passed before Grover finally had a buyer. Against great odds, Grover had managed to hold the ranch together for almost a decade in a stark and unyielding country. Now, finally having an exit strategy, he lacked only a good price for his flock. The offers, however, had not been attractive, and his friends counseled him to hold out a little longer. Surely he could sell at a decent profit. And so he waited. A week went by. Two weeks. Offers rose incrementally. Then, without warning, the bottom fell out of the market. With the closing date upon him, he had no choice but to sell at subpar prices. In September 1927, Grover transferred title to the ranch to Cressie Weatherby of San Angelo, Texas, for $2.85 an acre. Considering that similar land was going for $3.00 to as much as $5.00 an acre, what he got for it was, like the sheep, well below market value. But Grover had reached the end of his rope. It was time to get out.[40]

When Grover and Elsie and their four children crested the rim of the canyon for the last time, it was the end of a twenty-year saga for the Sutherlins. There were good years, to be sure. But between the many deaths they had endured, the natural attrition of the extended family, the years of hardships following Mart's death, and the market drop that prevented a more prosperous exit—there was little left in Pinto Canyon for the Sutherlins. As if to leave behind their troubles, and any memories that may have lingered, the family scattered to the four winds. Some ended up in Mississippi, some in Florida, some in California, some in Washington, some in Mexico. Most, however, eventually settled around El Paso and parts of southern New Mexico.[41]

If Grover had felt imprisoned in Pinto Canyon, perhaps it put a wanderlust in his being. For years the family remained in constant motion, like some itinerant clan. "Grover had a penchant for moving," Elsie later claimed. "Every time we got our farm in good shape, Grover wanted to move." They went first to Arizona, where he tried his hand running a horse ranch. When that, too, went bust, they moved to Sierra Blanca, eighty-six miles southeast of El Paso, and then to El Paso itself. He and Elsie finally ended up in Seminole, Texas, near the New Mexico state line, where they lived out the rest of their days tending a small farm.[42]

The Sutherlin experience in Pinto Canyon was unique only in its particulars. The outline was similar to the stories of hundreds of other homesteaders who came west for a fresh start but found success to be elusive. Even when the weather was agreeable and the market stable, even if predators could be held at bay and the neighbors kept in good standing, there were no guarantees. The bonanza that so many sought was like a shimmering mirage, something that could be seen but never seized.

∧∧∧

Until around 1914, nearly all the homesteaders in Pinto Canyon had been incoming. For most, however, the expected success never materialized. As a result, the decade that followed witnessed a slow exodus. As difficult as the early years had been, however, they would pale in comparison with the years to come. For this was before the Great Depression conspired with drought to bring even the hardiest ranchers to their knees. Before ethnic relations would founder, before José's most promising son would die at the hands of Texas Rangers. As a microcosm of the larger theater that was the Big Bend, Pinto Canyon's tortured backdrop fittingly reflected the harsh drama that was to play out within its folds.

6

Sangre del Cordero

Drought, Depression, and Texas Rangers

The Mexicans were intimidated by the whites at all times.
You never lifted an arm against them. [Pablo] should have
defended himself, but he didn't.

—Manny Perez

IT WAS A COOL spring morning as Gregorio Prieto motored up
the dirt road along Horse Creek to the ranch of Sim Weatherby, the
Prieto family's newest neighbor. He pulled up to the corrals where
the moving sea of woollybacks milled about behind the wooden
rails. Weatherby stared incredulously as Gregorio got out of the
truck and strode toward the corrals. Stepping up on the bottom rail,
Gregorio scanned the flock for his father's sheep. Soon he spotted
one of his ewes, opened the corral gate, and stepped inside. Wad-
ing through the animals, he grabbed a young ewe under the jaw as
he straddled her body. Sim stepped forward, yelling over the bleating animals, ordering him to let her go. But Gregorio only shook
his head, pointing to the prominent brand across her nose. "She's
mine," he yelled back.

Sim stiffened, his eyes narrowed. He told Gregorio the only thing
he was going to get was the business end of his gun if he didn't get
the hell out. With that, Sim started for the house. He wasn't going to
tolerate that kind of back talk, especially from a Mexican. But one of

his hired hands caught up to him. The Prietos were a respected family, the man whispered earnestly, and, besides, there were too many witnesses. Sim argued, but he knew the man was right. He turned back to Gregorio and yelled that he better take his animals and get off his property. But Gregorio was scarcely listening. He had never slowed in his task. He loaded his sheep into the back of his truck and latched the tailgate. A dust plume rose from behind his truck as he drove away. Sim stared after him. Not today, but someday, he probably thought as he watched Gregorio leave. He would show that cocky son of a bitch who was boss.[1]

By the end of the 1920s, the social fabric of Pinto Canyon had fundamentally changed and, with it, the peaceful dynamic that had existed between the four original families bound to a greater or lesser degree by debts of gratitude. All but one of the first run of homesteaders were gone: James Wilson, Albin O'Dell, O. C. Dowe, and the Sutherlins had all moved on. Of the four original families, only the Prietos remained. Unlike the others, they had persevered. But over the next decade, even the Prieto family would founder. Although José was better positioned to survive the drought and depression of the 1930s than his neighbors, he could not lessen the impact of the crisis on those around him. Nor could he control the consequences of his son Gregorio's actions, the fallout of which would bring tragedy upon his family. What neither aridity nor a crashed market could do, losing his youngest and most promising boy would do with surgical precision: break the spirit of a man.

The Prietos' Coming of Age

José's success had not been by luck alone, for he had done more than simply survive the first two decades of the century; he had thrived. His range savvy and well-honed frugality were among his greatest strengths. Although he'd received only the most basic of schooling, he held advanced degrees from the institute of the desert. He approached problems with an open mind and people with an open heart, moving through life with grace. In so doing, he'd largely avoided the plague of discrimination, having gained the respect of Hispanics and Anglos alike. But it was also true that he had enjoyed the good fortune of a large family, including four healthy sons who served as a labor pool during their years at home. Together, they'd been a ranching powerhouse. And although José Jr. never cared for the ranch life and Pablo would focus on his education, Victoriano and Gregorio were true cowboys who worked horses and livestock with competence and ease.[2]

But now José's boys were moving on. After marrying Martina Jaquez in 1924, Victoriano moved his family to Mexico. As he found work on various ranches, his family continued to grow. By 1930, the couple were living with their three sons and a daughter on Rancho de Álamos de Chavira outside Aldama, just north of Chihuahua City. Victoriano's brother, José Jr., would be the next to go, but he would choose a different path. Having a more mechanical bent than his brothers, he calmly set about tinkering with anything that needed fixing. José undoubtedly encouraged his son's interest, for the only thing more valuable on a ranch than a cowboy is a good mechanic. If there was roadwork that required a tractor or new spark plugs for the Model T, José Jr. was the man for the job. Around 1925, eager to escape the ranch life, he moved to Marfa and found work as a truck driver. About a year later, he married twenty-six-year-old Francisca Vasquez, who soon gave birth to their first son, Juan. Living in town, José Jr. would have had opportunities to further hone his mechanical skills. To him, rebuilding a carburetor or setting valve clearances was more fulfilling than herding sheep or goats could ever be—and in his genius for the nuts and bolts of the world, his passion shone.[3]

One Bad Seed

Gregorio didn't like white people and they didn't like him. It was racism on both sides.

—Manny Perez

Gregorio was not José Prieto's most humble son. Unlike his brothers, he did not share the calm temperament and quiet humility of his father. Nor had he known the poverty his family had risen from or endured the hardships that were routine for most Hispanics living in the borderlands. It was a disparity that grew starker during hard times. During the 1922 drought, a Red Cross survey reported some

Gregorio Prieto around
age seventeen, ca. 1921.
Photo courtesy of the
Prieto family collection.

fifty families living in poverty along the border. In José's hometown
of Ruidosa, the survey found one family of nine to be starving. By
contrast, Gregorio had never suffered for want of food. Or really
suffered much at all. Despite meager origins, José kept his family's
larders stocked and continued to gain ever greater financial stability.
The same year the Red Cross found such miserable conditions on
the border, José was busy purchasing additional land and applying
for US citizenship. As a result, Gregorio's life—relative to most—
had been privileged.[4]

That very privilege may have become his handicap, for it almost
surely went to his head. Gregorio was known for his swagger and
braggadocio, as well as for his womanizing. His marriage to Pilar
Nuñez from Ruidosa around 1928 notwithstanding, he always had
an eye for the ladies. And surely they also had an eye for him. Strut-
ting about in his button-down shirt, his pants tucked smartly inside

his embroidered boots, his hat casually cocked to one side, Gregorio cut quite a swath. If his philandering ways forced Pilar to look the other way, they did not escape the notice of others. In the wider community of predominantly devout Catholics, such behavior would not have put him in a favorable light.[5]

Gregorio could be a charmer, but he seemed devoid of the compassion that flowed through his father—a man who taught his children never to kill except for food or to protect the flock and never to run their horses unless there was a good reason to do so. But if Gregorio's horse acted up, throwing its head or bowing its back, Gregorio dug in with his spurs and whipped the horse savagely with the ends of his reins, forcing it straight up the nearest mountain until it was stumbling from exhaustion. And for Gregorio, unlike his father, to kill meant nothing. While out on horseback one day, he spotted a doe coming through a low saddle near the ranch house. Pulling his rifle from the scabbard, he shot it. When a second deer appeared, he shot that one too. And then—so the story goes—a third. Tiring of the game, he casually rode away, leaving the carcasses to the coyotes and vultures.[6]

Gregorio could also be quick with insults, often deriding his genial sisters as "worthless sheep." Because he would back down from no one, trouble always seemed to trail him. If confronted, he never hesitated to throw punches. Nor did he make any effort to hide his hatred for Anglos. Perhaps the sting of discrimination had left a chip on his shoulder, but anger always seemed to seethe just below the surface. By the time he was twenty-one, he had also started running afoul of the law. After landing in the county jail for aggravated assault, Gregorio pleaded guilty at his court hearing, and he was assessed a fine of twenty-five dollars in addition to time served.[7]

By the late 1920s, however, things seemed to be looking up for Gregorio. His marriage to Pilar coincided with his father's purchase of nine sections of adjacent land that included a house about a mile and a half northwest of the Prieto Ranch house. Built years before by Juan Nuñez on land purchased from the state by Juan's brother, Hilario, the unadorned but functional house was perched on a bench above the creek bed and offered a commanding view of Pinto Canyon. Four hundred yards southwest of the house, the two-track road crossed Pinto Creek for the last time before climbing steeply out of the canyon and winding around Cerro de la Cruz to continue west-ward toward Ruidosa and the hot springs. The house was a blessing, especially considering Pilar would soon give birth to the couple's only two children: Lydia in 1931 and Esperanza in 1934. Perhaps José hoped that the new house and starter herd of sheep and goats would encourage his son to be more responsible, that Gregorio might at last make the family proud. But if that was his hope, he would be sorely disappointed.[8]

Pablo at La Lydia

In 1927, José's youngest son would be the last to leave. José and Juanita were probably beaming with pride as they watched Pablo board the train bound for El Paso. For in leaving to attend high school, he would be the only one of their nine children to pursue his education. When the train finally pulled into the Union Depot in El Paso six hours later, Pablo stepped down from the passenger car dressed in his Sunday finest, eager to begin his studies at the Lydia Patterson Institute, where he would stay while receiving lessons in English, mathematics, science, and literature for the next several years. Founded in 1913, "La Lydia" was built to memorialize a prominent member of the Methodist Women's Missionary Society who had established schools for Hispanic boys in Segundo Barrio, a neighborhood in El Paso. The two-story brick building, commissioned in her honor and designed by the famous El Paso architect Henry C. Trost, housed one of very few institutions in the state where Hispanics could gain a higher education. Although the primary mission of the school was to educate boys seeking to minister to Hispanic Methodists, Pablo was, like many of his fifteen classmates, probably more interested in the educational opportunity than in becoming a man of the cloth. At that time, Presidio County schools, including the segregated Blackwell School in Marfa, offered classes only through the ninth grade. La Lydia picked up where other schools left off, allowing Spanish-speaking students to earn a high school degree, which gave them the option of attending college.[9]

Whatever Pablo's specific long-range plans, he did not seem destined for ranch life. Being ensconced in El Paso exposed him to a more cosmopolitan society than any he had ever seen. By all

Pablo Prieto around age twenty, ca. 1927. Photo courtesy of the Prieto family collection.

Pablo Prieto (*far right*) with classmates at the Lydia Patterson Institute in El Paso, ca. 1930. Photo courtesy of the Prieto family collection.

appearances, he fit in nicely. Photographs from his days at La Lydia show Pablo dressed to the nines, the corners of his handkerchief showing smartly from the pocket of his double-breasted jacket, his well-tended pompadour rising in a wave above his forehead. He is shown posing with other students in front of the brick façade of the building, standing beside a glimmering black 1930s sedan—one foot resting on the running board—and embracing his classmates, a boutonnière gracing the lapel of his jacket. Athletic and handsome, he undoubtedly drew attention from the ladies. There is no clue to his humble agrarian roots. Instead, he looks more like an academician—a sophisticated young Hispanic man poised to become a doctor or lawyer or scholar. Life was full of promise, his future bright.[10]

New Neighbors on Horse Creek

Old man Weatherby was pretty salty, and so was his son.

—Ted Gray

The same year Pablo left for school in El Paso, Sim and Cressie Weatherby purchased the old ten-section Sutherlin Ranch from Grover, who had probably been far more eager to leave than they were to arrive. The deed was placed in Cressie's name, but the purchase was paid for with a $25,000 loan from Dr. J. B. Chaffin, a San Angelo physician. Taken out on the eve of the Great Depression, it was a loan the Weatherbys would later regret. A rancher from the mesquite and live-oak country along the northern edge of the Edwards Plateau, Sim had always worked around sheep. Born in 1886 on the Williams Ranch in Brown County southeast of Abilene, he was the second child of Ora Bush and Benjamin Weatherby, a stockman from Mississippi. Sim's childhood was spent on their leased sheep ranch north of Sweetwater. In 1909, at the age of twenty-two, he married fifteen-year-old Cressie Bell Counts—a farmer's daughter from Erath County. For the next ten years they lived on a sheep ranch in Coke County before buying a house in San Angelo. By that time, Cressie had given birth to three sons—Harper, Theron, and James. Although Sim would run ranches near Big Lake, Del Rio, and as far west as Pinto Canyon, most of his ranching operations were within a hundred miles of San Angelo, where he and Cressie maintained a residence throughout their lives.[11]

Sim may not have spent much time in Presidio County before buying the ranch, but he had known about the Sutherlin Ranch for years. Taking advantage of drought-distressed ranchers pricing their sheep to sell quickly, Sim purchased a flock of old ewes from Mart Sutherlin as early as 1916, two years before Mart contracted the flu that ended his life and sent his family reeling. But any vestiges of the Sutherlin family's quiet struggle were soon to be overshadowed by bloodshed. For Sim would not be the congenial neighbor the Sutherlins had been. Instead, he would bring to the canyon a measure of strife unknown in the early years of settlement—even during the bloody Mexican Revolution. And although his tenure would be brief, it would change the course of history in Pinto Canyon.[12]

The fact is, Sim Weatherby was a known troublemaker. Although some claimed him to be only a bit salty, others saw him as an outright

James Sim Weatherby, ca. 1940. Photo courtesy of Kay Weatherby Ellis.

bully. By at least the early 1930s, his cantankerous nature was beginning to make headlines. In April 1934, Sim shot and nearly killed a
thirty-eight-year-old cashier at the Big Lake State Bank. Whatever
incited Sim's rancor, it was enough to cause him to pull his gun and
shoot the man in the chest, the bullet passing just inches from his
heart. Originally charged with assault with intent to murder, he never
served time, perhaps because the man survived. One year later Sim
was in the local newspapers again after one of his own ranch hands
brought suit against him for failure to pay wages he owed. The court
found in Eleno Guerrero's favor, allowing him to recover a judgment
for $82.50—two months' worth of wages—that the court agreed had
been unfairly withheld. Sim appealed, but the judgment stood. In
terms of bad publicity, Sim was on a roll. Between these two events,
he would find himself in the courtroom once again, this time on a
charge of attempted murder.[13]

Depression and Drought

**In 1933 to '34 it didn't rain, and everybody had to sell their
livestock.**

—Harper Weatherby

In 1929, the stock market crash on Wall Street ushered in one of
the most defining periods in US history, bringing the excesses of
the 1920s to a screeching halt. Declining job security, a downturn in
consumer buying, and loss of confidence in the stock market coupled with the foundering banking system together caused the US
economy to grind to a near standstill. But even as families lined
up outside soup kitchens across the country and unemployment
soared, there was little to indicate things were awry deep in Far West
Texas. The *Big Bend Sentinel* mentioned nothing of the stock market
fiasco, nothing of banks collapsing, nothing of growing unemployment lines in Dallas and Houston and El Paso. Instead, headlines
announced new construction on the Paisano Hotel, the largest and
most prestigious in Marfa to date. Like La Lydia in El Paso, it was
designed by the architect Henry Trost. The *Big Bend Sentinel* also

ran stories about the new Episcopal church under construction and announced that Fort D. A. Russell had been designated a permanent army post. In the first issue of the most trying decade in US history, the newspaper ran a headline seemingly detached from the growing national crisis: "1930 Slips in Very Quietly in Marfa."[14]

What no one could predict was that the quiet entry of 1930 also heralded a period that would see Presidio County cascade into despair. The last major economic depression, between 1893 and 1897, had long since faded from the collective memory. Recovering from the crash of 1921, commodity prices climbed rapidly and remained strong throughout the 1920s. Beef prices were the highest they'd been since before the war. Wool stood at slightly more than thirty-six cents per pound, and mohair reached an all-time high of seventy cents per pound. But having risen to such heights, their descent was even more precipitous. By 1932, both commodities had bottomed out, with wool cascading by more than 76 percent and mohair by an astounding 87 percent.[15]

Ranchers were especially hard hit, but they were not alone. The first few years of the decade would see the national income drop by almost 50 percent. Unemployment soared to a quarter of the national workforce. Banks fell like dominoes. By the end of 1932, more than fifty-seven hundred had failed across the nation. Hundreds of thousands of Americans were suddenly homeless. Some squatted in shanty towns along the outskirts of cities. Others wandered the countryside looking for work or hoboed over the iron rails. But those who stepped out of boxcars in Marfa were told to keep moving. By the summer of 1933, one out of ten people nationwide had become dependent on relief funds. Finally acknowledging the crisis, the *Big Bend Sentinel* cautioned that if relief funding ceased, "social disorder taking the form of robbery, thievery, etc. would soon be prevalent."[16]

In spite of the drop in agricultural prices, rainfall across the region had been ample. Between 1929 and 1931, precipitation hovered close to average. In 1932, the monsoon season alone brought nearly eight inches in Presidio, causing the Rio Grande to destroy riverside crops and wash away the customs house. Flooding around Marfa halted both rail and highway traffic, leaving the town marooned for days. Even so, the moisture was welcome news, the *Sentinel* opining that "prospects for fall and winter pastures are brighter than for many years."[17]

Prospects were momentarily bright, but after the 1932 monsoon the rain stopped. The remainder of the year brought less than a quarter of an inch in Presidio, and the first five months of 1933 brought none at all. By mid-June the *Sentinel* was reporting that ranges were drying up and that "ranchmen declare that rain is badly needed." A month later, with temperatures steadily rising, the *Sentinel* complained that the heat is "burning up tender grass that might be sprouting before it has had time to attain growth." Substantial rains failed to materialize. By the end of November, the total for the year at Presidio amounted to little more than two inches—less than a quarter of the annual average.[18]

The dryness persisted. During the first few months of 1934, Presidio received only a tenth of an inch. On 8 March an unseasonably strong cold front wedged southward, driving a raging dust storm before it. The air chilled as a gale force lifted the barren topsoil high into the sky. Already many ranchers were resorting to feed, their grass reserves depleted. By the beginning of June, an inch and a half had fallen in Presidio, but between the rains the sun beat down with renewed fervor, wringing the land dry. Newly sprouted grasses the moisture had coaxed out of the soil began to wither. Dust storms sandblasted what remained. Toward the end of June the *Sentinel* reported that "the entire section is sweltering under a heat wave that is said by old timers to have no parallel in their memories of weather chronicles."[19]

Presidio County was not alone. Through the summer of 1934, most of the country was gripped by drought, with all or parts of twenty-six states sweltering under cloudless skies. Crop failures across the Great Plains led to nationwide shortages of livestock feed. Waiting for better prices, many ranchers held their stock. But it was a grave mistake. Summer rains had not offset the dryness and heat. In some places, so much topsoil was lost that grass could no longer grow. Waterholes dried up. Springs ceased to flow. And then animals began to die. Hardened ranchers normally loath to ask for anything appealed to the federal government for help. Following a bitter fight in Congress, the Jones-Connally Act added beef to the list of commodities covered under Agricultural Adjustment Administration rules. Ranchers began lining up. In Marfa, representatives from ten area counties applied for relief under the Class A (highest priority) division of the Emergency Drought Relief Program. A relief office opened in early July in the Paisano Hotel to receive applications for assistance. Within two weeks the *Sentinel* announced that 1,884 cattle had been purchased in Presidio County alone.[20]

By 1934, beef prices were starting to edge out of their slump but still only bringing around four dollars per hundredweight—that is, if a buyer could be found at all. Feeder calves—young stock ready for the feedlots—might bring more if they were in good condition, but few were. The older cattle, often the skinniest and weakest, brought next to nothing. Meanwhile, distraught ranchers watched as their cattle grew gaunt, their eyes bugged out, and their skin tightened over bony ribcages like drum heads. There were too many animals for what little grass remained, too many animals and too little rain. Cowboys gathered the herds and cut out the weaker ones to be sold to the government. After submitting their applications to the relief office, the ranchers waited. If the application was accepted, a date was set for the appraisal. The offers were humbling—as little as a dollar for a calf or five dollars for a two-year-old, and never more than twenty dollars—even for a top cow. Ranchers could reject the offer, but few did. Once the agreement was signed, the herd was separated into two groups—those branded "R" were shipped to canneries to feed those on relief rolls. The rest were condemned—the ones the cowboys had to shoot on site before skinning them to provide proof of their death. Then the carcasses were burned. Columns of smoke marked the passing of the government agents from one ranch to the next, the dry summer air tainted by the smell of burning flesh. By the end of August, the federal government had purchased nearly five thousand head in Presidio County alone, about six hundred of which had been condemned.[21]

As bad as it was for cattle ranchers, sheep and goat ranchers— usually occupying more marginal land—had it far worse. Although they had a much weaker lobby in Washington than the cattle industry did, in early September Congress reluctantly added sheep and goats to the relief list, offering around $2.00 for sheep and $1.45 for Angora goats. By June 1935, more than 280,000 goats and more than a million sheep had been purchased nationwide. All those fit for consumption were sent to the Federal Surplus Relief Corporation. As with cattle, the rest were to be shot, skinned, and burned—but at a much higher rate than had been the case with cattle. An average of 75 percent were condemned nationwide. Between the relief program and those lost to the drought, the number of goats in Presidio County declined by nearly 20 percent and the number of sheep by almost half over the course of only five years.[22]

Modest rainfall in July and August 1934 briefly renewed hope. But after August, rain ceased altogether. In Presidio there would be no measurable precipitation recorded for the rest of the year. By early September, the *Sentinel* proclaimed 1934 to be the driest and hottest in seventy years. "The great West of the United States is passing through experiences that are terrible," it lamented. Already, some three thousand people were on direct relief in Presidio County—a full third of the county population. Around 270 unemployed men were placed on relief projects across the county, including repairing the road between Ruidosa and Porvenir and cleaning out the spring above Ruidosa.[23]

Meanwhile, in the midst of the heat and drought, of dry air and blinding sunlight and dying livestock and no money to be had, tensions mounted. It was a difficult time for even the best positioned of ranchers. But those below the Rimrock—where conditions were even more severe, where the buffer of more productive range didn't exist—had even fewer options. The Prietos and the Weatherbys and their neighbors were among the latter. In the hot and dry lowlands, the effects of the drought were magnified. Already marginal range was made more marginal still. What little forage remained was sparse and its hold on the earth tenuous.

By 1934, Sim Weatherby was only one of many ranchers across Presidio County unable to pay his property taxes—which amounted to less than seventy-five dollars. Like scores of others, he'd had to sell most of his stock. What income they brought him barely covered expenses. Operations were at a standstill. It was the kind of time when favors had to be called in, when it paid to be close to family, to be in good standing with your neighbors. But rather than forge alliances, Sim was busy collecting enemies, and foremost among them was the son of his closest neighbor.[24]

El Rinche

With few exceptions, the Ferguson Rangers composed a
sorry lot, and the effectiveness and hence the reputation of
the force plummeted. The fundamental flaw was simple
incompetence.

—Robert M. Utley, *Lone Star Lawmen*

Walter Fleetwood Hale Jr. wore his single-action army revolver with
pride, the bright nickel finish engraved with cartouche-like flourishes.
A gold band just above the carved pearl grip bore his inscribed surname, Hale. Known as "the gun that won the West," the .45 Colt had
been gifted to his father by the citizens of Belton, Texas, after he was
elected town marshal. He later wore the gun while serving as a Texas
Ranger. After retiring, he passed the revolver to his son.[25]

A large man with a round, fleshy face, his head topped with a
thatch of thin, straight red hair, Walter Jr. was not quite six feet tall
but weighed a full 240 pounds. At sixteen he'd married Pauline
Marie Lisenbe, and after stints in the oil field and as a city laborer, he
left his hometown of Belton to serve as a police officer in Waco and
later as city marshal of the tiny Central Texas town of Calvert. It was
there, in 1931, that Hale killed his first man.[26]

In the throes of the Great Depression, small-town banks in Texas
were easy targets for robbers. Bonnie and Clyde were only two of
many outlaws who chose the easier pickings in rural areas. The situation had grown bad enough that the Texas Bankers Association
began offering bounties. The signs they posted on bank storefronts
were unequivocal: "REWARD: Five thousand dollars for dead bank
robbers, not one cent for live ones." If the bounty deterred some robbers, it didn't deter them all. In early May, Marshal Hale received
a warning from the Waco chief of police that the Calvert bank was
to be hit. Hale posted watch across the street and began an extended wait. About a week later, Hale saw two men dressed in overalls
enter the bank. Through the bank window he could see one of the
men pull a pistol on the lone cashier. After forcing the cashier into
the vault, they scooped up about $7,000 in cash and made their get-
away. As they exited, Hale opened fire, but they were able to reach

their car. The men sped out of town amid a spray of bullets from Hale and other men alerted by the gunfire.[27]

As the robbers raced out of town, a bullet pierced the fuel tank, causing it to catch fire—a trail of flaming gasoline marked their path. They didn't make it far before screeching to a halt, abandoning the burning car, and running into the woods as Hale and the posse barreled after them. Hale later recounted that the robbers took cover behind a log and began shooting. Hale took aim with his .45 and shot one of the men through the head, killing him instantly. The other man, who'd had half of his face blown off during the escape, was captured. For killing the thief, Hale collected the $5,000 reward—more than a year's worth of wages during the lean years of the Depression. There was only one problem: some witnesses claimed the robber's weapons had been found in the burned-up car and that Hale had thus killed the robber in cold blood. The controversy that surrounded the killing forced Hale to resign two weeks later. He moved his wife and four children to Austin, where the reward money likely held them over comfortably until he enlisted with the Texas Rangers the next year.[28]

Walter Fleetwood Hale Jr. following the Calvert bank robbery in May 1931. Photo courtesy of Judy Baker Elliott on behalf of the Robertson County Historical Survey Committee.

Hale was fortunate that Miriam "Ma" Ferguson, the first woman governor of the state, had ordered a reorganization of the Texas Rangers. The Rangers' open support of her rival during the election had done little to endear her to the floundering Texas institution. With her election came their day of reckoning. The first thing she did was fire the entire Ranger force, replacing them with a completely new set of appointees, few of whom had much experience. Hale did have a law enforcement background, although not necessarily the kind of experience he needed. In 1933, Ranger Hale received assignment to Company A, stationed in Marfa under Capt. Jefferson E. Vaughan, who had gained something of a reputation during the Mexican Revolution. He was once called out for pistol-whipping a Mexican citizen on the streets of Valentine before throwing the man in jail. Even so, Vaughan engendered enough local trust to be elected five times as sheriff of Presidio County. Sitting broad-chested on top of his stout, paint-colored horse, his hat cocked over his left eye, Vaughan looked the part of a Texas Ranger.[29]

The Death of Pablo

It was obvious that the neighbors didn't want us to succeed. I was told by a former sheriff of the county that it was a certain big rancher adjacent to our place.

—Harper Weatherby

They never thought the Texas Rangers would come up and shoot their kids.

—Oscar Perez

By the summer of 1934, Walter Hale Jr. had been a Ranger in Presidio County for a year and a half—long enough to get his feet beneath him and gain some familiarity with the area and the people. With so many area residents destitute, much of his job involved investigations of petty theft. On 4 June he and Ranger Bob Burl were sent down to the Rio Grande to take reports of ranchers who claimed to be missing livestock. Later that evening the two met in Valentine, where they were to stay for the night.[30]

Hale was the first to arrive in the modest railroad town of some five hundred souls. As he was getting settled in at the boarding-house, a Model A Ford pickup pulled up with two men he recognized to be Sim Weatherby and his son Harper. Sim motioned Hale over and said he needed an officer: someone had cut his fence and stolen seven hundred head of goats off his ranch. And he had an idea of who had done it. His ranch adjoined the Prieto property. There was bad blood between them, he said, because they had wanted to buy his place and he had refused. Hale heard Sim out but explained that he was on another assignment and all he could do was report the theft when he returned to headquarters. Reluctantly, Sim and Harper turned back for Pinto Canyon.

After Burl arrived at the Valentine boardinghouse, the two Rang-ers took their supper and afterward sat on the porch talking. They were about ready to retire for the evening when they heard the dis-tant hum of an engine, gradually growing louder. They stepped out into the road to see two headlights racing toward them. Seconds later Sim's truck slid to a halt before them in a cloud of dust. Agitated and disheveled, Sim jumped out and, trying to catch his breath, explained they had been ambushed. Pointing to a bullet hole in the top of the truck cab, he said the Prietos were trying to kill them. Just as they drove by the Prieto house, Harper added, someone had opened fire. Hale and Bob exchanged glances. Goat theft was one thing, but attempted murder was something they could not ignore. Strapping on their guns, the two Rangers climbed into Sim's truck and the four men started off into the night.

They headed southwest out of Valentine across the Brite Ranch flats, speeding across broad Capote Draw toward Pinto Canyon almost forty miles distant. The engine chugged up hills and hummed down their backslope, two narrow headlight beams shining rigidly into the darkness. As they drove, a gauzy waning half moon began to rise above the eastern horizon. By the time they reached the can-yon rim, it was nearing midnight, the moon already starting to cast shadows.[31]

After the steep descent into the canyon, the truck crossed Pin-to Creek—barely flowing—and continued westward, the mass of Chinati Peak silhouetted against the southern sky. As they neared Gregorio's house, Sim pulled to the side of the road and cut the engine. The four men walked down the Pinto Canyon road paralleling the creek bed until reaching the steep two-track road that led up to Gregorio's. The house was dark; everything was very quiet. Hale motioned Harper to the back door to see that no one escaped. Then, posting Sim to the right corner of the house and Bob to the left, he approached the door and rapped sharply—a sound that pierced the silence of the night so abruptly it caused Hale to jump. After a moment, a faint light showed under the door. He knocked again.

Inside, Gregorio was jarred awake by the knocking. Sitting up wearily, he asked who was there. Hale answered, saying he was with the Texas Rangers and that they were there to investigate a stray bullet that had hit the Weatherbys' truck. There were just a few questions he wanted to ask. Aggravated, Gregorio said he knew nothing about it. In any case, it was late and they should come back another time. But Hale insisted. If Gregorio would just come out for a minute, he said, they would be on their way. For several moments there was silence. Then the inside light went out. Seconds later the door was jerked open with such force it slammed against the inside wall. Gregorio stood in the doorway, blinking into the darkness. He could barely make out three men in the dim moonlight. Then, from the corner of his eye, he saw the man on the left raising a gun to his shoulder.

Gregorio spun around to take cover just as Sim's shotgun exploded—a flash of light, a thunderous roar that shattered the silence, echoing off the walls of the canyon. Gregorio flew forward, landing face down inside the open doorway. Pilar screamed and, reaching for his outstretched hand, pulled him inside before slamming the door shut and bolting it. Baby Esperanza and her sister, Lydia, wailed from their bed. Except for the gasping of Gregorio, desperate to regain his breath, there were no other sounds. When Pilar looked out the window again, the men were gone.

The air was filled with the acrid smell of gunpowder as Pilar struggled to light a lantern. Gregorio was splayed out on the floor in a pool of blood. His back burned as if on fire, peppered with birdshot. He gasped for air, all but life itself having been knocked out of him. As his breathing gradually slowed, three-year-old Lydia held a shaking tin of water to her father's lips. He drank slowly, wincing with each gulp. "Don't cry, *mija*," he told her. "Everything will be okay." He coughed, the convulsion sending electric-like jolts through his spine. He was hurt, but he didn't know how badly. He wasn't sure he wanted to know.

Outside, the four men retreated to the creek bed, where they huddled in the darkness to formulate a plan. Hale didn't like the situation. They didn't know how badly Gregorio was hurt or how many men were in the house. What he did know was that the adobe walls were so thick that the place might as well be a fortress. He wanted backup. He asked Burl to return to Marfa to pick up his regular partner, Charlie Curry—someone, he claimed, who could think as fast as any man alive. Burl left, and ten minutes later they heard the Model A sputter to life and listened as the sound of the engine gradually grew fainter until it topped the rim and could be heard no more. It would take at least three hours before he returned with Charlie, maybe longer. The men stretched out on creekside boulders to wait. The moon had risen high in the sky, casting crooked shadows across the canyon floor and skylighting the gently swaying branches of ocotillo, snaky and spiked and stiff in the night. Somewhere overhead a nighthawk peened.

Inside, Pilar knew she needed to get help. Gregorio was flat out on the floor, his breath a harsh rasp in the silence. She stood by the back door and crossed herself before stepping outside, clutching Lydia, who lay sleeping in her arms. For a minute she stood in the darkness. When she was sure it was clear, she started down the footpath to José Prieto's house. Normally she would have a lantern. Normally she would not be walking at night, when rattlesnakes lie waiting in dark shadows. She prayed as she walked that none would be on the trail. She quickened her pace, but it was more than a mile and a half distant. It seemed she had walked a dozen by the time she arrived. Breathing heavily, she opened the door and stepped inside.

She found Pablo alone. He was home from La Lydia for the three-month summer break from his studies. He wearily sat up in bed while she explained what had happened. Blinking away the sleep from his eyes, he rose and quickly dressed. On the way out, he grabbed his hat and a .30–30 Winchester from behind the door. He levered the action to see that a cartridge was in the chamber and then closed it. He scanned briefly around the room for extra cartridges but could find none. They set out down the trail, Pablo setting the pace with Pilar close behind. They walked fast, without stopping, until they reached the house. Inside they found Gregorio still sprawled on the floor, breathing uneasily. Esperanza had fallen asleep beside him.

Dawn broke to find Hale, Sim, and Harper still waiting by the creek. As the first rays of light hit Chinati Peak, they looked up the road to see a lone Hispanic shepherd driving a flock of sheep out of a pen. He herded them down the road toward the men before suddenly turning them back the way they had come. A minute later he turned them back again. Sim looked at Hale. That was where they had been fired on last night, he said. The herder must be trying to cover the tracks. Motioning for Weatherby to follow, Hale edged up the road, keeping just out of sight. When they got close, Hale stepped out into the road, his pistol drawn.

The herder threw his hands up, his eyes widening as they settled on Hale's badge—a symbol that struck terror into Hispanic residents of the borderlands. El Rinche.[32] Just then they heard a vehicle approaching and looked up to see Weatherby's truck returning from Marfa. The truck rolled to a stop, and Burl and Curry stepped out. Together they marched the shepherd up to Gregorio's house and sent him inside to see if Gregorio was dead or alive and to urge the Prietos to surrender. He stood at the door and called out in Spanish. The door opened and he was let inside. Minutes passed. An hour. Finally, the shepherd stepped outside. He said Gregorio was in bad shape, but the Prietos would not come out and they would not surrender. Hale and Curry looked at each other. If they waited for the sheriff, the Prietos might escape. Besides, they reasoned, several local deputies were Hispanic and probably friends of the family. Hale brought his hand down to rest on the butt of his pistol and turned to Curry. "I'm going in," he said. Curry looked up and grinned. "Let's go."

The two men approached the house slowly, keeping their eyes on the window. Hale stepped up on the porch and rapped on the door. "Hale of the Texas Rangers!" he yelled. "Open up!" From inside they heard a woman respond, but they could not make out what she said. Just then the outside door flew open as an interior door was slammed shut and bolted. Hale started for the door, but Curry pushed him aside. Larger even than Hale's 240 pounds, he crashed through the door. Pablo looked up from where he sat beside Gregorio. Before he could react, Hale leveled his pistol and fired, blowing a hole through the left side of Pablo's body. Pilar screamed from the corner of the kitchen, baby Lydia wailing at her breast. Curry trained his gun on her. "Put down the baby!" he ordered. But she only cried out in Spanish and clutched Lydia tighter. He glared at her for what seemed an eternity before finally lowering his gun.[33]

Hale walked over to where Pablo lay slumped over on the ground and prodded his lifeless body with the toe of his boot. He then turned and toed Gregorio in the ribs. No response. Searching the house, they found a .30–30 Winchester, a 12-gauge pump shotgun, and a .30–06 Springfield bolt action rifle. All were loaded. Forty-eight live rounds lay on the table in the front room. None had been fired. An hour later Sheriff Joe Bunton arrived and took depositions from Hale and Curry. Shortly after they left, Dr. John Peterson arrived from Marfa to find Gregorio alive but in critical condition. For Pablo, there was nothing to be done except the paperwork. Principal cause of death: gunshot, left side. External cause: homicide.

Peterson drove Gregorio to his clinic in Marfa, where he removed more than a hundred pellets of birdshot from his back. Gregorio was damned lucky. Had Weatherby been any closer, it would have blown a hole through his body. As it was, his back would be forever peppered with scars, but his life had been spared. Gregorio would recover quickly, but there were rumors the Rangers would come back to finish him off, that he wasn't safe at the clinic. Some said he wasn't safe anywhere at all.[34]

José's Pistoleros

The four-door Model T sputtered and bucked up the Pinto Canyon road toward Marfa, a plume of white dust billowing behind it. The car was full of men: José Prieto's *pistoleros*—men from the ranch, relatives from Ruidosa—men ready to put their lives on the line to protect the Prietos. All but José himself were armed and loaded, the muzzles of their rifles sticking out the windows, the car so bristling with guns that from a distance it looked like a metal porcupine. José Prieto sat in the passenger seat holding on as they ran pell-mell over potholes and stones, his face stone cold, eyes red from the dust and the anguish of losing his son to a violent and senseless death—the one son who had escaped a life of labor, who would navigate the world of the learned, the one who would have made the Prieto name ascendant. *Never again,* he thought to himself. *Those Rinches will not kill any more of my boys.* He winced at his own rage, a feeling as unnatural to him as a suit of armor.[35]

The car spread a great cloud of dust as it sped into Marfa, the chassis canting awkwardly as they skidded around corners. Men stepped out of the street, women gathered their children about them. News had spread like wildfire. Everyone knew they were gunning for Hale and Weatherby. The car pulled abruptly to a stop in front of the jailhouse. José got out and walked to the door while the men waited in the car, eager for what lay ahead. But once inside, Sheriff Bunton only shook his head. He told José that Gregorio was in no danger and that Weatherby and the Ranger were under the protection of the law. There was nothing to be done. It was a matter for the courts. The state would prosecute. Justice would be served.

Headlines in the *Big Bend Sentinel* following Pablo Prieto's murder. Courtesy of the *Big Bend Sentinel.*

José emerged from the jailhouse, his eyes cast down, a look of defeat across his face. He shuffled back to the car and got in and, for a moment, just sat. His men wanted blood, they wanted revenge. But the anger had gone out of José; only grief remained. Violence was not the solution. It had never been a solution. It had only caused misery and would only cause more of the same. The men grumbled as the driver turned the wheel of the car and motored away, the porcupine following its own faint tracks back out of town.

Pablo's Funeral

On Wednesday, 6 June, the small group of mourners gathered around José and Juanita at the eastern edge of the Merced Cemetery in Marfa as the priest sprinkled Pablo's cold body with holy water and recited the final prayer before they closed the casket and lowered it into the freshly dug grave. José had probably not thought of owning a town cemetery plot. Buying it was an unwanted expense during tight times, a loss appended to a loss. But it seemed appropriate for his most promising son. José could easily have scratched out a plot at the ranch that would have suited him and Juanita—or in a simple rock cairn grave in the family plot on a hill outside Ruidosa. Perhaps he felt a town plot was more fitting for their erudite and sophisticated son. In any case, a plot in the town cemetery was one small way to honor Pablo's unrealized future. It was one that had been unfairly taken, as if the odds against him had not been stacked high enough already.[36]

Seventeen days shy of his twenty-seventh birthday, Pablo was the first in his family, indeed of his entire lineage, to have sought a high school diploma. He was soon to graduate from the Lydia Patterson Institute as a man of learning, as easy in a suit as he had been in cowboy boots in his youth. From La Lydia, for all anyone knew, the world was his oyster. But at the wrong place at the wrong time, the contingencies of life bore down on him in a senseless breach of fairness. For the trouble that Gregorio had drawn to the family, it was the most innocent of them all who paid the price. Fate, in this case, was the master of irony.

The Trial

On Wednesday, 8 August 1934, two months after Pablo's death, it was standing room only at the Presidio County Courthouse as the bailiff called for all to rise. Judge C. R. Sutton entered the chamber and sat at his bench. The twelve Anglo jurors were sworn in by the court clerk and seated. The attorney for the defense, E. B. Quinn, sat beside Walter Fleetwood Hale Jr., who was wearing his Colt revolver—an allowance the judge had made based on rumors that the Prietos would kill him if he was let free. "[It was] probably the first and only time in the history of trials of any kind in this or any state," the *Big Bend Sentinel* reported, that "the accused, Ranger Hale, carried firearms in the courtroom . . . [and] wore it during the hearing of the witnesses."[37]

District Attorney Roy R. Priest, representing the state against Hale, rose and argued that the jury, consisting solely of Anglos, would not be impartial in deciding the case, that it was not, in fact, a jury of the peers of the victims but only of the defense. Quinn countered that the jury had been carefully selected according to established procedure and would in no way be biased against the plaintiff. In the end, the jury selection was allowed to stand. The first witness was Sheriff Bunton, who claimed he had arrived at the scene of the shooting to find Gregorio seriously wounded and Pablo dead. He testified that several guns were found inside the house, although none had been fired. Dr. Peterson appeared next, confirming that Pablo had died as the result of a gunshot wound.[38]

When the trial resumed the next morning, Pilar was called up to testify. When asked to identify the man who killed Pablo, she pointed a shaky finger at Hale. He shifted in his seat, his jacket opening just enough to see the pearl handle grips of his revolver. The district attorney asked her to recount the sequence of events. She said that when the Rangers came to investigate the stray bullet that had struck Weatherby's truck, Gregorio claimed he knew nothing about it and told them to come back another time. But the men persisted, she continued, and when Gregorio finally went to the door, he was shot. The Rangers left but returned in the morning and broke through the door of the house. There they found Pablo sitting by the side of Gregorio, whereupon they shot and killed him. There was also one additional detail: neither Gregorio nor Pablo had been armed.[39]

Finally called to testify, Hale said that when Gregorio came to the door, he had a gun in his hand and that "the shooting of Gregorio, by someone in [his] party, then occurred." Later in the morning, Hale continued, they returned to the house and were again refused admittance. They broke through the door and found Pablo sitting on the floor with a gun in his hand, which forced Hale to shoot in self-defense. Other witnesses were called. Area ranchers said that Sim Weatherby had openly accused the Prietos of stealing his sheep and goats. Others took the stand to say the Prietos were an honest and upstanding family and that accusations of theft were preposterous. The closing arguments hinged on a single point of contention. The prosecution maintained that although there were guns in the house, Pablo had been unarmed when he was killed. The defense countered that because Pablo had been armed, the shooting had occurred in self-defense, that Ranger Hale had only been fulfilling his duties as an officer. In the end, it was simply Hale's word against that of the Prieto family. On Friday at four thirty in the afternoon, after three days of testimony, the case was turned over to the jury.[40]

Twenty-four hours later, on Saturday, 11 August, the jury foreman, J. A. Williams, rose before the hushed courtroom and announced the verdict: "We, the jury, find the defendant not guilty." The courtroom erupted in protest as the Prietos and their supporters stared in disbelief at the faces of the jurors. Hale stood triumphantly and, after vigorously shaking the hand of his attorney, strode out of the crowded courtroom amid the clamor to where his wife waited in their car, packed and ready to go. There was only one option left him. His assignment in Presidio County, similar to his duty as city marshal of Calvert, had come to an abrupt end. If one of the Prietos didn't go after him, someone else would. Things had gotten too hot for Hale.[41]

"El diablo nos está llevando"
(Things are going to hell)

Perhaps the drought and the Depression were making everyone crazy. Perhaps it was hunger, or booze. But all hell seemed to be breaking loose, with ethnic tensions soaring higher than at any time since the Mexican Revolution. Five days before the trial, a ranch hand named Antonio Carrasco had killed his boss and his boss's wife at their ranch between Valentine and Van Horn before torching their house and fleeing. Just as the Hale trial got under way, headlines in the *Big Bend Sentinel* read, "Mexican Maniac Runs Wild, Attacks Three Girls, Winds Up in Jail Here." Marfa resident Pedro Hernandez, supposedly under the influence of marijuana, was said to have gone berserk and attacked three Hispanic girls. After the girls fled to the church for protection, he then attacked the priest who had rushed to their defense. In other stories, Gregina Conterrez pleaded guilty to aggravated assault, while Alfonso Bustillos was convicted of assault with intent to murder. Such headlines certainly wouldn't have helped the Prietos' case, in which the most striking difference between the plaintiffs and the defendant was the color of their skin.[42]

Walter F. Hale Jr.'s judgment of acquittal. Courtesy of the Presidio County District Court, Marfa, Texas.

The spike in ethnic strife was part of a much larger trend. Reeling from economic disorientation, many Americans sought a convenient scapegoat—one they readily found in the Mexican-American community. Being already on the lower rungs of society, its members had been especially hard hit during the Depression. US attitudes and policies only added to their suffering. Following a spate of anti-Mexican rhetoric and laws enacted to restrict immigration and force deportation, the federal government repatriated around a million people during the 1930s in a series of massive roundups. During the winter of 1931 a huge corral across from the customs house in Juárez held more than two thousand *repatriados* huddled and starving. Women swarmed around the warehouses picking up beans that had fallen out of gunnysacks being carried inside. In the trickle-down of suffering, Hispanics nearly always found themselves holding the short end of the stick.[43]

The Aftermath

Everyone began to carry guns. Distrust that had relaxed in the waning years of the Mexican Revolution was abruptly revived as old prejudices returned. Bartolo Villanueva, one of José's hired hands—a man who vowed to protect the Prietos at all costs—oiled his rifle and strapped on the bandolier that he would wear for months to come—one he'd worn fighting with Pancho Villa during the revolution. He became José's personal bodyguard, always on the lookout for the Weatherbys, for the Rinches. When someone arrived at the ranch, he watched down the road to see if anyone followed. In the quiet evenings after work, he strained to hear the sound of distant motors. Awakened in the night, he listened for footsteps. Nothing more would be allowed to happen to the Prietos. Bartolo would see to it personally.[44]

The surge of suspicion and prejudice in the aftermath of the trial prompted Sheriff Bunton to write to the adjutant general of Texas, Henry Hutchings, requesting the entire Ranger company be removed from the area "for causing a great deal of [bad] feelings." Although Hutchings did not comply with the request, he did transfer Hale to Austin, where he served out the rest of his Ranger term under Capt. D. Estill Hamer. Pablo's murder had been only one of a string of missteps by the Texas Rangers during one of their dark-est hours—a murky time in a long and controversial history. Ma Ferguson had dealt with the Rangers as political pawns and made appointments with little regard for their level of experience, or their integrity. As the effectiveness and reputation of the "Ferguson Rangers" plunged to new lows, the effects of her meddling became increasingly evident.[45]

Meanwhile, Hispanics in Marfa and along the border lived in con-stant fear of the Rangers. History had proven the Rangers could get away with cold-blooded murder—as they had at Porvenir in 1918, so they had with Pablo. For the Prietos and Pilar Nuñez's family, the fear had grown intensely personal. Rumors ran through the town and countryside and among the hamlets of the borderlands. Some believed the Rinches were out to get them all. "They were like mad dogs," recalled one of Pilar's cousins. José had lost one son and nearly lost another for reasons he couldn't grasp. The court system had failed to deliver justice. Anglos seemed above the law. It was clear that if his family was to be protected, they would have to do it themselves. He couldn't bear the thought of losing another son to such violence. And he knew Gregorio, for one, would seek revenge. With Victoriano already in Mexico, José prepared to send his remain-ing sons to join him. Perhaps in Mexico they would be out of danger. Perhaps there they would be allowed to live their lives without fear of being shotgunned by bitter neighbors or killed by Texas Rangers. One thing seemed certain: they were no longer safe in Texas.[46]

The Weatherbys' Exit

Under the pretense of having exhausted the available pool of impartial jurors but more likely because of mounting unrest, Sim Weatherby's defense attorney moved for a change of venue. Judge Sutton agreed and sent the case to Upton County, more than a hun-dred miles away in the declining oil boomtown of Rankin. Lack-ing local context, the case held little interest. When it was brought before the district judge on 24 September, it was summarily dis-missed. It was the second time in six months that Sim had gotten away with attempted murder. If he'd felt justified in having shot

Gregorio in the back—an act that most would consider cowardly—certainly now he felt completely vindicated. Sometimes, he may have reasoned, violence pays.[47]

Nevertheless, it had been a tough run for the Weatherbys. After borrowing $25,000 to purchase the ranch in 1927, Sim had spent a small fortune having a new netwire fence built around its periphery. If he had a couple of good years before the financial crash, they were likely offset by the cost of getting the ranch up and running. Added to the expense of fencing and water improvements was the never-ending war against lions and eagles and wolves and bears—a problem made worse by having their ranch neighboring the Chinati Mountains, which were essentially a refuge for wild animals.[48]

Sim's eldest son, Harper, did his best to help. He bought a pack of hunting dogs and, during the six or so years he lived there, killed two bears and thirty-four lions—one of which nearly cost him his life. The lion lay on a mountain ledge, its foreleg caught in the steel jaws of a leghold trap, the draghook (attached by a chain to the trap) hopelessly tangled in brush. Having no gun, Harper climbed to a small bluff just above the cat and began hurling rocks the size of cantaloupes, slowly pummeling the cat to death in a series of skull-crushing blows. Lifting a large rock to deliver the coup de grâce, his footing gave way, sending him careening toward the wounded animal. Landing just beyond the cat's reach, however, he was able to scramble quickly away.[49]

In 1928, Sim was able to pay off the balance on the ranch, thus removing the Sutherlins' lien on the property. But he'd been unable to pay down the loan from Chaffin. In 1936, he got a second loan—this time for $40,000—secured against the ranch and due in only five years. But the very next year he failed to pay his property taxes, a breach of the terms of the loan. By 1940, Weatherby's fate seemed about sealed. Sim made a couple of desperate attempts to avoid losing the ranch, first by leasing the mineral rights and then by attempting to transfer title to his sons. But Chaffin had already called in the loan. In the transfer to the sons, the deed admitted it was contingent upon a "claim which is being asserted by Dr. J. B. Chaffin of Tom Green County, Texas, but which claim is not a just or legal claim against said land."[50]

Apparently Chaffin's claim was legal enough. Only a month after gifting the ranch to his boys, Sim signed it over to Chaffin in exchange for the cancellation "of one certain note for forty thousand dollars . . . none of said principal having been paid." It may have been for the best. For one thing, he'd probably become estranged from many area residents after the incident with Gregorio. Sim was also probably confident he would never turn a profit below the Rimrock. The same rocky, cougar-infested scrublands that had failed the Sutherlins also failed the Weatherbys. The land had long since lost its ability to support enough livestock to carry a family, and it probably had been in decline since the first decade of ranching there. Intensive grazing without pause for so many years, first by Mart and then by Grover, who had few options to do otherwise, had been done at a price. The situation was clear enough by 1934 that Sim leased a ranch south of Odessa and sent Harper to run it. But in losing the Pinto Canyon ranch, they also gave up the Odessa ranch, quitting Far West Texas for good. Sim and Cressie returned to San Angelo, where he would continue to ranch until his death at the young age of fifty-four, when he suffered a massive stroke. He lingered in Shannon Hospital for seven days, before death finally claimed him on June 5, 1941. In terms of time, their brief twelve years in the canyon was short. But in terms of their effect on the Prieto family, the Weatherbys had left a legacy of heartache and grief.[51]

The Drought Breaks

By the beginning of 1935, the drought seemed to be breaking, with 1.36 inches falling in Presidio between January and March. But it was a false promise. The next two months brought virtually no rain, and in early May an epic dust storm scoured the region. "High winds swept the worst dust storm of the season across the highland country Wednesday afternoon and night," announced the *Sentinel*. "Visibility was as little as one hundred feet at times. . . . Cattle, already harassed by lack of green feed, suffered still further. Ranchmen were grave about the drouth situation." Meanwhile, hundreds of farmers across the Rio Grande in Mexico were facing starvation. With no water in the river with which to irrigate crops and their government impotent to help, they had nowhere to turn. One desperate farmer just above Candelaria put everything on the line by planting his crop in the drying mud of the river bed, the delicate sprouts waiting to be washed away by the next flood.[52]

Rainfall finally returned in June with an inch and a quarter falling in Presidio. "The best in five years soaked the Big Bend area this week," reported the *Sentinel*, and "ranchmen generally felt better than they have for several years with good grass assured and the price of beef better than it has been in years." The drought began to reverse itself. By August, rains had become widespread, with Marfa receiving 8.25 inches in that month alone. Pinto Creek began to flow again, and the Rio Grande rose to flood stage at Ruidosa. At last the *Sentinel* announced that the drought had broken: "This year, for the first time in five years, enough rain has fallen, over the greater part of the drought-stricken areas, to bring the stored up groundwater supply up to a level where it can be reached by the roots of farm crops." After 1935, rainfall increased almost every year for the remainder of the decade, but overall recovery was slow. For all but one of fourteen months between July 1934 and August 1935, the region had been locked in severe drought—long enough that much of the damage was irreversible. In some places there was nothing left to water.[53]

∧∧∧

The 1930s was a period of incredible hardship for millions across the country. But in Pinto Canyon, ranchers shared the added burden of being in a drought-prone and marginal region to begin with. For the Prietos, the drought and Depression had been only the backdrop to an even greater tragedy. They could not have known that the departure of the Wilsons and the Sutherlins would also bring about an end to an earlier period of harmony. Gregorio's recklessness, combined with the Weatherbys' hostility, all fueled by drought and debt, pushed things to the breaking point. The situation had become a powder keg.

Pablo's death served up a devastating blow to José and Juanita's spirit. To suffer the additional indignity of seeing the people of their own county—a supposed jury of their peers—exonerate the murderer, seemed like the ultimate betrayal. The expansiveness that had attended José's ascendancy began to close. By that time, he had already bought the last section of land he would ever own in the United States. Out of fear of losing more sons to violence, he would send them south—away from the dreaded Rinches to areas less prone to discrimination. But Mexico would not turn out to be the safe haven José hoped; freeing them from one threat only opened the door to others. Although his three remaining boys would spend the next decade in Mexico, only two would return.

7

Madre Patria

Mexico, Family, and Ranching in the 1940s

Donde es mi tierra, donde me vaya bien? (Where is my
country, where things go well?)

—old Mexican proverb

THE RED 1948 Dodge pickup stalled midslope on the *cuesta* as it
climbed out of Pinto Canyon. Juan Morales hit the brakes just in time
to keep the truck from rolling back. "Piedra!" he yelled. The passenger door swung open, and Sue Prieto jumped out, ran to the edge
of the road, picked up a rock the size of a melon, chocked the rear
tire, and jumped back in. Juan shifted the truck into gear, revved the
engine, and popped the clutch. The truck lunged forward—the rear
tires spraying a plume of rocks and dust rearward—and climbed
another twenty yards before it stalled again. "Piedra!" yelled Juan.
Sue jumped out again and found another rock and chocked the tire
and got back inside. Juan popped the clutch. The tires spun, dust
and gravel sprayed, and the truck charged forward again. From stall
to stall—sometimes as many as five or six—they slowly climbed
their way out of the canyon in increments until they finally reached
the rim where they could motor on, less encumbered by gravity than
before.[1]

Juan Morales was the Prieto family's most recent addition. And
of José's three sons-in-law, he was by far the most eccentric. With

his practical jokes and riotous laughter, he easily became a perennial favorite of the family. But Juan was also a skilled cowboy who would provide much-needed help on the ranch at a time when José needed it the most. After sending his sons to Mexico for protection, he'd lost a major source of labor. His daughters took up the slack for a time. But as they began to marry, their husbands provided a new labor pool—some of whom fit nicely into the Prieto family and some who did not. Juan Morales and Manuel Perez easily won José's respect, and he, in turn, portioned out sections of his ranch for them to run as their own. But Juan Benavidez was different. The tension that developed between the two men prompted Juan to strike out on his own. Only through his legendary tenacity and a dogged work ethic was he able to build a ranch for himself, and he did so without having been given the leg up the others received. Meanwhile, Mexico did not prove to be the refuge for his sons José had hoped. Instead, their stay would be marred by violence as senseless as what had befallen Pablo. By the time the last of the Prieto sons returned to the United States, the bridges they left in Mexico had burned to ash.

A Decade of Prosperity

The 1940s proved a prosperous counterpoint to the hardships of the 1930s. Following the end of the drought, both rainfall and market prices trended upward. Averaging slightly more than eight inches a year during the 1930s, annual rainfall in Presidio County during the 1940s would average more than eleven inches. The summer of 1941 was the wettest on record, with Presidio logging more than twenty-three inches of rain—considerably more than twice the average. And following this strong start, rainfall hovered at near or above average for the rest of the decade, including a few notable deluges. In October 1944 a fifteen-foot-high wall of water roared down Cibolo Creek after two days and nights of continuous rain. After the flash flood, homes, businesses, roadways, and—for the fifth consecutive year—crops around Presidio were left in ruin. Although the floods left widespread damage, the heavy rains spawning them served to revitalize the grasslands.[2]

Agricultural prices also remained strong throughout the 1940s. Following 1938, the value of commodities climbed dramatically.

World War II finally did what Roosevelt's New Deal had struggled to do—restore the American economy. The ripple effects of the wartime economic boom extended to nearly every industry in the nation, not the least of which was agriculture, providing food and fiber to feed and clothe the troops. Between 1938 and 1942, lamb prices rose by 60 percent and wool prices doubled and held steady throughout the decade. Mohair followed the same trajectory but trended higher than wool, as it had since the early 1920s, averaging around 44 percent higher between 1940 and 1946.[3]

During World War II, the US government set prices and bought all domestic wool and continued to do so through 1947. Without such controls, the price would have risen far higher than it did, but by preventing a postwar crash the controls helped stabilize the market. Mohair was largely riding on the coattails of wool. After initially placing restrictions on civilian use of both wool and mohair, the government deregulated mohair in 1942. Even so, because mohair served as a replacement for wool in civilian markets, prices remained strong. The Wool Act of 1947 enabled the Commodity Credit Corporation to buy wool at 42.3 cents per pound to ensure that the national supply of wool would remain steady. Two years later, mohair was also added to the list. But such price supports did nothing to prevent the growing trend away from animal fibers in favor of new synthetic fibers.[4]

Even with the rainfall and strong economy, problems began to plague stockraisers in many parts of the country during the 1940s, most notably rising costs. The military draft and urban migration created severe labor shortages, causing ranch wages to double. By 1945, some five hundred thousand Texans had migrated from rural counties to join the industrial workforce—a loss that was never rectified. By 1950, for the first time in state history, more Texans lived in cities than in the countryside. Ranchers streamlined operations to make up the difference. Replacing men with machinery allowed some to hire fewer workers, but the price tag made it an expensive solution. Meanwhile, wartime rationing and reduced imports pinched supplies of basic food items, rubber, and gasoline, even as routine ranch equipment was dramatically increasing in price. By 1948, wholesale prices had more than doubled since the beginning of the decade, having risen at an average rate of 8.2 percent per year.

For many Texas sheep and goat raisers, the combined effects of labor shortages and rising operating costs forced them out of the business. The number of sheep on Texas ranches fell by nearly a quarter, from 11 million in 1943 to fewer than 6.5 million by 1949.[5]

Ranchers in the Big Bend were far less affected—at least partly due to the availability of cheap labor from Mexico. Another factor was rangeland that could support either cattle or sheep—a flexibility not available in many other parts of the state. If it had been, sheep and goat raisers elsewhere might have transitioned to cattle as a less labor intensive alternative. Lacking such options, many south-central Texas operations folded. By contrast, in the Big Bend, sheep numbers increased eightfold over the course of the decade. Presidio County had fewer than 26,000 sheep in 1940; by 1950 it had more than 206,000. As existing sheep and goat ranchers stocked up, many cattle ranchers transitioned to take advantage of wartime demand. The surge in sheep was attended by a 34 percent decline in cattle and a 22 percent decline in goats over the same period.[6]

As the county became increasingly devoted to sheep, infrastructure soon followed. In 1942, the Highland Sheep and Goat Raisers Association was founded in Marfa, and the following year Harper Rawlings built the Marfa Wool and Mohair Warehouse, with a capac-

Working sheep at the Prieto Ranch headquarters. Photo courtesy of the Prieto family collection.

ity of more than a million pounds. Although cattle numbers would rise again during the 1960s, the beef industry in the region would never fully recover. In 1945, the esteemed Highland Hereford Breeders Association held the final Highland Fair and Feeder Sale. The glory days of the Highland cattle kingdom had passed.[7]

The meteoric rise in the number of sheep in Presidio County during the 1940s did more than eclipse the cattle industry. It also triggered a spike in predator populations. Even though mountain lions, wolves, and even coyotes were known to take down an occasional calf, lambs and kid goats were pursued by a far wider range of predators, including bobcats and eagles. In response, nearly forty federally employed trappers were assigned to the area, killing around 4,000 animals in 1941 alone, including 3,000 coyotes, 935 bobcats, 24 mountain lions, and a wolf. But when it came to golden eagles, the predator war reached epic proportions. As the Big Bend sheep and goat raisers' greatest nemesis, golden eagles wreaked havoc on area ranches. One Marfa rancher claimed to have lost eight hundred lambs in only two months. To combat the problem, ranchers from Brewster, Presidio, and Jeff Davis Counties organized the Eagle Club in 1942 and hired pilot J. O. Casparis of Alpine to wage aerial warfare against the birds. It proved incredibly effective. Over the next six years, Casparis shotgunned nearly 5,000 eagles—a feat that renowned Texas naturalist Roy Bedichek lamented had no equal. But Casparis was not alone. Until the practice was banned twenty years later, an estimated 20,000 eagles were killed from aircraft across the Southwest in the most prolific state-sponsored slaughter of raptors ever known.[8]

The Prieto Boys in Mexico

My grandfather sent them to Mexico to be safe, and they weren't.

—Ida Benavidez-Taulbee

Following Pablo Prieto's murder and the exoneration of Hale, fear and anger peaked among Hispanics in the borderlands, especially toward the dreaded *pinche Rinches*. Some believed the Rangers

intended to run the Prietos—and any other "Mexican troublemak-
ers"—out of the region. If José ever had faith in the US justice system,
surely those days had passed. Unable to offer his sons safety in the
United States, he turned to the *madre patria*—the Prietos' homeland.
By 1930, Victoriano was living and working on Rancho de Alamo
north of Chihuahua City. With four children—Francisco, Josefina,
Jesús, and Domingo—he and Martina were one of six families living on the ranch. The ranch headquarters was like a village. Now
thirty-eight years old, Victoriano was older than all but six of the
residents—making him a senior employee and probably one of the
ranch's top hands. It was only natural that José would think of Mexico and his son Victoriano, who had carved out a life for himself and
his family over the course of the last decade. There was no reason
Gregorio and José Jr. couldn't do the same. But José didn't want to
leave it to chance. He needed his sons to be safe, but he also wanted
them to be successful—and what he knew about success in his own
life had always come from ranching.[9]

Likely within days of Gregorio's release from Dr. Peterson's clinic,
José was pressing cash into his hands and those of José Jr. for gas,
food, a few months' rent. And, with that, they packed up their families and turned south into Mexico to join their brother. It would
have been a long drive down rough dirt roads to get there. Arriving at Aldama, they were probably met by Victoriano, who would
have ushered them to his house at the ranch. Before long, however,
both men moved their families into the nearby hamlet of El Carrizo,
where José Jr. found work as a well driller. After Gregorio's money
ran out and his credit line from his brothers had been exhausted, he
found work cowboying on nearby ranches.[10]

Right-Hand Woman

**Mother was the only lady who would go out with grandpa . . .
she was his right-hand man.**

—Manny Perez

With his sons down in Mexico, José had lost his principal source
of labor. He had hired up when needed and usually kept a man or

two from Ruidosa or Barrancas, but otherwise his sons had provided much of the year-round labor the ranch required, for many hands made light work. But with his sons no longer available, José relied increasingly on his wife and daughters, even as their own availability diminished. Petra was the first to marry, leaving only Sotera, Frances, Lupe, and Jane. While they all possessed the skills and could make a hand in a pinch, not all of them enjoyed the work nor were they all equally proficient. Of the four, Lupe and Frances helped him the most. Frances especially enjoyed working beside her father, didn't mind long days in the saddle, and could ride, build fence, shoe horses, and shear a goat as well as any of her brothers. Now it was her time to shine, stepping in to fill the void, becoming José's right-hand woman.[11]

José Prieto and his daughter Frances, ca. 1935. Photo courtesy of the Prieto family collection.

In fact, it was not unusual for women to take up ranch work. Although boys tended to receive more training in such undertakings than girls, society was less formal below the Rimrock, and so was the division of labor by gender. Because marginal profits left little to spend on hired help, many remained reliant solely on family members to operate their farms and ranches. The standard division of labor was one born of genetic distinctions such as greater muscle mass and more aggressive tendencies that gave men an advantage when it came to hard manual labor and the rough-and-tumble nature of ranch work. But you used what you had, which gave some families a distinct advantage over others. That was at least part of the reason James Wilson never attained the success that José did. Having four sons, José had been lucky. But relying on their help was now a thing of the past.

Juan Benavidez and a Different Style of Success

It was his way or the highway.

—Teresa Benavidez

Well before José sent his sons to Mexico, his daughters had already started to fly the coop. In 1929, Petra became the first of José's daughters to leave when she married Juan Benavidez. Born in 1905 in Sanderson, Texas, to Paula Maldonado and Blas Benavidez, Juan spent his early years helping his father with his shearing business. Along with a crew of men, Blas and Juan traveled the seasonal shearing circuit with either a truck or trailer fitted with drive belts and pulleys that ran mechanical clippers, typically using the powertrain of the truck. From the Sanderson area, they worked eastward toward Dryden and westward at least as far as Valentine, where the nearby Sierra Vieja lowlands were the local epicenter for sheep and goat raising. It was while shearing at the Prieto Ranch that Juan first met Petra. If they dated at all, it was a brief courtship. The two were wed in the Catholic church in Marfa the day after Christmas, after which Juan took his new wife back to his father's ranch outside of Sanderson. Over the next three years, Petra gave birth to their only two children—Benjamin in 1931 and Ida the following year.[12]

Juan Benavidez with an Angora goat, ca. 1940. Photo courtesy of Ida Benavidez-Taulbee.

José was probably excited to receive his first son-in-law into the family, especially one who was a professional shearer. Regardless, it would be several years before Juan started working for José. In 1936, Juan and Petra moved into the old Nuñez house on the Prieto Ranch, where Gregorio and Pilar had lived before leaving for Mexico. In exchange for Juan's help, José offered to go halves on the kid crop—the standard unit of exchange to family members, one that allowed his children and their spouses to start herds of their own. But Juan had only worked for a year or so before their relationship began to sour. Juan claimed José reneged on the deal and kept all the kids for himself. If so, it would have been uncharacteristic of him—and he may have had a good reason for doing so. Regardless, the event served to confuse Frances, who had never seen her father angry at anyone.[13]

Juan decided to strike out on his own. In 1940, finding nearby land for sale by the Denison and Pacific Railway Company, he paid the first installment on three sections—two of which abutted José's original homestead section at El Tanque de la Ese and were only about four and a half miles by road from the Prieto ranch house.

The third was on the northern flanks of Sierra Parda about four miles to the south. Although his were all patented, nonstate sections having no homestead requirement, Juan wasted no time in making improvements. On a flat piece of ground along an upper tributary to Arroyo Escondido, he and a friend threw together a makeshift shack out of scavenged lumber and corrugated sheet metal. Thin shafts of light streaked through the empty nail holes. If it rained, buckets were placed accordingly. In the meantime, Juan started saving money in earnest by shearing sheep and goats on local ranches, all the while slowly building up his own goat herd. From such meager beginnings, his ranch would eventually become a successful and enduring operation.[14]

The Benavidez family had lived in the tin shed for only a few months before Juan started building a more permanent adobe house. Meanwhile, as money became available, Juan slowly expanded his ranch. Over the next few decades he would purchase a total of eight sections, which formed a nearly contiguous block around his headquarters. Extending from Pinto Creek northward almost to the Chinati Hot Springs, most of his land was rolling chino grass hills that ran along the front of the Sierra Vieja. As he came into more land and livestock, his shearing business slowed until finally he was exclusively working his own herd—one that eventually swelled to some three thousand sheep and five thousand goats. His was a success born of ambition and a good deal of goat sense. At a glance he knew a good mohair goat from a bad one. Those not up to his standard were sold for meat or simply eaten at home. And despite a lack of education, he had an easy way with numbers. By threes, he could count a herd of sheep or goats with rapid speed, his final tally almost always right on the money.[15]

Most of the year Juan worked alone or with the aid of Ben or Ida. But during shearing time Juan hired extra help. One of those hired hands was Chano Porras, from Barrancas. He was a master of his work, able to shear as many as 125 animals a day, but like most men Juan hired, Chano was an illegal worker. The cost of hiring him was low, but it was always a calculated trade-off against the risk of being caught. He minimized it by always having one man as a lookout. As the shearers bent to their tasks at the "six-handle" stationary shearing machine Juan set up at headquarters, the lookout scrambled to the top of a light pole near the entrance to the ranch and literally posted watch. If the lookout spotted the Border Patrol, he signaled the men below, who promptly dropped their clippers and ran to the creek to hide. Not that the officers didn't notice the corral full of sheep and the abandoned clippers. But with no men to apprehend, it became a cat-and-mouse game. Fortunately for Juan, enforcement was more relaxed in those days. Even so, there was still an occasional raid—and associated fines. As a result, it paid to take precautions.[16]

Even with undocumented workers, labor costs mounted—from a dollar a day to three dollars a day per person or more, leaving Juan reluctant to use outside help. Instead the burden fell increasingly on his family. For Ben and Ida, that meant childhoods lacking much

Benavidez family at the ranch. From left: Juan, Petra, Ida, and Ben. Photo courtesy of Ida Benavidez-Taulbee.

of anything resembling a social life. When his children were old enough to attend school, Juan purchased property in Marfa, where the children could live during the week. But as soon as they were let out on Friday, Juan was waiting with the truck full of supplies, ready to take the children home to the ranch. On weekends, when other children might be riding bikes or playing *las escondidas* (hide and seek) or *la patada del bote* (kick the can), Ben and Ida were fixing water gaps, pulling wells, leathering pumps, and working sheep and goats. "Our life was real rough," recalled Ida. "The wool and mohair we took to Marfa Wool and Mohair. If it sold, we sold it. If not, it was there in the warehouse until someone would buy it. So my dad would go to the grocery store and charge it and charge it and charge it until he got money to pay."[17]

Although Petra spent part of every Sunday at an altar she arranged in a corner of the house in accordance with her Catholic upbringing, Juan had no interest in religion, which he considered mere superstition. Unwilling to allow religious traditions to cramp his style, he treated Christmas and Easter like any other day. They might slaughter a pig or a goat for dinner but only after a full day of work. And there was never a tree or any presents. In fact, on some occasions Juan would take things away. Ida's favorite horse was a jet-black gelding loyal to her and her alone. If anyone else tried to mount him, they were likely to be bitten for their effort. Because Ida was too short to step into the stirrup and mount him, the gelding would stretch out his legs and thus lower his body so she could clamber up. In the evenings after dinner, he would wait by the screen door for mesquite beans and scraps of tortillas Ida had saved. But despite her undying devotion to the horse, one day without warning Juan sold it to a deer hunter who wanted it for his son. Ida begged and cried, but Juan never let sentiment cloud a business decision. They needed the money, and that's all there was to it.[18]

Indeed, Juan was famously tight-fisted. Some might say miserly. Unless there was no other option, he refused credit so as to avoid having to make interest payments. Instead, he carried a wad of cash in his pocket for purchases, including, at times, enough even for big-ticket items such as a ranch truck. By the same impulse, nothing was bought that could be grown. They kept chickens and hogs and tended a garden that kept them in chiles, squash, carrots, tomatoes, and the occasional watermelon. But aside from *carne asada* and *chicharrones* from slaughtering a pig around Christmas, the only meat consumed was from the nonproducing animals of their Angora goat herd, with occasional "mountain oysters"—the testicles of male Angoras—saved after the castrating of young billies in the spring. His children would not know the taste of beef until after leaving the ranch. He likewise disdained fruit or candy as extravagances. Over the protests of his children and grandchildren, he would claim he had grown up eating little more than pinto beans and *fideo*—thin Mexican noodles—and had turned out just fine. Food was for nourishment, not for fun.[19]

According to some stories, Juan's famous frugality occasionally crossed the line. Since one of his sections was surrounded by the Prieto Ranch, he sometimes had to herd his animals through José's pastures. If some of José's goats got mixed in with his along the way, they might not be returned. One of José's sons-in-law recalled finding where a goat from the Prieto Ranch had been killed beside the road. He followed the trail of blood all the way to the kitchen door of Juan's house. Recklessly, Juan once advised his brother-in-law in a moment of blind self-incrimination that a rancher should never eat his own animals when he could just as easily eat those of his neighbors. But José disdained conflict. Even when he believed a theft had taken place, he refused to confront Juan. Instead he reinforced in those around him the value of honesty. "José believed in the straight road," recalled his grandson Oscar Perez. "He was a very, very honest man. If one of your animals got out, he would make sure it got back to you."[20]

If it was true that Juan sometimes failed to return José's goats, he almost certainly felt justified in doing so. Remembering the kid crop he believed José owed him, in his eyes it may have been a simple matter of settling a debt. But he also clearly harbored resentment. Whenever someone mentioned a Prieto in conversation, Juan commonly erupted with expletives. Petra—a calm counterpoint to Juan's periodic fulminations—always remained quiet and reserved. "I never heard her yell," recalled her granddaughter Teresa. "She never yelled at my grandfather. She never yelled at us." More inclined to laugh, Petra's good humor and even-keeled temperament helped balance out her explosive husband. "No one else could have put up with my grandfather," claimed Teresa. Still, even those who saw him as unscrupulous found it hard to argue that Juan Benavidez had been anything but successful.[21]

Manuel Perez, the Sharecropper's Son

In 1936, José's third daughter, Frances, married Manuel Perez. The son of Petra Vasquez and José Perez, Manuel was born at Ochoa, about twenty-five miles southeast of Ruidosa, in 1911, on the cusp of the Mexican Revolution. Prior to immigrating to the United States, the family had lived on a ranch near Peguis Canyon, some twenty miles west of Ojinaga, where they were sitting targets for Pancho Villa, who was ranging about the countryside seeking commissary for his troops. Family lore holds that one day Villa and about two hundred of his men descended on the Perez family's ranch, pitched camp, and ordered all the livestock seized to feed his troops. Two weeks later Villa's column, many on Perez horses, marched on, but not before recruiting José's fourteen-year-old son to his ranks. When José protested, Villa simply said it was *por la causa* and that he was lucky they didn't take more. Manuel's older brother remained with the Villistas for two months before escaping on foot back to the ranch.[22]

José had his son back, but the family had lost everything. They packed up what they could carry and moved across the Rio Grande to Texas, just like a million others who fled Mexico during the revolution. Having fifteen children, however, José probably chose to avoid the official crossing at Presidio and the eighteen-dollar-per-person visa charge and consular fee. They most likely crossed at Vado de Piedra or perhaps closer to where the family settled in the tiny farming community of Ochoa, ten miles northwest of Presidio. Ochoa was not an old town; it had gradually developed around the farm of Esteban Ochoa, grandson of Presidio pioneer Ben Leaton. By the time of the Presidio Valley irrigation and cotton boom starting around 1914, the community had swelled to some seventy-five families.[23]

The Mexican Revolution helped populate the town. José was likely one of the hundred or so refugees that Esteban Ochoa hired to construct an irrigation ditch for his farm. Like most of the other families, the Perez family stayed on as sharecroppers. Whatever José was able to grow on his farm, he could keep one-third, the balance going to Esteban. Although Ochoa got a school around 1911, education was never an option for young Manuel. Instead, his father insisted that he work to help support the family. Every day he headed into the Chinatis leading a pack burro that he loaded with wood he gathered in the mountains to be exchanged for beans, coffee, sugar, and shoes. He didn't care for the work, but nothing was worse than his father's temper. One day Manuel returned to find the milk cows had gotten out and trampled the corn seedlings. It did not matter he was not to blame. José whipped him with a pair of horse reins. But it was sometimes far worse. Manuel carried two knots in his back throughout his life after being beaten with an ax handle. That may have been the reason he never lifted a finger in anger toward his own children. He'd seen enough violence for one lifetime.[24]

Manuel left home as soon as he was able. Perhaps it was because of the memories of his childhood, but Manuel would never farm again. The work he enjoyed most was on horseback, but he took whatever

Manuel Perez in Pinto Canyon, ca. 1945. Photo courtesy of the Prieto family collection.

work he could find. During the Great Depression, Manuel was one of dozens of workers on the county relief rolls who built stone culverts and embankments on Pinto Canyon Road. It was around that time that Manuel met Frances, probably at a dance in Ruidosa. They married in Marfa in June 1936 and had their first child, Ruben, in 1937. By that time Manuel was working at nearby Rancho Berrendo, but around 1941 he began working for José Prieto.[25]

For weeks at a time, Manuel and José would camp back in the mountains, leaving Frances with Juanita at the ranch house. Once a week or so they rode back to the house to resupply, filling panniers with coffee, beans, and tortillas, sometimes a bit of dried meat, and lard that Juanita rendered from lambs. Manuel could have done without lard; the very smell of it set him retching. Instead, he supplemented their diet with wild game. One day while scouting for game, he sat down on a ledge for a smoke. As he rolled his cigarette, he casually looked down to see a big buck just below him, completely unaware of his presence. He set down his tobacco and quietly grabbed a large rock. Taking careful aim, he hurled it downward, hitting the deer in the head, killing it. Back at camp as the two men set to skinning it, José scratched his head as he looked for the bullet hole. Finally, he asked where he shot it, as a wide grin spread across Manuel's face.[26]

Before long, Manuel and Frances moved into the old Nuñez house that would later carry the Perez name. They were glad to have a place of their own, for by that time, the family had grown. Carlos was born in 1939 and Manuel Jr., or "Manny," followed a few years later. Although the house was an upgrade, it was still relatively primitive. Water had to be hauled up from Pinto Creek—provided it was running. If not, water had to be hauled in barrels from nearby Ojo Acebuche or from the Ruidosa Hot Springs five miles to the northwest. If one traveled as far as the hot springs, it only made sense to wash clothes and, if time allowed, take a bath. While her three boys splashed in the water down by the creek, Frances bent to her washboard beside the small waterwheel that turned lazily just above the bathhouse.[27]

Manuel could not have picked a better time to start ranching on his own. It was a time when the market was strong and the rains fell with ease, allowing him to make more money than he had ever seen.

One year Manuel did particularly well. After paying José a year's worth of rent, he bought a new truck and new beds for the family. For the children, it was a Christmas to beat all—with new clothes, store-bought toys, and even tricycles. Still, as her mother had done throughout her life, Frances supplemented their income by making and selling *asadero*—cheese made from goat's milk. It was the boys' job to collect the hard yellow berries of the prickly *trompillo*, or silverleaf nightshade. The crushed berries caused the curd to separate from the whey. After skimming out the curds, Frances cooked them slowly on a *comal* or skillet, fashioning them into flat cheese "tortillas." These she sold door-to-door in Marfa, and the sales of those cheeses, as well as lard and kid goats for *cabrito*, usually paid for the family's in-town purchases.[28]

Juan Morales, a Horse of a Different Color

He never grew up. He was always a kid.

—Oscar Perez

Although Sue was José and Juanita Prieto's eldest daughter, she would be the second to last to marry. Her younger sister, Lupe, would wait longer, but not by much; her marriage to Marfa business owner Adam Miller followed Sue's by only a few months. Age forty-two at the time of her marriage, Sue was well outside the bridal norms of that era. She may have been comfortable living with her parents, and certainly Juanita appreciated her help around the house. Or perhaps she had been holding out for just the right man. If so, she finally found him in one who became the most beloved of the Prieto family—and more of a handful than Sue could ever have predicted.[29]

His name was Juan Palomino Morales. Wearing a mischievous grin, as if it were a matter of great pride, he explained that his name meant "Johnny horse feedbag." Born in 1899, he was the eldest of nine children raised by Ruidosa farmer Leonardo Morales and his wife, Porfiria Orona. Although Juan attended the Ruidosa School, he disliked it so much that he once tossed the scent gland of a skunk into the woodstove, causing the suspension of classes for several

Juan Morales at Rancho Alamito, ca. 1945. Photo courtesy of the Prieto family collection.

days. Juan much preferred the cowboy's life. As a young man, he worked on ranches around Ruidosa and above the Rimrock—the Brite Ranch and the old Cibolo Creek Ranch. At sixteen he married Guillerma Hinojos, from the quicksilver mining village of Terlingua about eighty miles to the southeast. But their union did not last. In 1935 Juan filed for and was granted a divorce, the court finding his allegations—likely infidelity—reasonably proven.[30]

Probably having met at a dance in Ruidosa, Juan Morales and Sue Prieto married in August 1940 at the Prieto house in Marfa. A short, stocky man with distinctively Indian features, Juan was probably the most fashion conscious of any of his peers in the borderlands. His white hat was always tilted just so, his clothes clean and pressed. But on days he went to town, Juan dressed to the nines—a new pearl-snap western shirt, ironed blue jeans with the cuffs rolled up to accommodate his short legs, a leather belt with his name tooled on the back, and an ornate belt buckle. Although short of stature and bearing a slight stutter, Juan was quick to laugh, mischievous, and had a gift for holding forth on the comic exploits of his past.[31]

Juan had been working for José for some time before they decided it would be good to have a camp house near the back southwestern

A view of Rancho Alamito toward the west. Photo courtesy of the Prieto family collection.

corner of the ranch. Using local stonemasons, José had a rock house built just below the toeslope of Sierra Parda. With the house, a set of pens, and a bit of cross fencing, a third ranch was created out of the larger Prieto place—what Juan called Rancho Alamito. As he had with Manuel, José set Juan up with enough sheep and goats to get started and charged a reasonable and flexible yearly rent. It would be Juan and Sue's home for the next twenty-five years.[32]

Sited below the craggy cliffs of Sierra Parda, the new house was arguably in the most scenic location on the ranch—but also the most inaccessible. If the place seemed ill-suited to the most out-going of the extended Prieto family, it was an isolation that Juan would later use to great advantage. Though their home was remote, for the many years they were there Juan and Sue maintained a tradition of extending invitations to the children of friends, relatives, and neighbors to stay at the ranch, often for weeks or months at a time. Unable to have children of their own, theirs was a door many young boys passed through over the years to work at feeding chickens, collecting eggs, milking goats, pulling weeds, and gather-

Juan Morales with two Angora billies. Photo courtesy of the Prieto family collection.

ing firewood. But Juan also taught them the skills of the cowboy's trade—how to ride a horse and round up livestock, fix fence, shear sheep and goats. And then there was Juan's gauntlet of practical jokes, for Juan was also an incurable prankster.[33]

His antics were legendary. Anyone who'd spent time around Juan could rattle off their favorite Juan Morales story, which usually involved being the butt of his jokes. If no one was home when Juan passed by the Perez house, for example, he couldn't help but sneak inside and make certain adjustments, like swapping out sugar for salt. If he found an unoccupied vehicle, it was hard for him to resist the urge to jack it up and place blocks under the rear axle, leaving the rear tires off the ground by half an inch. The next driver would be left in a state of utter confusion as the rear wheels spun impotently, the vehicle still as a stone. But most of the time it was the young and gullible who became the hapless victims of his shenanigans. One of his favorites was the coin-in-the-funnel game. He would have children balance a quarter on their forehead, and the challenge was for them to tilt their head down and drop the coin into a funnel wedged in their pants. If they made it, Juan told them, they could keep the money. As the boy concentrated on the quarter, probably thinking about the candy he might buy with the prize, Juan would empty a glass of water down the funnel. One boy who spent the night at the ranch was bewildered to find his foot wouldn't fit into his boot the next morning. There was something inside. When he turned it upside down, a dead rattlesnake fell out. Just out of view, Juan could always be found shaking in quiet laughter.[34]

If he could find no better targets, Juan's hired help would do in a pinch. He once had a man from Mexico name Lache, a highly devoted employee if not exceptionally bright. Feeling his mischievous urges rising, one day Juan told Lache that Pancho Villa was going to attack the ranch and they would have to protect it.

"What do you want me to do?" asked Lache, very much concerned.

"I want you to hide and post watch," replied Juan, who then directed Lache to an old pumpjack with a tarp over it and told him to get under the tarp and not make a sound. After Lache had hidden himself, Juan aimed the water sprinkler toward the tarp and turned it on. Later, long after dark, Juan started shooting his rifle into the air and yelling "Viva Villa!" The next day, he opened the tarp to reveal a moist and frightened Lache, still faithfully posting guard. He said he heard Villa, but never saw him because it had rained all night long. Juan once asked another worker if he'd ever been on a Ferris wheel. When the man said no, Juan tied him to a tractor wheel and sent it careening down the road. But certainly one of his most ambitious jokes was the time he convinced two of his workers to get married, that it wasn't right for them to sleep in the same bunkhouse otherwise. When they reluctantly agreed, he dressed one of the men in Sue's clothing, had the other don his best outfit, while Juan—Bible in hand—officiated the wedding. If Juan could find no ready victims for his pranks, he'd improvise. One day the Perez family arrived at the Rancho Alamito to find Juan's horse painted blue. After the laughter died down, they asked why he would do such a thing. Juan just shrugged and said he simply wanted a horse of a different color.[35]

Manuel Perez's youngest son, Manny, was particularly close to his uncle Juan—who became his *padrino* (godfather), much to the envy of other boys who vied for Juan's attention. For months, Manny stayed with Juan and Sue, his absence probably easing the burden for Frances. And despite his youthful proclivities, Juan proved to be a capable role model for impressionable young Manny, for aside from his juvenile sense of humor, Juan was also an exceptionally hard worker. Although he "was a short man," recalled Manny, "to me he stood seven feet tall." By the first light of dawn, the pair would set out, Juan riding high on a red horse named Enmascarado (masked one) for the white blaze on its head, and Manny dwarfed by his mount—a palomino named Baby. Riding across the long *cordones* (rolling hills) spanning the northwestern part of the ranch, the two checked water gaps, dropped off salt blocks, and checked on the sheep and goats scattered around the ranch.[36]

Roundup was an especially exciting time for Manny. With a crew of cowboys, Juan and Manny would ride out around three in the morning, their food and bedrolls tied behind their saddles. Sunrise found them high up on the flanks of Sierra Parda. When it finally grew too steep for the horses, they would set out on foot, one man leading the horses back down to the bottom. The rest scaled the steep slopes until reaching the summit, where Juan maintained a remote camp—little more than a modified rock shelter stocked with a few canned goods and some dried jerky. The next day the crew would awaken early and begin herding the goats down the mountain by foot, rolling boulders down the mountain to drive the animals before them. Once gathered near the toeslope, they remounted their horses and drove their herd the rest of the way to the corrals.[37]

During the long, lazy days of summer between lambing and shearing and episodic floods that blew out the water gaps, Juan found time for more leisurely pursuits. One of the drainages off the western flanks of Parda was known as La Mota, and it had a cold freshwater spring that ran over bedrock, creating a waterfall. On a nearby tree hung a can with a long wire handle Juan could use to dip water out for a drink without having to dismount. A long strand of barbed wire stretched between two trees served as a rack to dry venison out of reach of scavengers. Passing by, Juan would grab a piece of jerky from the line and hand it to Manny. With the summer heat shimmering in waves over the desert flats below, he and Juan would tie their horses, hang their clothes in a tree, and sit beneath the waterfall, letting the cool water drain the heat out of their bodies.[38]

For entertainment, Juan built a small goat-roping arena beside the corrals. Using a pair of jumper cables and a record player, Juan jerry-rigged an outdoor PA system, and between deejaying songs, he served as rodeo announcer. As young Manny readied his loop, Juan held forth with his best radio voice: "And now, ladies and gentlemen, from the Alamitos Ranch is Manny Perez on his horse Baby!" Taking swigs from his burlap-covered bottle of sotol, Juan continued with the announcements, which became ever more animated as they echoed off the rocky slopes of Sierra Parda. Of course, there were times when Juan twisted off a bit more than usual. Despite serving as a part-time deputy sheriff at parties in Ruidosa where he was supposed to maintain order, Juan was often the drunkest and rowdiest person there. It had the effect of keeping fellow revelers more concerned about Juan's state of mind than getting busted themselves, there being few things as unsettling as a drunk with a badge and a gun. Later, if Juan was unable to keep his truck on the road, he simply put Manny in his lap to do the steering. But arriving at home safely was only part of the problem. For he still had to find a way to explain things to Sue as she stood waiting at the door with her arms crossed and lips pursed.[39]

Sue and Juan Morales at one of his famous barbecues. Photo courtesy of the Prieto family collection.

But if Juan had a weakness for partying, he also had a weakness for the less fortunate—the young and old, infirm and homeless. When he wasn't looking to play a joke on someone, he was usually thinking of a way to help them. Rarely did he make a trip to Marfa without leaving behind a kid goat for some *viejita* or for a family of hungry children. Kids flocked to his side as he handed out nickels and dimes so they could buy a piece of candy. When one old man, Gilberto Martinez from Ruidosa, had finished work for José or Manuel, Juan took him in at his ranch for another month or so, in exchange for nothing more than the old man's company and having to endure Juan's relentless practical jokes.[40]

El Rancho del Pando

By the early 1940s, José Prieto was probably confident his sons would remain in Mexico. Or perhaps it had been his plan all along. But around 1943 he made a trip to El Carrizo, where he and Victoriano scouted for ranches to buy. They eventually found what they were looking for about twenty-five miles northeast of Chihuahua City and just west of El Carrizo, along the eastern flanks of the Sierra del Cuervo (Crow Mountains). At only three sections, El Rancho del Pando (meaning something like "the bent-over-backward ranch") was not large, but it could serve as a base of operations his sons could call their own.[41]

José remained long enough to complete the transfer of title and to purchase a herd of cattle to stock the ranch. He also made it clear to his three sons this ranch was theirs to share and build on as they could. It would also constitute the bulk of their inheritance, meaning Pinto Canyon Ranch would go to their sisters. The herd was to be divided into four equal parts, with José and each son retaining an equal share. Being the eldest, the most experienced in ranch work, and having the largest family, it made sense to leave Victoriano in charge. Since he also would tend José's cattle, he was to receive 50 percent of that calf crop on top of his own.[42]

Being one hundred miles south-southwest of Ojinaga, Victoriano and his family could visit Pinto Canyon only infrequently. Still, they made the journey at least once between 1940 and 1946. Less encumbered by children, José and Juanita made more frequent trips to visit

Map showing location of El Carrizo and the Prieto Ranch.. Map by D. Keller.
Used with permission. Copyright © 2018 Esri, ArcMAP, National Geographic,
DeLorme, HERE, UNEPWCMC, USGS, NASA, ESA, METI, NRCAN, GEBCO,
NOAA, increment P Corp., and the GIS user community. All rights reserved.

their sons. In 1943 the *Big Bend Sentinel* noted, "The José Prietos have
returned from Chihuahua, Mexico, where they visited their three
sons on the Carrizo Ranch." It may have been the same trip that
Manuel Perez's oldest son, Ruben, recalled when he got in trouble
for smashing some of the watermelons growing in Victoriano and
Martina's huge garden.[43]

With a world war raging overseas, José Prieto was doubly glad
his sons were not one of the ten million Americans drafted into
service. A pacifist at heart, he could not condone his sons fighting
wars—perhaps even less so in support of a country whose judicial
system had failed him. In having sent his sons to Mexico to protect
them from the Texas Rangers, he had also protected them from the
war. But Gregorio might have been better off drafted, for trouble
would follow him regardless of where he went. As if his drinking
and carousing hadn't tarnished his reputation enough, he damaged
it further by killing a man in cold blood.

One day Gregorio walked into a cantina in El Carrizo and, stepping up to a man passed out at the bar, pressed the barrel of his revolver against the sleeping man's head and blew his brains out. There was undoubtedly bad blood between them, but Gregorio's cowardly act landed him in prison in Chihuahua City, where he would remain for the next year. Meanwhile, without Gregorio to provide for them, Pilar and his two daughters would have been destitute had Victoriano not taken them in. But if Gregorio was ever grateful for his brother's help, it never showed. Instead, he acted as if it was a thing owed him, as if all good things to come his way had been divinely ordained.[44]

If José hoped the Mexico ranch might encourage his sons to work together and manage their affairs equitably, he would be sorely disappointed. Instead, Gregorio complained bitterly that Victoriano had received preferential treatment from their father. Taut as a drumhead, he was primed to erupt. In Gregorio's eyes, the two long-eared hound dog puppies that José had given to Victoriano's family proved they had been favored. One day while pitching hay in the barn, perhaps obsessing over the insult he felt, Gregorio in a sudden rage speared one of the puppies with the pitchfork. The injured puppy was still alive, but Victoriano was furious. This time Gregorio had really crossed the line. Squaring off, the normally mild-mannered Victoriano ordered his brother to take the dog to the vet to get it stitched up. And then to never return.[45]

Ranching in the 1940s

The 1940s would prove to be the last great decade for ranching in Pinto Canyon and the Sierra Vieja lowlands—pastures bursting with sheep and goats, hillsides a moving sea of white wool and mohair. Ample moisture and strong markets buttressed the industry like never before. And problems arising in other parts of the state were less acute in the Big Bend, the ruggedness precluding the degree of mechanization occurring elsewhere, for with the exception of farm trucks, 4×4 surplus Jeeps, and mechanical wool clippers, ranching in the lowlands remained much the same as it always had. In a word, primitive.

Sheep and goat raising remained labor intensive and followed a rigid schedule that corresponded to the seasons and the life cycle of livestock. Warm Texas weather typically allowed operators to shear twice yearly without fear of cold stress. The work itself was done by specialized itinerant shearing crews, called *tasinques*. Ranchers in Pinto Canyon and the Sierra Vieja typically hired crews from the Sanderson area—the epicenter of sheep and goat raising west of the Pecos. Crews arrived completely self-contained. The shearing rig was a truck or trailer custom-built to accommodate ten sets of mechanical clippers, called "hands," which were belt run, typically off the drive shaft of the truck. Crews were headed by a *capitán*, who oversaw the work, owned and maintained the equipment, kept the crew fed and sheltered, and arranged the contracts with individual ranchers. The *tasinques* were typically paid half the amount given the *capitán* for each sheep or goat shorn, with tokens or even stamped pieces of tin serving as legal tender within the outfit.[46]

Shearing was an orchestrated effort. The workers first sorted the animals by age and gender. They readied the equipment, greased and oiled the gears and couplings, and sharpened the shearing blades. They swept the shearing surface, usually a concrete slab,

Shearing crew clipping mohair at Rancho Alamito. Photo courtesy of the Prieto family collection.

clean. When all was ready, they started the truck engine, with the belt drive engaged, and the operation suddenly sprang to life with a great racket of rotating pulleys and squeaking belts and the mechanical clatter of a score of clippers. The shearers dragged the animals backward by their front or rear legs to the clipper stations. They balanced the sheep on their haunches and supported them with their thighs, whereas with goats, the shearers laid the animals on the floor with their legs tied. Once positioned, the animals usually became docile, as if in a trance. The shearers removed the belly wool first, then the left flank, rump, neck, and left foreleg, followed by long "blows" (clipper strokes) from the flank to the top of the head, finishing with the right foreleg, flank, and rump. From daybreak to dark, shearers labored over their animals, often at the rate of a hundred or more per day.[47]

At the back of the truck the *capitán* sat before an iron disk. After wiping the disk clean, he spread glue across it, followed by a powder-like grit. Once the glue dried, cementing the grit in place, he put the belt around the vertical shaft and engaged the drive, causing the disk to spin like a record player. One by one, the *capitán* took his collection of dull shearing combs and cutters and pressed them onto the disk, moving them forward and back from the center to the outer edge, sharpening the teeth in a spray of sparks, stopping occasionally to examine his progress. Once he had sharpened the combs and cutters, he set them on a leather pad to cool before returning them to the shearers, who traded them out for dull ones. Other men stood at a table ready with twine to tie up each fleece (or in the case of mohair, simply to gather it) before placing it in an eight- to ten-foot-long burlap sack suspended from a metal hoop stand. When the sacks were about half full, a boy was recruited to climb inside and stomp the wool or mohair down, packing it tightly. Once full, the sacks—weighing around 350 pounds each—were sewn shut with a long needle, marked according to age and grade of the fiber, and stacked in preparation for transport to town.[48]

Because time was money and scheduling was tight, the workers maintained a fast and furious pace. With eight to ten shearers going full tilt, an outfit could process a thousand or more sheep and goats per day. Depending on the size of the flock or herd, the shearing was usually finished in a matter of days. As soon as the last animal was

Manuel Perez (*far right*) sitting on a wool sack with his shearing crew and his sons Ruben (*second from left*), Carlos, and Mannie (*to his right*). Photo courtesy of the Prieto family collection.

shorn, the crew packed up and moved to the next ranch to start the process all over again. In their wake, flocks of dehaired sheep and goats were turned out upon the hillsides, threads of white following narrow trails through the mariola and acacia. Springtime shearings always carried risk. One year in April, José had just finished shearing when a late-season norther blew in. Lacking their fleece, three hundred of his animals perished in the cold.[49]

Following shearing, the animals were typically treated with insecticide. Biting and sucking lice, sheep keds, bot flies, mange and scab mites, blowflies, and ticks can all infest sheep and goats. Left untreated, they cause not only a decline in the quantity and quality of the fleece or hair but sometimes even the death of the animal. To prevent infestations, about two weeks to a month after shearing, all the animals were dipped in a medicinal solution. The technique was not particularly old. The first dipping vat in the state was built in 1865 by pioneer Texas sheep rancher George W. Kendall to control scab. After a bit of trial and error, Kendall had a stonemason construct a vat that soon became the standard for sheep and goat dipping across the country.[50]

The dipping vat itself usually consisted of a long and narrow but deep concrete trough set in the ground, with a ramp at the far end that allowed the animals to climb out. Beyond the ramp was a "dripping floor," or a concrete slab sloped to direct the solution back into the vat. The deep side or "front" of the vat was connected to a narrow chute that opened to a small pen at the back. Two men stood on either side of the vat. As the sheep were driven forward, each man grabbed one and dropped it into the solution. Meanwhile, two more men used long poles bearing curved metal bars to push the animal's head under the liquid. The sheep and goats emerged at the other side, the solution dripping from their bodies onto the sloped platform.[51]

If shearing was burdensome, lambing and kidding could be far worse—a period sometimes described as a "month long hell of worry and toil." Ewes (female sheep) and does (female goats) typically gave birth in mid-April, after the risk of freezing weather had passed. Shortly beforehand, the animals were brought closer to headquarters so they could be watched and protected from predators. The *hijadores* (lambers) helped with any difficult births, encouraged indifferent mothers to nurse, cared for orphaned or injured animals, and posted watch for predators. Lambing season was the herds' most vulnerable time of year, and protecting lambs and kids required vigilance. Being the second greatest source of income after wool and mohair, the lambs and kids had to survive if the ranch was to be a financial success.[52]

A week or two after birth, lambs and kids were castrated, both to produce better meat as well as to ensure that females would be bred only by choice sires. The lamb or kid then received an earmark—a mark unique to the ranch and registered with the county—cut in one or both ears. By slightly modifying the mark each year, the rancher could tell the age of any one sheep. José Prieto's earmark was a broad, shallow U-shaped cut on both ears just below the tip. In addition, he branded a single line across the animal's nose. At the end of the process, boys were put to work swabbing the fresh wounds with *tecole*, a homemade black smear made of pine tar, iodine, and creosotebush extract, said to smell so foul that "no self-respecting fly would come close to it."[53]

As the weather grew cooler, bucks were turned out with the nannies and the rams with the ewes. The date was critical in order to control the timing of births. Since sheep and goats gestate over a period of five months, breeding them in November means that lambing would begin in April. After about a month, the bucks and rams were removed. Better able to defend themselves against predators, they were turned out on the higher, more distant pastures. The nannies and ewes, on the other hand, were turned out on lower pastures closer to headquarters so it would be easier to watch over them. At last, for a few months, life on the ranch relaxed until it was shearing and lambing time again.[54]

Trouble in Mexico

Aside from the troubles with his brother, Victoriano was enjoying greater stability and success than ever before, thanks to the improved beef market brought by the war as well as the more agreeable weather conditions. Even so, problems remained. Rancho del Pando was adjoined by another ranch owned by the infamous Cano family. Chico Cano had acquired the ranch sometime before 1920—likely from the proceeds of his banditry or from smuggling Mexican cattle to US buyers. At the urging of his older brother, he retired there in the waning days of the revolution. After the US Cavalry ambushed him and his men in the spring of 1919, an event that left him with a bullet wound in his back, his days on the front lines were numbered. He turned away from the fighting and banditry that had dominated his life over the previous decade and settled into a quieter life.[55]

Because Chico Cano and his wife were unable to have children of their own, Chico's brother José—who had been prolific—urged the couple to adopt two of his own sons: Chico Jr., born in 1917, and Nicolas, born in 1926. Although Chico's bandit days were behind him, his drinking days were not, which soon became a problem. One day, his head swimming from sotol, Cano convinced a young boy to balance a bottle on top of his head. Cano took aim and fired but missed the bottle. The boy fell dead. Remorseful, Cano gave up the bottle for good. But fate seemed to pursue him. In 1941 Cano was thrown from his horse and suffered a serious head injury. He received treatment in a Juárez hospital for several weeks and convalesced at a friend's house for months afterward. But he was never the same. Chico lingered in a compromised state until August 1943, when he died of stomach cancer in Cedillos, Chihuahua.[56]

By the time Chico Cano died, whatever money he'd set aside was gone. Lacking the funds to pay for his father's funeral, Chico Jr. asked Victoriano for a $400 loan. Poor though he was, in the manner of his father he found the money to lend him. But nearly three years later, the loan was still outstanding. In March 1946, while in El Carrizo getting a permit to market his calves in Juárez, Victoriano ran into Chico Jr. and asked about the money, which he needed for his trip. Chico said he had the money but needed to go back to the ranch to get it. Victoriano drove Chico while his brother Nicolas followed in their truck. A few miles short of the ranch, the engine died on Victoriano's truck. Nicolas left to get a chain so they could tow him back to the house. He returned with Gregorio a short while later, by which time Victoriano and Chico had succeeded at getting the truck started. As night fell, Gregorio and Nicolas drove on, with Chico and Victoriano following behind. But upon nearing the ranch, Gregorio could no longer see Victoriano's headlights. The truck had stopped. Nicolas turned the truck around, and they had not gone far when they came upon Victoriano's truck in the bar ditch. Inside, Victoriano lay slumped over the wheel, the cab awash in blood. He had been stabbed seven times, the slashes on his hands betraying his desperate attempt to defend himself. Inside the glove box sat a revolver, just out of reach. Gregorio strained to see where Chico had gone, but it was too dark. He had disappeared into the night.[57]

By the time the coroner came with the sheriff, Chico had emerged from the darkness. He admitted having killed Victoriano but claimed it was self-defense, although he bore no wounds. The trial was brief, the court finding him guilty of murder. Like Gregorio before him, he was thrown in prison in Chihuahua City. Nevertheless, Victoriano's three oldest boys wanted revenge. Martina realized that if they stayed on the ranch, one of her boys was bound to kill Nicolas or be killed trying. Taking only a wheat mill and their personal effects, they left the ranch for good. The family resettled in the nearby town of San Pedro, where the boys found work in the fields and used their

wheat mill to make flour. But their pay was barely enough to support the family. If their finances had been tight before Victoriano's
death, they were now nearly destitute, the pain of loss accentuated
by poverty.[58]

Although Martina knew Gregorio was a scoundrel, she likely
hoped he would help her family the way they helped his when he
was in prison. But that was not what he did. Instead, with his brother out of the way, Gregorio took it upon himself to sell Rancho del
Pando, along with all the livestock—thus liquidating the Prietos'
assets in Mexico. The proceeds he kept for himself, giving nothing
to Victoriano's struggling family. If he felt justified, he may have reasoned that this move somehow evened the score—the unfair advantage given Victoriano by José, the fight they'd had after he injured
the puppy. Or perhaps fairness didn't even occur to him. The money
from the ranch would provide a comfortable living for him and his
family, allowing him to pay down his debts. Or fines levied against
him. One thing was certain—Gregorio felt no obligation to anyone
but himself.[59]

The Second Prieto Diaspora

Within a few years of Victoriano's murder, the Prietos slowly began to
filter back into the United States. Mexico had lost its luster. The place
they had gone seeking refuge had turned against them. Once again,
as it had been with José's family in the 1860s, the Prietos would turn
their backs on Mexico, even if their migrations trickled out over the
span of a decade. In 1949, Victoriano's daughter, Guadalupe, crossed
at Presidio, where she was received by José and Juanita. Over the next
several years, the rest of Victoriano's family followed. Most ended up
in Pecos, where they found jobs as migrant farm workers. José Jr. and
his wife Francisca also gradually migrated northward, moving for
a time to Coyame, where, in 1943, she gave birth to their youngest
son, Joe. By 1947 José Jr. was working as an auto mechanic in Ojinaga, where their second son, Filemon, died, probably from leuke-
mia. Finally, in 1952, José Jr. moved back to Marfa ahead of his family,
probably in response to pleas from his father, who was increasingly
in need of help around the ranch. Three years later, José Jr. brought
his family to Marfa.[60]

Gregorio would be the last to leave Mexico. Maybe he'd finally soured on the country or felt that greater opportunity awaited in the United States. More likely he'd run out of money, having blown through the proceeds from the sale of the ranch and livestock, and was out of options, or at least those options that didn't involve work. He set his sights on other prospects. His family still owned the ranch in Pinto Canyon, and with his last remaining brother being uninterested in ranching, he probably saw opportunity. Although Gregorio well knew the Pinto Canyon ranch was supposed to go to his sisters, certainly he could find a way back in to stake his claim.

∧∧∧

Pinto Canyon ranchers could not have known the 1940s would be the last decade of relative prosperity. World War II finally brought the country out of the Great Depression, assuring a strong market to feed and clothe the troops. In spite of government controls, ranchers were guaranteed good prices for their products. The spike in wool prices caused sheep and goat numbers to outpace cattle in Presidio County for the first time in sixty years, while ample precipitation allowed ranchers to stock the land to capacity. But the war also created labor shortages across the country. José Prieto suffered his own shortage after sending his sons to Mexico. Two of his sons-in-law helped ease the crunch, and José carved out portions of his ranch for each of them. But strained relations with his third son-in-law, Juan Benavidez, caused a quiet feud that simmered for years—a rift that divided a family most believed was indivisible. Meanwhile, Mexico was not the panacea José had hoped it would be. In exchange for an act of kindness—a loan to a neighbor in need—Victoriano was repaid with a violent death. It may have been circumstance. It may have been chance alone. The fact remains that there was one common thread between the death of Pablo and the death of Victoriano, and that was Gregorio.

8

Tierra Seca

A Parched Land for Lost Dreams

The sheep at the end would just cry because the water would never get to them. And then they started dying. It was the one time that Grampa cried.

—Manny Perez

And many a boy would become a man before the land was green again.

—Elmer Kelton, *The Time It Never Rained*

THE WINDMILL blades turned lazily in the morning sun. A trickle of water pulsed out of the pipe with each rise of the sucker rod. At ten o'clock, it was already almost a hundred degrees in Pinto Canyon. Sweat dripped off the brim of Manuel Perez's hat as he dipped water out of the tank and transferred it to the several fifty-five gallon barrels in the back of his pickup. He squinted up at the Chinatis, seared by the sun, a great angry orb of light and heat that burned away the clouds and dried the springs and scorched the grass to a crisp. When the barrels were filled almost to the top, he drove slowly down the road, trying to keep the water from sloshing out. Pulling up to where the bleating goats stood in front of the concrete trough, dry as a bone, he could hear their cries grow more shrill. As he began to pour water into the trough, they rushed in a mass frenzy to drink, the older and weaker ones pushed aside to cry in agony, unable to reach the water before it was gone. Others were already too weak to care, and some of them began to lie down in the dust and die. Tears rolled down Manuel's face as he closed the tailgate of his truck. Like some level of hell, this nightmare had descended upon them. And there was nothing to be done about it and no end in sight.[1]

The seven-year drought of the 1950s placed a stranglehold on ranching operations in Pinto Canyon unlike any other drought of the century. It not only crippled the industry but also caused irreparable damage to the land itself—a loss of topsoil and a quickening of desertification from which the canyon would not recover. As a result, the decade would mark the end of large-scale ranching in Pinto Canyon and the Sierra Vieja lowlands. The drought had taken too much, and as more ranches failed or converted from sheep and goats to cattle, the logistics of ranching the lowlands became increasingly constricted, the economics ever less profitable. Arriving at the end of the 1940s, the Shelys would be the last family to ranch in Pinto Canyon. But their introduction would be anything but easy. Seven years of heat and aridity served to dispel any ambitions of turning a major profit. At least not on livestock. As the drought intensified, and for years after it finally broke, Pinto Canyon ranchers had to find other ways to make ends meet—whether that was hunting, candelilla, sotol, or mining. Any profit that could be squeezed from the land allowed whatever tenuous hold they had to last just a little bit longer. But for the Prietos, it was only borrowed time. After sixty years in Pinto Canyon, José's ranching days were coming to an end.

The Shelys and a Dry Awakening

In February 1948, the old Wilson Ranch was sold to Terry W. Shely, a rancher from near Marathon, about eighty miles to the east. Born in 1890 in Brackettville, Texas, to Esther and Terry M. Shely, Terry, along with his brother Charley, spent their formative years on the family sheep and cattle ranch in Kinney County. But around 1915 the Shelys pulled up stakes and headed west in a covered wagon. The family settled in far eastern Brewster County on a ranch south of Tesnus—a tiny rail siding and station stop on the Southern Pacific Railroad some twenty miles southeast of Marathon. Terry Jr. was serving as postmaster at Tesnus when he was drafted into service during World War I. The end of the war found him safely back at the ranch, although he had suffered permanent hearing damage from a shell that exploded near him.[2]

Within a couple of years of his return, Terry had married Belle Clark in Marathon, and in 1927 she gave birth to their only child, Fred. He came of age on the family ranch. Not until he joined the merchant marines during World War II did he catch his first glimpse of the world beyond. Soon after returning, he married Kathryn Hegelund, who'd recently moved from Arizona with her family. By that time, his father had started scouting for other land. Terry first heard about the Pinto Canyon Ranch from his friend James Logan, an Alpine rancher and feed broker, who offered to swap ranches. Seeking an exit from the partnership with his brother, Terry accepted.[3]

By the time the Shelys first saw the Pinto Canyon Ranch, it was not the same ranch that James Wilson had sold some twenty years before. For one thing, it had passed through three owners since Wilson sold it to J. H. Young in 1927, and it had more than tripled in size. After Young died—within three short years of signing the deed—Leroy Cleveland, son of pioneer W. H. Cleveland, bought the ranch and ran it for another five years before his divorce forced the sale of the land again. In 1936, Jim Watts—who had married

Terry Shely as a young man at the Tesnus Ranch. Photo courtesy of Terry Jean Shely-Allen and Nancy Shely-Costantino.

James Wilson's daughter Mamie, and who had acquired the old George Sutherlin place in 1928—purchased the ranch, bringing it to a total of twelve sections. But in 1943, he bought another spread north of Marfa and sold the Pinto Canyon Ranch to James Logan.[4]

Logan had been in the region for only three years when he purchased the Pinto Canyon Ranch. Born in 1907 in a railroad section house in Coahoma, just east of Big Spring, Texas, he found work in a feed store in Colorado City shortly after leaving high school. In 1932 he married Jo Avis Askey, whose father mentored young James in the ranching business. After their first son, Daniel, was beset with asthma, the family relocated to the drier climate of Alpine around 1940. As their son's health improved, James opened a mohair and feed business before purchasing the Pinto Canyon Ranch from Jim Watts in 1943.[5]

While overseeing his feed and mohair business in Alpine, James left the day-to-day ranch operation to Bill Elms, who lived with his family at the old George Sutherlin house. Having relocated to Presidio County from Real County around 1940, Elms later became known for his award-winning registered Angora goats. With the help of mostly Mexican cowboys—some illegals, others operating under the Bracero Program (a series of agreements between the United States and Mexico aimed at filling labor shortages stemming from World War II)—Elms oversaw the several thousand Angora goats ranging across the ranch.[6]

In the five short years James Logan owned the Pinto Canyon Ranch, he spent a good bit of his time in the county clerk's office handling the details of purchasing nineteen additional sections to the north and west, effectively tripling the size of the ranch. It fit with his penchant for trading and land deals. After five years, maybe he'd run out of land to buy. Or perhaps the commute from his home base in Alpine had grown tiresome. Regardless, selling the ranch was also a chance to help out a friend who was ready for change. In 1948, James traded the now eighteen-thousand-acre Pinto Canyon Ranch to Terry Shely in exchange for his part of the Tesnus Ranch and assumption of two outstanding notes on the property.[7]

In the early spring of 1948, Terry and Belle Shely moved into the old Wilson house and Fred and Kathryn Shely moved into Mart Sutherlin's old house on Horse Creek, where Grover Sutherlin had lived with his family for so many years until he sold it to the Weatherbys, who held it until Dr. Chaffin called in the loan in 1940. So it was from the Chaffins that Fred leased the house and the rest of the fifty-five-hundred-acre ranch. Although only leasing the Chaffin place, Fred and Kathryn bought a half interest in the Pinto Canyon Ranch. In addition to the land, the livestock and the ranch equipment—amounting to two half-ton pickups, a tractor, and a horse trailer—were likewise owned in partnership, with each of the four holding a quarter interest.[8]

Terry and Fred operated the two neighboring ranches independently but frequently worked side by side, especially during roundups and shearing time. They almost always depended on mules to get around, though sometimes in pastures that were too steep or rough they were forced to carry out their work on foot. Ranch roads were few in number, and the ones that did exist were hell on tires. In any case, riding a mule was often faster and easier—a five-mile mule ride between the two ranch houses was closer to a ten-mile trip by truck. As with other lowland ranchers, the Shelys ran mostly Angora goats and a lesser number of sheep, allowing more efficient use of the range. To further diversify, they also kept a small herd of cattle scattered around the less rugged pastures.[9]

Fred and Kathryn's house lay just below Chinati Peak, which rose abruptly about fifty yards beyond the ranch house. Kathryn gave birth to two daughters in the five years after they moved in: Terry Jean in 1951 and Nancy in 1953. Located more than three miles from the county road through Pinto Canyon, the family was more isolated than they would ever be. "It was a very remote place to live, but we didn't know any different. I loved it," Terry Jean recalled. Life was routine if demanding, with the familiar yearly cycle of lambing and roundups and shearing and the never-ending repair of fences and patching of leaks in waterlines and troughs. The remoteness also meant life remained relatively primitive. Electric lines would not reach the canyon until 1954. In the meantime, Terry and Fred invested in wind-powered electric generators, which, along with a bank of glass-encased Delco batteries, provided enough electricity to run the lights, a few household items, and their radios, which on good days could pick up as many as three stations. The 250,000-watt "Border Blaster" station XERF out of Ciudad Acuña across from Del Rio

Fred Shely on horseback in Pinto Canyon. Photo courtesy of Terry Jean Shely-Allen and Nancy Shely-Costantino.

came in loudest. In between playing the hillbilly and gospel music that made up the station format, disk jockeys hawked cherished items like authentic autographed pictures of "Jesus H. Christ."[10]

For years, the Shelys' world was contained within the confines of Pinto Canyon. It was a world populated with livestock and family and the men who worked for them, many of whom were almost like extended family. One family of brothers, the Serranos, offered a full range of services—the older ones working livestock while the younger ones tended the garden and did chores around the house, awaiting the day they would get their turn in the saddle. Another, Sabino Hernandez, worked for Fred for years and became something of a guardian to the girls. On rare occasions when the Border Patrol showed up and the men scattered, the kids scattered with them, sensing the urgency if not fully understanding the reason for their flight. In the evenings, their workers joined the family to play baseball in the yard, and on holidays they shared meals. Isolation and proximity created bonds that transcended the strictly segregated divisions in town. Instead, ranches often had an equalizing effect so that men and women were judged more according to their personality or skill level than their ethnicity.[11]

During this time the Shelys' main connection to the outside world was through the postman, Juan Prieto—son of José Prieto's brother Francisco—who became a fixture of the ranching community below the Rimrock. After Juan finally got the Marfa-Ruidosa-Candelaria mail route he had long hoped for, there was only one problem—he didn't own a vehicle. But he knew someone who could help. As many in the extended Prieto family had done before him, Juan approached his uncle. After listening to Juan's predicament, José offered him a loan that allowed him to buy a brand-new 1947 Chevy pickup, thus launching a career that would support him and his family for more than thirty years.[12]

Ranchers along the Pinto Canyon road always anticipated Juan's arrival, for his services extended far beyond those of an average postman. In the morning before leaving Marfa, Juan stocked up on groceries to distribute along the route, each family receiving the coffee, block ice, or gasoline they had ordered. He also provided job placement and transportation services. Before roundups, ranch owners could tell Juan how many men they needed and for how long. At Ruidosa he hand-picked the men he believed to be the best match and then delivered them: $1.50 to ride in the back of the truck or $2 in the front. Along with his delivery and shuttle service, for many Juan also served as their primary connection to the outside world. Whether by bringing newspapers or by sharing town gossip, he relayed stories from Marfa to Candelaria and all points in between. If a dance was to be held at the old barracks in Ruidosa or if someone's house in Marfa burned to the ground, more than likely Juan was the one to tell you about it.[13]

Pinto Canyon on the Big Screen

In 1950, Pinto Canyon became the setting for part of writer and director Alan Le May's feature-length film *High Lonesome*. Opening scenes were shot from the Sierra Vieja Rimrock and others from several nearby ranches on the Marfa Plain. A number of scenes, however—including the climactic gunfight—were shot at the old

The Horse Creek schoolhouse in a still from the movie *High Lonesome*. *High Lonesome* (1950) ©Le May–Templeton Pictures.

Horse Creek schoolhouse, outfitted to look like a general store. The schedule was tight. With only twenty-six days allotted for filming, the crew worked at a furious pace. It required twelve people just to operate the equipment—eight to run the massive Technicolor camera and four more to handle the sound effects.[14]

The Pinto Canyon scenes were shot first, in early January. In addition to the twelve actors flown in from Hollywood—including Chill Wills and Jack Elam—sixty local residents were cast as extras, including the Shely family. Cast as a "psychological Western drama," the movie tells the story of a drifter, played by John Drew Barrymore—son of legendary stage and screen actor John Barrymore. Although the younger Barrymore was not the actor his father had been, the movie was a modest success, especially locally. The film debuted in Marfa in September 1950, breaking all prior attendance records at the Palace Theater. Over the two days it was in town, an estimated twenty-five hundred people—about 70 percent of the entire population of Marfa—viewed the movie. Locals were dazzled by the richly saturated color and the "beautiful results in many of the scenic

shots," although the film did not effect the boom in tourism predicted by the *Big Bend Sentinel*. It did, however, set a precedent for local
production; *Giant* was filmed outside Marfa in 1956 and received
widespread critical acclaim, including the Academy Award for Best
Director.[15]

The Seven-Year Drought

**It was so hot that the lizards had to spit on their feet to get
across the rocks.**

—**Adam Miller**

On the cusp of the worst drought in the twentieth century, the ranges of the Big Bend were bursting at the seams. Thanks to the strong
market and ample precipitation of the previous decade, stockmen
had increased their herds to historic levels. By 1945, the county was
stocked at nearly twice what it had averaged over the past four and
a half decades—more livestock than it ever had or would ever have
again. By 1950 that number had already started to taper off. But with
nearly 35,000 head of cattle, 16,000 goats, and 206,000 sheep, it was
still far beyond its historic carrying capacity. As they approached the
most severe and protracted drought anyone could remember, ranchers found themselves overextended and vulnerable at the worst possible time.[16]

The drought began in 1951 and would last an unprecedented
seven years—twice as long as the drought of the 1930s. According
to tree-ring data, which shows precipitation trends over long periods of time, it was the worst drought in the Big Bend in more than
three hundred years. Although no single year during the 1950s was
as dry as 1934—the worst on record according to instrument-based
data—the sheer length of it made the cumulative effect much greater. Reaching far beyond the confines of the Big Bend, the drought
shape-shifted and migrated, expanded and contracted, intensified
and abated at different times in different regions so that, over the
entire course of the disaster, it affected nearly every corner of the
continental United States. Originating in western New Mexico and
eastern Arizona, it drifted southeastward to southwestern Texas. By

1953, it had expanded from the Big Bend northeastward across Texas and Oklahoma. The following year it intensified and spread across most of the central United States. But 1956 was the worst year of them all, with the entire south-central part of the country locked in severe drought. It was not as widespread as in 1934, but the year was preceded by five years that were almost as dry. As a result, the protracted aridity made the decade as great a hardship as the 1940s had been one of prosperity.[17]

In the Sierra Vieja region, the drought was markedly worse below the Rimrock. Over the course of the seven-year drought, Valentine averaged 69 percent of normal annual precipitation whereas Candelaria averaged only 59 percent and Presidio, a mere 51. Considering that anything below 75 percent can spell disaster for ranching operations, it made for a dire situation. The driest year in Pinto Canyon was probably 1953, when rainfall at Candelaria amounted to a little more than four inches, about 34 percent of average. But underscoring the uneven distribution of rainfall, Valentine and Presidio experienced the worst of the drought in 1956, when rainfall totaled six inches (45 percent of average) and less than two (17 percent of average), respectively.[18]

Ranching during the Drought

The drought descended early in 1951 with a dry spring and uneven rainfall throughout the monsoon months of summer. Temperatures soared, with Presidio registering record-breaking highs of 114 degrees F in both May and August. After October, no rain fell at all. By the end of the year, Candelaria totals were nearly five inches below average. The following year offered a slight reprieve, but the area still received only about 20 percent of the average monsoonal rainfall while temperatures soared as high as 112 degrees F. By the second year of the drought, conditions were growing serious. Grass failed to germinate. Shrubs were beginning to wither.[19]

José Prieto and his men began burning spines off prickly pears and the leaves off sotol plants using blowtorches. After the men split the seared sotol heads open with hand axes, the sheep and goats rushed in to eat the nutritious heart. But restricted to the flanks and slopes of the Chinatis, sotol was a finite and temporary fix. As the

The Prieto Ranch house in 1953. Photo courtesy of David and Ann Amsbury.

last of the grass was eaten down to the roots and the prickly pear and sotol laid to waste, the sheep and goats began to eat the woody parts of shrubs as well as weeds they normally avoided. Some even ate the bright yellow flowers and silvery stems of the toxic *telempacate*, which gradually weakened the animals until at last they stood trembling, arching their backs, salivating a green froth as they slowly died from the poison. By the latter part of the 1950s, Presidio County ranchers were losing as much as $100,000 worth of sheep a year due to their animals' ingestion of toxic plants.[20]

As the drought progressed, dirt tanks that José had constructed in arroyos around the ranch dried out until the clay bottoms cracked into pieces cupped skyward like hundreds of tiny plates. Normally reliable seeps and springs slowed to a trickle and then ceased to flow altogether. As surface water became unavailable, José and his sons-in-law were forced to rely on groundwater alone. But with only four wells on the ranch and no system of pipelines, the men spent long days filling fifty-five-gallon barrels from the windmills beside their houses and hauling water around the ranch by truck. A two-hundred-gallon airplane fuel tank Manuel bought from Fred Shely certainly helped. But as days dragged on under the searing sun, there was no way to keep up. The animals started to die. The emotional effect was devastating. "The worst part was the tear in my dad's eyes," recalled Manny Perez, "because he was my hero."[21]

As dry as the first two years proved to be, they were only a prelude. After March 1953, not a drop fell for the next two months, setting the stage for a perfect storm. In the first week of May, as temperatures soared well above the century mark in the lowlands, blast-furnace winds gusted up to 60 miles per hour, raising dust that darkened the skies and scoured already barren rangelands. The *Big Bend Sentinel* pronounced it the "worst battering in the last thirty years" and noted "serious loss of top soil." The summer monsoon season failed to bring relief, delivering only about 30 percent of the normal rainfall. The temperature of the sunbaked earth rose under shimmering heat waves—the barren soil sterilized beneath the burning sun. Even when rain did fall, it could not penetrate the hardpacked earth. Instead, water flowed away, taking with it what little topsoil was left. As sheet-wash removed soil slowly, in layers, rills grew into arroyos that cut downward, turning the horizontal suddenly vertical and stealing the rangeland rain by rain, acre by acre, until in some places nothing but bedrock remained. By the end of the year, the area had a nearly eight-inch deficit, down by more than twenty inches since the drought began.[22]

Ranchers had little choice but to destock. Most had already sold the sheep and goats that would sell. The ones that remained might have been placed on feed, but the labor and expense of doing so was great. Many had to rely only on what the ranch could provide, which was increasingly meager. The number of livestock in the county plunged precipitously. Between 1950 and 1959, sheep numbers in Presidio County fell by 70 percent—from 206,000 in 1950 to 61,000 by 1959. Meanwhile, cattle numbers continued to decline, in a trend that had begun during the war as producers turned to sheep. It amounted to the greatest drop in livestock in the county since the birth of the industry. The only livestock class that increased was goats—a reflection of their ability to better utilize the ravaged ranges, and they increased by only a modest 32 percent—far from a replacement for the tens of thousands of animals that were being shipped out of the county as fast as trucks and trains could carry them.[23]

Had the market collapsed like it did during the Great Depression, things could have been even worse. Instead, wool received an unexpected and unprecedented boost when the US military ordered massive quantities of wool for the winter campaign of the Korean War. The resulting boom caused prices on wool exports to rise 130 percent between 1950 and 1951. If not for the ceiling imposed by the federal government, those prices would have climbed further still. But the windfall was to be brief. By 1952, prices had returned to their 1950 level, followed by a gradual decline through the remainder of the decade. Fortunately, declining prices were offset by a new government subsidy. Determined to reduce dependence on foreign fibers and avoid future shortages, Congress declared wool a "strategic material" with the National Wool Act of 1954. The more wool a rancher produced, the more compensation received. Because mohair was considered an alternate fiber, it tagged along for the ride. The act could not stem the natural decline in domestic wool production, but it did help stabilize the industry and partially offset the growing gap between ranch income and expenses.[24]

The Dimming Light

José Prieto might have weathered the drought of the 1950s just as he had weathered droughts before, but by now he had been ranching in Pinto Canyon for more than half a century—through years of rain and drought, booms and busts. Throughout he had persevered. But now, in addition to the drought, his own body was beginning to fail. For years he had refused to let his fading eyesight slow him down, but it was no longer something he could ignore. Driving back to the ranch one day with his daughter Jane, José drove off the road and into the bar ditch. He hadn't seen the curve. Realizing his driving days had come to an end, he turned to Jane and, looking at her through cloudy eyes, asked her to take over. He never drove again. If his pride suffered, no one could have known. For he took it in stride, remaining the kindly, smiling man he had always been, not allowing the dimming light in his eyes to dampen the spirit in his heart.[25]

Even so, José's world continued to close. In the late summer of 1952, Juanita was admitted to the Hill Clinic in Alpine for a pene-

Juanita Prieto holding a kid goat on the Prieto Ranch in the 1950s. Photo courtesy of the Prieto family collection.

trating pain deep in her gut and a flagging appetite that had caused weight loss. Her health had been sliding for some time. Her life-long habit of smoking unfiltered cigarettes, as well as her weight, age, and a heart condition, had taken a toll. By the time she entered the clinic, the whites of her eyes had already started to turn yellow. During the examination, Dr. Malone Hill found a tumor on her pancreas. She had only been at the clinic three days when she died at the age of seventy-eight from advanced pancreatic cancer. Services were held the next evening at St. Mary's Catholic Church, and she was laid to rest at the Merced Cemetery in Marfa, in the family plot José had purchased nearly two decades earlier for Pablo. Aside from his son's grave and headstone, the plot was empty. Through fading eyes, José must have contemplated the empty space beside Juanita's grave where his own body would soon be laid to rest.[26]

José Prieto Leaves Pinto Canyon

José's final days on the ranch were slow ones. The loss of Juanita, his partner and steadfast companion for more than sixty years, his slow-ing gait, and his failing eyesight, compounded by the most intense drought of the century, finally did what no other event of his life-time could do—force him to leave the ranch and the only life he had ever known. By 1956, with the drought growing worse by the day, José had gone all but completely blind. He tapped his last mornings down to the corrals with his walking stick and, leaning against the railing, asked questions of his young grandson, Adam Miller. Were the kid goats still in the house trap? Was the float valve in the water trough still leaking? "It seemed like he could see," recalled Adam. But the fact that he could not, and the weight of the world closing in around him, all pointed toward the only logical conclusion.[27]

Jane helped him pack what meager belongings he needed and drove him to El Paso, where he could be closer to the doctors he was beginning to call on with ever greater frequency. There, on the east end of town, less than a mile north of Juárez, José purchased a mod-est frame house with a tiny yard—one of hundreds of new cookie-cutter homes being constructed along the outskirts of the town. It was a world apart from anything he had known. Still, he mapped

José Prieto with one of his rams in the 1950s, shortly before he left the ranch. Photo courtesy of the Prieto family.

out his new surroundings in his mind well enough to where he could get around without help. Soon after they got settled in, Jane called on her niece, Ida—Juan Benavidez's daughter—to help take care of her grandfather. José's smile broadened when Ida took him for walks around the neighborhood, but otherwise he remained the quiet, content man she had always known him to be.[28]

José's last days he spent alone in the backyard, where he had ample time to reflect upon the past eighty years of his life—back to his earliest memories along the Río San Pedro in Mexico, of immigrating to the United States with his family when he was eight, and of coming of age along the banks of the muddy Rio Grande in Ruidosa. He may have thought about his first paid jobs cowboying and shearing above the Rimrock, of his marriage to Juanita and their first home, at El Tanque de la Ese, of meeting James Wilson and the first sections of land he purchased from the state, of long days loose herding goats, and of even longer days when it came time to shear. He might have recalled the birth of his nine children and the simple joy they brought to his life and his agony following the death of his sons, of the windfall years of the 1920s and 1940s, and the hardships visited on the family in the 1930s and 1950s. And, finally, of the triad of events that finally forced him out of the life he had built: failing eyesight, losing Juanita, and a drought worse than anything he had ever witnessed. A teetotaler his whole life, José learned to enjoy a single glass of *aguardiente* (loosely translated as "firewater") in the evenings and two cigarettes before bed—as if he were performing some kind of ritual before nightfall brought the darkness of the outer world to merge with that of his inner world, where his dreams for a time returned to him a life of color and light.[29]

The Drought Persists

Meanwhile, the drought continued to deepen. Following the driest year in Pinto Canyon, 1954 offered something of a reprieve when the monsoon season partially made up the deficit: in May, the *Big Bend Sentinel* announced that more than an inch of rain had fallen in the canyon. Receiving twice the normal rainfall in June prompted Juan Morales to walk into the offices of the *Big Bend Sentinel* bearing tufts of grass "thirty inches high above the roots" as a result of the summer rains. But even as the newspaper reported that the hills near Candelaria were greener than they had been for years, it would also be the hottest year of the decade. Although conditions were dramatically better than 1953—with Candelaria logging 81 percent of the average precipitation—after October, no rain would fall again for six months.[30]

The drought respected no boundaries, and things were just as bad south of the border. By the summer of 1955, conditions were so dry on his ranch in Mexico that O. C. Dowe reported losing six to seven head of cattle daily from starvation. But the following year would be the worst of all. After a wet January, rainfall diminished. July teased some moisture from the sky, but once again the monsoon failed to deliver, with Candelaria receiving only 39 percent of average. In Presidio it was far worse, the town receiving less than two inches for the entire year. In an appeal for government assistance to stabilize feed prices, the Highland Sheep and Goat Raisers Association offered a dire appraisal. "Unless this larger aid and price freeze goes into immediate effect on government aid feeds," noted the resolution, "the livestock industry in these disaster drought areas will cease to exist within the next six months."[31]

The association's claim may have been exaggerated, but it wasn't far from the truth. After six years of drought, optimism was hard to muster. The summer months of 1957 saw a return of broiling temperatures, with Presidio setting an all-time high of 117 degrees F in June. Meanwhile, nine out of twelve months brought below average rainfall at Candelaria, and the rain that did fall came when it was least needed. The distress selling of livestock, which had been nearly constant since the drought began, glutted markets, which guaranteed poor prices. Between 1950 and 1954, the value of live sheep had fallen by 33 percent and cattle prices by more than half.[32]

Adapting to Dry Times

As flocks dwindled and the wool and mohair clip bottomed out, economically distressed ranchers in Pinto Canyon faced losing everything unless they could find other sources of income. In this effort, they were more fortunate than most—for the land here contained a more diversified flora and fauna and a wider array of geological exposures, which offered a broader suite of possibilities to low-land ranchers, than did other areas. Out of necessity, many began to explore different strategies—some that that were legal and some that were not. Resources that had not been previously exploited or were secondary to the ranching operation became primary. In some cases, they would remain so even after the rains returned.

Of the various economic lifelines tested during the 1950s, hunting was one of the most lucrative and has proven the most enduring. Although ranchers had been charging to hunt wild game for years, fee hunting became a much more important source of income. During the drought of the 1930s, personal advertisements in the *Big Bend Sentinel* declared area ranches off limits to trespass hunters and threatened that "violators would be prosecuted to the full extent of the law." The hunger that attended the Depression likely triggered a marked increase in trespassing and pot hunting, which was certainly part of the problem. Another factor was the drought itself. Like domestic livestock, mule deer and other game populations began to decline. As they did, their scarcity increased their market value. Advertisements for hunting leases became commonplace, ushering in a new source of revenue that would eventually come to overshadow the livestock industry itself.[33]

Since hunting leases only sell the opportunity to hunt and do not guarantee a quarry, this new source of revenue held promise of profits with little cost to the rancher. It also capitalized on a previously unrealized—and largely renewable—resource. It was a prospect few ranchers were able to pass up, certainly not the entrepreneurial Juan Morales, who began running ads in the *Big Bend Sentinel* at least as early as 1956: "Deer hunters: have ideal location for hunting in Pinto Canyon on Ruidosa road. Will haul deer to your car. Only $50." The Shelys also began to take hunters and, if they were lucky, Fred might cook them a steak. But if clients expected to be pampered or laze away the day away in a deer blind, they may have been disappointed to discover they would have to hoof it if they wanted to bag a mule deer. By the early 1970s, Larry Godfrey, outdoor editor for *The Eagle* in Bryan–College Station, was among a group of men who tried. His stories about hunting in Pinto Canyon were filled with tales of woe as much as they were about hunting. The trips were evaluated by the degree of suffering—number of flat tires, scrapes, cuts, and near-death experiences—as much as by the number of bucks taken. One of his fellow hunters, after falling down and losing his eyeglasses and nearly—according to him—his life, put his hands together and prayed, "Lord, if you'll help me down without killing myself, I'll never climb these [mountains] again." During years visiting hunters were lucky enough to take down a deer, Fred might offer one of their off-season goat punchers to help pack out their quarry. If he didn't, they might have wished they'd hired Juan Morales.[34]

Cooking Candelilla

Growing throughout the lower reaches of the Chihuahua Desert, candelilla—meaning "little candle" in Spanish—is a small, shrubby plant that produces a high-quality wax believed to be an adaptation to arid conditions. Inconspicuous in appearance, the plant occurs as a low-growing clump of upright, slender, cylindrical stems found most often on rocky or gravel limestone hills. Utilized for centuries by *curanderos* (traditional Mexican folk healers) for kidney ailments and venereal disease, candelilla was later subject to refinements in processing that made it commercially available. The wax was used extensively during the world wars, primarily as waterproofing for tents and ammunition.[35]

In 1911, a chemist from Monterrey, Mexico, developed an effective and efficient processing method that launched industrial candelilla wax production. That same year, the first wax factories in the region were built in the far eastern Big Bend. The largest was built at Glenn Springs, south of the Chisos Mountains in what is now Big Bend National Park. Surrounded by thousands of acres of candelilla plants, the Glenn Springs factory was considerably larger than earlier such operations, and it included a water storage system, a boiler room with towering smokestacks, and six large extracting vats. Because processing was labor intensive, a small community developed around it. It contained a well-stocked general store, a large house for the foreman, and several dozen huts for the workers and their families. There were smaller local factories as well, including one near Pinto Canyon owned by O. C. Dowe. Two Alpine men started another in 1927 on the Rio Grande about six miles outside Ruidosa.[36]

Although early operations used wooden vats and required boilers to make steam, by 1935 most operations were switching to metal vats that could be partially buried and fired directly. Use of these metal vats also allowed operators to be more mobile so that, when the candelilla supply was exhausted in one area, the *candelilleros* simply loaded the entire operation onto a truck and drove to another location. This newer, simpler setup also made it affordable for smaller operations, and the industry thus expanded into nearly every niche where candelilla grew, including remote areas with limited quantities.[37]

Those who could were not long in embracing the industry, and many in the lowlands started wax operations, most often using men out of Mexico to run them. Wax was also frequently smuggled out of Mexico. In 1937, the Mexican government placed a heavy export tax on candelilla wax, sparking a vigorous cross-border smuggling trade. Although it was illegal to take it out of Mexico without paying the tax, once it was across the border US law imposed no restrictions. But being of poorer quality—often containing impurities such as ash and sediment—this wax was mixed with cleaner domestic wax, which fetched higher prices in Presidio or Alpine.[38]

The process was simple, if laborious. *Arrieros* (gatherers) pulled the plants from the ground and packed them to the processing area on burros. Workers piled the plants beside a metal half drum until enough had been gathered to make a run. Filling the vat with about two hundred to three hundred gallons of water, the *candelilleros* set to work. They charged the firebox with wood or pitchfork loads of dried, spent candelilla plants, then ignited them. As the water heated, the men heaped approximately five hundred pounds of fresh candelilla into the vat, stomping the *yerba* (herb) down to water level, then pulling closed the two-hinged metal grates over the vat and clamping them shut. The men measured out sulfuric acid and added it to the mix. As the water boiled, the wax separated from the stems and rose to the surface in a brown, frothy foam they skimmed out with an *espumador*, a colander-like spoon, sieving the raw wax, or *cerote*, into a barrel. The men continued firing the vats until they ran out of candelilla or daylight. After the wax dried, the *candelilleros* removed it from the barrel, broke it into angular chunks with a hammerstone, and stored it in burlap sacks.[39]

Small wax operations sprang up across the Sierra Vieja lowlands —at least four or more in Pinto Canyon itself. José Prieto and all three of his sons-in-law had their own. These temporary "wax camps" were little more than flat pieces of ground near a source of water. Aside from the wax vat, crude brush shelters completed the camp. Springs with nearby rock shelters were even better. For the men, the days were long and the work was hard and dangerous. At José Prieto's operation about a mile up the Pinto Canyon road from his house, Adam Miller remembered, one of the men fell into a vat of boiling acid water, leaving him scarred for life.[40]

Juan Morales's Sotoleros

Unlike candelilla, whose production was legal, the making and selling of sotol—the distilled juice of the sotol plant—carried significant legal risk, as well as a far greater profit margin. Such opportunity did not escape the notice of Juan Morales. His official position with the sheriff's department notwithstanding, as early as 1950 Juan was overseeing his own sotol business, supervising the production and personally seeing to its distribution.

Like an upturned green mop, its long, slender, drooping leaves splayed outward from the main trunk, the sotol plant has been important to desert people for millennia. The woody flower stalks, reaching heights of up to sixteen feet, were perfect for brush shelters, spear shafts, and fireboards. The long, toothed leaves were woven into baskets, matting, and sandals. Most of the biomass of sotol, however, resides within the fleshy basal trunk. As a result, this starchy *cabeza* (head) served as an important food prehistorically, and it eventually became the raw material for making fermented beverages.[41]

By at least thirteen hundred years ago, the practice of fermenting semisucculent plants such as sotol and agave had already been well established in the Americas. Archeologists believe that the people of Casas Grandes (Paquimé) in northwestern Chihuahua, Mexico, used a series of rock-lined ovens in the systematic production of such fermented beverages. Although large-scale distillation of fermented drinks awaited the Spanish invasion, fermented agave and sotol would ultimately form the basis of the most common distilled liquors of northern Mexico. By the early twentieth century, sotol already had a long and storied history as the alcohol of the *campesinos*. But the real heyday of sotol came after the passage of the Volstead Act of 1919, when demand skyrocketed, making the Prohibition era in the United States the age of sotol in the borderlands. As a counterpart to the sweet-mash corn liquor of the Deep South, sotol was essentially the Mexican version of moonshine. Any day of the week you could drive to a certain tree outside Ruidosa, honk your horn, and a man would appear from the brush to take your order—typically in gallon increments.[42]

Prohibition had long since ended by the time Juan Morales got into the business, but sotol remained the go-to alcohol in the borderlands because it was a cheaper—and often more potent—alternative to legal liquor. Just as Juan was launching his sotol business, he and Sue were hosting five-year-old Jerry Lujan, whose family lived a few doors down from a house José Prieto had purchased in Marfa. Whenever Jerry saw Juan's red 1948 Dodge pickup pull up in front of the Prieto house, he could barely contain his excitement. One day, Juan invited Jerry to come stay at the ranch. After gaining his mother's reluctant approval, Jerry jumped into the truck with Juan and Sue, ready for his big adventures. It was still dark the next morning when Juan shook the young boy awake. "It's time to get up," he said. "When you live on a ranch, there's no time to waste!" Blinking the sleep away, Jerry groped about blindly for his clothes, only to find the legs of his pants filled with canned goods and his shirtsleeves tied in knots. Meanwhile, Juan sat just around the corner, listening to "El Gallito Madrugador" on XELO out of Juárez, his shoulders convulsing with laughter.[43]

After Jerry got his wardrobe straightened out, Juan asked him if he would like to go see some Indians up in the mountains. Imagining stoic braves in warbonnets bearing lances and shields, the young boy's eyes widened in fascination. Taking a pack mule loaded with goods, they headed southeast from the ranch house before turning into a narrow canyon in which erosion had carved out a series of small rock shelters where people were living. There, Juan began distributing food and clothing to the men, some of whom had brought their families. Meanwhile, Jerry was greatly disappointed that Juan's "Indians" looked and dressed no differently than Mexicans. They rode farther up the canyon to an even larger cave where an older man lived—the *patrón* of the *sotoleros*. Handing the man a bag of provisions, Juan discussed the plans ahead, then told him they would return in a week.[44]

By the time of their visit, Juan's *sotoleros* had already been hard at work setting up the *viñata* (sotol still). They had dug a deep pit and lined it with rocks before constructing a firebox of dry-laid stone to heat the copper *peroles* (kettles), leaving an opening through which they could charge it with wood. With the oven and firebox constructed, the men spread out below the canyon to collect the largest and hardest pieces of wood they could find, with preference given hardwoods such as mesquite, oak, and acacia.[45]

A few days after their trip to the *viñata*, *el patrón* appeared at Juan's house and said everything was ready. Jerry watched wide-eyed as the old man sat down in the shade of the porch. Producing a small wooden box from his pocket, he slid open the lid and took out a piece of cotton and a piece of flint. He then crushed a bit of dried grass in his hand and placed it inside the box. Striking the flint against a steel striker, the sparks landed on the bit of grass in the box, producing a wisp of smoke. The old man lifted the box up to his mouth and blew until the grass bits burst into flame. Placing the cotton inside the box, the flame grew brighter as he lifted it to light his hand-rolled cigarette. Then, blowing out the flame, he replaced the contents of the box, slid the lid shut, and returned it to his pocket.[46]

As Juan's truck bounced down the dirt road toward Ruidosa, Jerry craned his neck to see over the dashboard. Just east of town, he stopped one last time to glass for *la migra* then descended into an arroyo, cut the engine, stepped outside, and gave a loud whistle. Men suddenly appeared, coming from behind bushes and boulders, each bearing a large sotol *cabeza*, the leaves already hacked away. Moving back and forth from their hidden cache, within a half hour they had the bed of Juan's truck full of *cabezas* piled high against the railings. The next morning, Juan corralled his string of mules and fitted each with a packsaddle. Then he loaded four *cabezas* into each of the burlap sacks he had and, tying two sacks together, lashed them over the packsaddles. When the load was secured, Juan and Jerry led the sotol mule train and wound their way back into the narrow canyon and the waiting *viñata*.[47]

The *sotoleros* piled wood into the pit until the flames leaped skyward in a wall of intense heat. They fed the fire for several hours
until the rocks glowed red, the heat radiating out like a furnace. They
then filled the pit with three hundred *cabezas*, followed by a layer of
yucca and sotol leaves, then a thick layer of dirt shoveled over the
top. For the next two days they waited, occasionally adding water,
through a small hole dug into the center of the mound, to keep the
cabezas from drying out. Meanwhile, deep inside the pit, the long,
steady heat slowly broke down the bitter woody pulp into a sweet,
soft flesh. Finally, the men shoveled off the dirt and removed the
cabezas before chopping and pressing them to collect the juice, which
they directed into vats. They added water to this mix and then waited for the mash to ferment.[48]

After three days, the mash had turned a dark amber color and
was starting to bubble. The *sotoleros* poured the living liquid into the
copper *perole*, then pressed the copper cap over the collar. Lighting
the wood beneath the *perole*, they fed the firebox continually until
the mash reached 174 degrees F, causing the lighter alcohol to rise
out of the water as a vapor. As it traveled through the *pipote*, sub-
merged in a barrel of cold water, the vapor cooled, which caused it
to condense. From the *pipote*'s end the *vino* began to drip. This they
collected for a second cooking to further concentrate the alcohol,
so that what dribbled out of the *pipote* the second time was pure
sotol.[49]

Jerry watched transfixed as *el patrón* handed Juan his first taste.
As Juan brought the tin cup back down, his face had turned red,
his eyes wide as saucers. For a moment he didn't say a word. Then,
a smile spread across his face as he nodded to the old man. After
filling several five-gallon tin containers and tying them to his pack
mule, they headed back to the ranch, where Juan proceeded to get
gloriously drunk. The next day, nursing his hangover and harried
by a frustrated Sue, the three of them drove to Marfa, the five-
gallon containers safely stowed in the back. Later, Jerry walked to
the Prieto house to find a worried Sue, who asked if he would go
find Juan. She suggested Catarino's Bar, a few blocks away, as a
good place to start. When Jerry entered, he found Juan and the bar-
tender carefully funneling the clear liquid from the tin containers
into empty glass bottles. Curious, Jerry settled in to watch. Once

the bottles were filled, the men at the bar began ordering rounds of Juan's fiery sotol. It seemed only minutes before one of the men became very drunk and, staggering up to Juan, drew back and punched him in the face. Quick on his feet, Juan backed up and threw three punches in quick succession, knocking the man out cold.[50]

If Sue was fully aware of Juan's sotol business, she had learned to look the other way. Because she was less tolerant of his own drinking, Juan learned to be sly. For one thing, he routinely stashed sotol around the ranch. Some days he packed a lunch and a single glass and told Sue that he was going out to check fences. Instead, he met his neighbor Augustín Nuñez at the fence, where they proceeded to "have a hell of a party," recalled Adam Miller, Juan's nephew. Following an afternoon of drinking and storytelling, the men would slur their goodbyes and start for their respective houses. Half dozing in the saddle, Juan's head slowly bobbed as his horse followed the trail home.[51]

The Perez Family's Departure

As the drought intensified over the course of seven years, fewer and fewer ranchers were able to stay in business. Although the division of José's ranch created parcels of nearly equal size, they contained different features. Juan Morales and José Prieto both had springs on their ranches whereas Manuel Perez did not. He relied instead on about three and a half miles of Pinto Creek, which sometimes flowed and sometimes didn't. The land between the three parcels also differed significantly. Juan and José's ranches both wrapped around the lower flanks and toeslope of the Chinati Mountains. With a northern aspect, the slopes were cooler and moister than the open country, generally offered better grazing and a more diverse browse menu, and tended to be more resilient during drought conditions. But aside from sections within the canyon itself, Manuel had mostly rugged hills and *cordones*, the eroded bolson deposits that extended westward toward the river. Being lower in elevation and farther from the Chinatis, they were also more susceptible to drought.[52]

Manuel tried his best to hold on to his livestock and the ranch. But as the drought dragged on and animals began to die, he was left with few choices. He looked for other work to tide him over, hoping to resume ranching once the rains returned. He couldn't have known how long the wait would be. Fortunately, an opening with the US Department of Agriculture as a quarantine enforcement inspector, commonly known as a "river rider," allowed him to remain close by. In 1947 an outbreak of hoof-and-mouth disease in Mexico had set off a furious effort to contain it. An extremely infectious and sometimes fatal virus-borne illness that affects cloven-hoofed animals, hoof-and-mouth causes a high fever and blisters on the feet and mouth of affected animals. To prevent the spread of the disease into the United States, the federal government paid men to patrol the Mexican border on horseback. If they found any infected trespass livestock, it was their duty to kill them and burn the carcasses. Every day for about a year, Manuel rode from Candelaria to a point about halfway to Ruidosa, where he met a fellow river rider, Clay Polk, coming from the other direction. There the two men would rest in the shade, eat their lunches, and enjoy a smoke before setting out for their return trip.[53]

Manuel might have stayed with the job, but it wasn't in the cards. After his position ended he found himself with few options but to leave Pinto Canyon for good. Over the next few years the family moved several times, until Manuel found work that suited him. After working in a feed gin in Marfa and a cotton gin in Balmorhea, they moved to the ranch of Ben Gearhart outside Valentine, and Manuel cowboyed for a few years more. But around 1958 he got a job at the Hueco Ranch near El Paso, and he would remain there until his retirement.[54]

For young Manny, leaving Pinto Canyon also meant his time with Juan Morales had come to an end. One day, shortly before he had to leave the Alamitos Ranch for good, Juan pulled over by the ranch gate and got out of his truck. As they gazed eastward across the ranch toward Sierra Parda, Juan turned to his young nephew. "I'm gonna tell you something," he said. "I've got an egg for you. But it's an egg you can't eat, I can't eat, and Sue can't eat. But someday, that egg will be yours." Only later would Manny come to understand what his *padrino*'s words meant and how fate prevented him from making good on his promise. It was a long drive to the bus station in Alpine the day that Manny had to leave to join his family. As he boarded the bus, Juan became uncharacteristically flustered. As if not knowing what else to do, he removed his wristwatch and handed it to the young boy. "I want you to have this," he said.[55]

The Shelys Move to Marfa

The Shelys arrived in Pinto Canyon at a difficult time. The wet years and strong market of the 1940s prepared them poorly for the dry years ahead. What success they enjoyed was probably limited to their first few years in the canyon, for after that the skies forgot to rain, the clouds burning away beneath a searing sun. "There was no grass. It was just awful," recalled Fred's daughter, Terry Jean. The degree to which the drought had become part of their lives struck home one year as Fred drove his family to vacation in Montana. As they headed northward and crossed the plains of Wyoming and Montana, the color palette began to change. Peering out the car window, their youngest daughter, Nancy, was bewildered. Finally, she asked what all the green stuff was on the ground. Had it been tawny and much more sparse, she might have recognized it as grass.[56]

Although Terry and Belle lived within sight of the Pinto Canyon road, Fred and Kathryn in the old Sutherlin house up Horse Creek were far enough away that their isolation could be absolute for weeks at a time. Once or twice a month, the Shelys drove into Marfa to stock up on groceries. But otherwise they remained on the ranch, the days long and full of light, the evenings often spent beside the radio or playing dominoes to pass the time. But, as most ranch families eventually do, around 1958 the Shelys moved to town so Terry Jean could attend school. At first Fred rented an apartment in Marfa, but they soon bought town lots and built a house. After years of isolation in Pinto Canyon, Fred and Kathryn enjoyed living a more socially active lifestyle. Outgoing and civic minded, over the years Fred became increasingly involved with his kids' sporting activities when not serving on the school board or the Marfa City Council.[57]

The Drought Breaks

About the time the Shelys left Pinto Canyon, the drought finally began to ease. In 1958 Candelaria logged more than eighteen inches of rain for the year, more than six inches above average. Four and a half inches fell in September alone, causing widespread flooding. Three Marfans on the way to the hot springs became stranded in Pinto Canyon, forcing them to spend the night in their vehicle. Despite the abundant rain, however, the effects of the drought lingered—not the least of which was the psychological impact. As late as April of the next year, when Ruidosa received more than half an inch of rain, the *Sentinel* reported that "the rain gave new hope to ranchers looking for an end to long years of drouth, understocked ranches, and fading bank balances." With drought conditions having become routine, few wanted to risk an optimistic tone. It would not be until July 1960, after a five-inch rain, that the newspaper staff cautiously announced that "observers have stated the long drought is now broken."[58]

The rains of 1958 brought renewed hope but could not erase the lasting effects of the worst drought in modern history. They could not make up for months of aridity and broiling sunlight when grass grew brittle and finally rattled away in the wind, as dust lifted high into the sky and thirsty sheep and goats milled and cried. When trains could barely carry cattle, sheep, and goats out—or feed in—fast enough. When those who did not or could not find a market for their animals found instead foreclosure notes in the mailbox, the smell of death heavy in the air.

After the Grass Was Gone

As bad as the drought's effects were on the ranching industry, those wrought upon the land may be the most enduring. Grasses and forbs died, shrubs withered. Even hundred-year-old cottonwood trees became stressed to the point where they stopped reproducing or perished altogether. Without vegetation, the soil baked under the desert sun. Over time, even drought-tolerant soil crusts—a living mantle made of soil bacteria and fungi—begin to die. The loss of crust reduces the organic content of the soil and interferes with nutrient cycling, thus impoverishing fertility and undermining structure. Less resilient, it is more susceptible to erosion—the winds that blow it and the downpours that wash it away. Amid a background of stepwise deterioration caused by intensive and uninterrupted grazing, droughts like those of the 1930s and 1950s triggered sudden and precipitous declines that were largely irreversible.[59]

The carrying capacity of the land, as measured by the number of animal units per acre or—in the case of the desert Southwest—the number of acres per animal unit, is an imperfect but nevertheless illuminating indicator of land productivity. In 1950, before the drought arrived, there were more than eighty thousand animal units in Presidio County, amounting to one for every twenty-four acres. In 1959, after the drought had ended, that number had cascaded to thirty-five thousand animal units, amounting to one animal for every sixty-two acres. Although that figure would decline to as low as thirty-eight acres per animal in 1974, never again would there be as many grazing animals as there were in 1950, before the drought punched the air out of the regional ranching industry. Many things changed, but perhaps nothing so much as the lasting impacts to the land; if that land is desert, it is slow to recover, if it recovers at all.[60]

It had been fifty years since Manny last saw the Morales Ranch, the home of his cherished *padrino* Juan, where he was told about the egg that would never be his. Many things were just the way he remembered them—the house, the rocks Sue placed in the concrete border, the windmill and water tank, the rise of rocky Sierra Parda to the southeast. And yet, something didn't seem right. He found himself staring at the ground beneath his feet. Where as a child, Sue had him rake the driveway between the house and the barn before one of Juan's feasts or rodeos, now there was nothing left to rake—only bare rock exposed to the heaving sky.[61]

A Matter of Inheritance

As the end of life drew near, José Prieto deeded all nineteen sections of his Pinto Canyon ranch to his daughters in shares, undivided. His sons had already received their inheritance years earlier with the ranch in Mexico. Never mind that Gregorio had squandered his portion. Now it was his daughters' turn. Because Jane never married, having devoted years of her life attending to her father, she received an undivided half. The others—Sue, Petra, Frances, and Lupe—were to receive the remaining half, divided equally between them. The terms were straightforward. To each he bequeathed the ranch, "for and in consideration of the natural love and affection I have and bear for my daughters." José may also have felt that the ranch would be better off in their hands. One thing was clear—he wanted his daughters to keep the ranch intact.[62]

Gregorio's Return

In 1958, Gregorio finally left Mexico for good. Having lived there for the last quarter century and gotten a good feel for the possibilities—or lack thereof—he may have believed the United States had more to offer. Most likely, he'd run out of options. The fact remained: in Mexico, he was a known criminal who had committed murder. Perhaps back in the United States he could get a clean start. Knowing his sisters had inherited the ranch, he might already have been scheming to turn the situation to his advantage. With his last remaining brother lacking interest in ranching and Gregorio lacking any respect for his sisters, he would soon strong-arm his way to being manager of the Prieto ranch lands—a development that would ultimately cause the Prieto family a loss greater than any of them could have imagined.[63]

$$\sim\!\sim\!\sim$$

The drought of the 1950s largely sounded the death knell to ranching in the lowlands and left Pinto Canyon—like most of the ranges across the Big Bend—more impoverished than ever. As the heat and aridity dragged on year after year, as if it were the new normal, ranchers adjusted as best they could, mostly by liquidating stock and finding other sources of income. Hunting, candelilla, and sotol helped, but ultimately they were not enough. The Shelys, who would be the last family to ranch in Pinto Canyon, arrived shortly before the drought delivered a harsh welcome. And just as drought presaged an uncertain future for the new arrivals, it conspired to bring José Prieto's long ranching career to a close, forcing him to quit the only life he'd ever known.

9

Tierra Perdida

A Lost Land

Ranches don't fail, families fail.

—Chip Love

It was the lawyers that stole that land from us, and that real estate man.

—Manny Perez

THE PIPER Super Cub appeared suddenly over the Rimrock of Pinto Canyon like a giant metal bird, the nose pitched downward, diving at a furious pace. The engine raced as it gained speed, letting gravity do the work. Ten-year-old Adam Miller froze in his tracks, believing the aircraft was surely going to crash before his eyes. Then, moments before impact, the small plane leveled out, not ten feet above the ground. The landing gear raked through a stand of mesquite trees, one branch tearing a six-inch gash in the leading edge of the left wing. And then the plane was climbing again. Sharply banking, it made a complete circle a hundred feet above the ground before leveling out and—finally—slowly descending. A puff of dust flew up where the wheels hit the narrow dirt runway. The plane taxied wildly for a moment over the rocky ground before making a sharp 90° turn and coming to an abrupt stop thirty feet in front of the boy. The engine idled for a few seconds longer and then sputtered to a stop. The door flipped open, and Fred Shely clambered out, sporting a stained cowboy hat and pearl-snap western shirt, a wry grin across his face.

"You got any duct tape?" he asked the boy.

"No, sir," Adam replied.

"Come over here then and help me find some." After a few minutes of rummaging through the back of the plane, they found a roll beneath one of the seats. Picking up a yucca stalk from the ground, he broke it into a stick a foot long. He then pulled off a long piece of tape and taped the stick over the gash in the wing. With four additional strips of tape, the patch was complete.

"Okay, let's go," he said. José Prieto's grandson Adam was terrified of the airplane, but he dutifully climbed into the passenger seat and was trying to find the seat belt when Fred turned the key and the propeller buzzed to life. The plane lurched forward, turned, and finally headed straight down the runway. Fred pulled the accelerator full throttle. The tiny airplane taxied stiffly over the rocks and gopher mounds. Nearing the end of the runway, with not even thirty feet to spare, the tires lifted off the ground. The plane climbed skyward, sweeping the tips of ocotillos at the end of the runway. They were flying. The plane banked right, curving around Needle Peak, and then leveled out. The boy realized he had been holding on to the seat in a death grip. He breathed, trying to relax. In front of him, a nail suspended from a string swayed to and fro. "What is that for?" asked Adam. "That's to see if I'm flying level," said Fred, laughing, as they banked again, turning south. Chinati Peak loomed ahead, above the narrow canyon where they were to check on his Angora goats in the back pasture of Fred's ranch.[1]

Fred Shely's legendary flights were a sign of changing times in Pinto Canyon. Although much remained the same, newer technology like airplanes made travel faster across the expansiveness of the region and allowed some—like Fred Shely—to live two lives, one in town and the other at the ranch. Airplanes compressed the commute from town and hastened the never-ending inspections of stock, water, and fences. Although the outlay was substantial and workers still had to commute to on-the-ground work sites by foot or mule, the investment ultimately saved much wear and tear on the more grounded ranch vehicles.

While the 1960s were far wetter than the 1950s had been, they only barely allowed the ranching industry to stabilize following the major upset of the seven-year drought. For the Shelys, the decade brought new promise as they successfully launched a mining operation that provided income the drought couldn't touch. Meanwhile, after Gregorio assumed management of the Prieto Ranch, the trouble he towed came home to roost. Mishandling funds and neglecting to disburse the income equitably among his siblings would cause a long-standing rift in the family to broaden into a chasm. After a lawsuit was filed in district court, the fate of the Prieto Ranch would hang precariously in the balance.

When the long drought finally broke in 1958, it did so in a mighty drenching, suggesting that the weather had finally righted itself so that ranchers could get back to business as usual. But the 1960s—while significantly wetter than the 1950s—were still dry, far too dry to allow the range to fully recover. In Candelaria, the average annual precipitation for the decade came to slightly more than nine inches a year—about 20 percent below average, and in half of those years the figure dropped to more than 30 percent below average. Several years were even as bad as those of the 1950s. Only one year—1968, with more than fifteen inches falling in Candelaria—was notably wet.[2]

The market proved slightly more agreeable. Over the course of the decade, wool prices held steady, averaging around forty-six cents per pound. The bad news was that in 1960 the Pentagon dropped wool from the list of strategically critical materials. Following the Korean War, the US military had adopted synthetic fibers for uniforms, causing wool to fall by the wayside. The National Wool Act of 1954 had paid ranchers the difference when prices fell below the "incentive level." But now that cushion had been removed. Although subsidies persisted in some form for another twenty-five years, they were only a temporary buttress to a dying industry. The future for mohair looked bleaker still. After 1964, the mohair market took a nosedive such that for the first time since 1932, the value was on par with that of wool instead of hovering above it. The latter trend was especially worrisome for ranchers in Pinto Canyon, where mohair had long been the mainstay.[3]

More worrisome yet was the passage of a 1962 amendment to the Bald Eagle Protection Act of 1940 that put an end to the indiscrim-

inate killing of golden eagles. For the previous two decades, sheep
and goat raisers had waged aerial warfare against the largest bird of
prey in North America, which also happened to be the top predator
of lambs and kids. Pilot J. O. Casparis of Alpine claimed to have
killed at least eight thousand. Although one study concluded that
less than 2 percent of lamb and kid losses were due to eagle predation, ranchers in the Big Bend weren't buying it. Game warden Ray
Williams of Alpine asserted that every eagle killed saved between
ten and thirty lambs. More than a decade after the prohibition went
into effect, Fred Shely claimed he lost around eight hundred head
of sheep and goats in a single year. "[The new law] is bad for us in
the sheep and goat business because eagles kill lambs and kids," he
said. "They say they don't, but they do."[4]

The Pinto Canyon Mines

The uncooperative climate, coupled with the new restrictions that
left livestock more susceptible to predation, underscored the wis-
dom of exploring other options. If the Shelys had not considered
the potential for mineral wealth on their lands after first arriving,
by now they were likely thinking economic diversification might
be a good idea. There was ample precedent. The earliest successful
commercial mining in the region took place in the southern reaches
of the Chinati Mountains. The Presidio Mine, which gave birth to
the town of Shafter, was by far the most successful, producing thirty
million ounces of silver and other metals between 1888 and 1942.
Other less successful attempts at mining silver occurred at the Loma
Plata Mine, some ten miles northwest of Pinto Canyon, intermittently between 1928 and 1949. During the 1920s, at least two different
companies tried to capitalize on meager nitrate deposits around
Candelaria, with disastrous results. Despite these limited successes
beyond the Shafter area, however, new discoveries and new markets episodically rekindled interest in the mineral resources of the
region.[5]

Starting in 1952, the Shelys sold oil and gas leases to Cities Service
Oil Company, Humble Oil, and a man named Jerry Raisch, although
none led to more than preliminary exploration. Then, during the
mid-1950s, a small army of graduate students from the University

of Texas renewed hope in mineral deposits across the Sierra Vieja. During two long hot summers in 1953 and 1954, doctoral student David Amsbury mapped and described the geologic structure of Pinto Canyon. Among the minerals of economic interest he reported were uranium deposits on the north side of Cerro Hueco—a vertically jointed rhyolitic intrusion containing abundant natural cavities. Around 1958, Terry and Fred Shely had the section platted into mining claims, and they used dynamite and bulldozers to expose the deposit in a 6-by-8-foot-deep prospect, which they excavated for about 140 feet into the side of the mountain. Subsequent analysis proved the ore to be of medium grade—about 0.34 percent triuranium octoxide (U3 08). It was good enough to warrant further exploration. The bigger question was the extent of the deposits. In an effort to find out, Fred bought his own Geiger counter, and in 1975 the Shelys leased the claim to Meeker and Company for about six years. Perhaps imagining a bonanza, a trapper named J. A. Stewart, residing across the road in the old George Sutherlin house, stockpiled about two hundred tons of the radioactive ore in the old teacher's residence beside the Horse Creek schoolhouse. Lacking commercial quantities, however, the mine never grew beyond a prospect.[6]

Fortunately for the Shelys, uranium was not the only mineral of interest Amsbury identified. Another one was perlite, which had been used commercially only since the 1940s. Perlite is an amorphous volcanic glass with a high water content. When the mineral is heated above 1,500 degrees F, the water trapped inside vaporizes, causing the rock to expand to as much as sixteen times the original size, making it less dense and lighter in weight by volume. Because of these unique properties, expanded perlite has found use in a range of applications, notably as a lightweight aggregate in plasters and concrete, as drilling mud, and as insulation, among others. Estimated to be as much as four hundred feet thick, the perlite deposit was situated several hundred feet above the canyon floor, on a mountain slope just south of the Pinto Canyon road. Around 1961, the Shelys leased the claim to Perlite Producers, Inc., which got the operation off the ground. Being a surface exposure, the mineral-bearing rock was simply blasted and bulldozed out as an open-pit operation. The ore was then loaded and trucked to Marfa, where a refining plant managed by Paul Wofford crushed, sorted, and washed the perlite before shipping it off for further processing, including the cooking process that would cause the rock to expand. The *Big Bend Sentinel* enthused that the deposit was estimated to be a twenty-five-year supply and that one million cubic yards of it were expected to be used in the Lake Amistad dam being built about 350 miles down–river on the Rio Grande.[7]

Under different circumstances, the perlite mine might have been viable. But the remoteness and difficulty of getting the ore out of the canyon proved problematic and costly. To take the long way downriver to Presidio was too far. Climbing out of the canyon on the dirt road, on the other hand, was hell on trucks and even worse on tires. In 1965 the company moved the refining plant to the mine in Pinto Canyon to consolidate operations. The partially processed ore was then trucked directly to loading elevators beside the railroad. That practice increased the efficiency of the operation, but it was not enough. As it was with most of the earlier mining efforts in the region, deposits were never abundant enough to justify the trouble of getting machinery in and the ore out. The lowlands have always seemed to resist such enterprise.[8]

Fox in the Henhouse

There's one in every family. He was the one.

—Manny Perez

Although the Prieto Ranch remained largely destocked during the drought of the 1950s, following a few years of decent rain the range had recovered enough that leasing the grazing rights became an option. In 1960, the family leased the ranch to Jesse Vizcaino of Marfa—the most unlikely of candidates. Born in 1927 to Julio Vizcaino and Carmen Jiner, for years he'd run the family's department store in Marfa. Having enjoyed financial success with his town business, he was finally able to pursue a long-standing dream. Despite having never run stock, Jesse purchased a herd of Angora goats, hired a foreman, and began commuting to Pinto Canyon on weekends. It was satisfying enough that only five years later he would buy a ranch of his own about ten miles north of Pinto Canyon, and he would operate it for the next twenty-three years.[9]

In the meantime, Gregorio Prieto had positioned himself to be manager of the Prieto Ranch. It was a good option for Gregorio, who did not have to do much except collect the rent, pay the taxes, and oversee the hunting operation. Juan Morales would have been better for the job, however. Already there and being intimately familiar with the ranch, he would have been a good fit. More importantly, he could be trusted. Had it been put to a vote, he would have won handily. Instead, over the protests of his sisters, Gregorio assumed management on authority of his overbearing nature. Because he was bullish and prone to outbursts, no one had the nerve to challenge him. He tyrannized his sisters, sometimes leaving them in tears. Having seen his mother, Frances, reduced to sobs one too many times, Manny finally confronted him. "Every time you come, you make my mother cry," he said. "I want you to stay away." There was no love lost between them. "He didn't like me, and I didn't like him," recalled Manny. And the fact remained that placing Gregorio in charge of family property was akin to putting a fox in charge of the henhouse.[10]

A Triad of Tragedy

The 1960s were troubled years for the Prietos. It was a decade distinguished by a series of deaths that began to whittle the family away. After returning from Mexico, José Prieto Jr. moved his family from Marfa to Odessa, where he opened his own auto repair shop, realizing a long-standing ambition. But whatever satisfaction he enjoyed was brief. In mid-July 1961, he crawled under a car to adjust the brakes. There should have been an axle stand. Instead, only a small screw-type jack supported the car. Perhaps he pulled too hard on his wrench, but the car slipped off the jack, crushing his head and chest and killing him instantly. He was found by two Odessa schoolgirls walking by shortly after the accident. An unexpected death dealt a tragic blow to the family. But at least it had not been the result of a gunshot or stabbing, like that visited upon his brothers.[11]

Only a year and a half after José Jr. died, José Prieto lay on his deathbed at home in El Paso. His health had been sliding for about a month because of advanced arteriosclerosis. His bearded and bespectacled doctor, Armando Garcia Cantu—one of the last in the city to make house calls—had been visiting with greater regularity. Known for his tenderness, especially toward the elderly, the doctor was likely fond of José—a man who even at the end of his life never wavered from being his gracious self. On 24 February 1963, at the age of ninety-six, José Prieto died of a stroke—some tiny piece of unstable plaque having blocked the flow of blood to his brain. He might have lived a bit longer. But after a life spanning nearly a century, little could change the fact that his time was near. If nothing else, his tired body was deserving of rest.[12]

Having been raised Catholic, José would be buried in the Catholic manner. Even so, his spiritual leanings may have sprung from more mystical traditions. Although José never spoke openly about religious matters, Frances used to watch curiously when her father left the house without explanation and walked toward the mountains. One time she followed him, just out of sight. He did not walk far but entered a small side canyon just out of view of the house and stopped. Frances's eyes widened as her father slowly raised his arms and turned his face to the sky. She strained to listen. But if something was said, she could not hear what it was.[13]

The Prieto family assembled at St. Mary's Catholic Church in Marfa as they had so many times before—his five daughters and last remaining son, his twenty-five grandchildren, and his fourteen great-grandchildren. He had been loved by them all, and, because he was the patriarch of the family, his passing held great significance. After mass, the mourners rose and sang "In paradisum" as José's grandsons carried his casket out of the church. Afterward, at the Merced Cemetery, in the plot that José himself had purchased thirty years earlier and where he had stood twice before, to see a son and his wife buried, the family gathered around as his casket was lowered into place at Juanita's left side.[14]

Modest and humble throughout his life, José had nonetheless been a legend in his own way. Transcending the barriers of his ethnicity and the confines of his socioeconomic standing, he had been a source of support—financial and otherwise—not only to his wife and children but also to his brother and cousins and nephews, as well as his neighbors. A tenacious work ethic coupled with his innate business sense allowed him to assemble a ranch far grander—and wealth far greater—than anything his childhood had prepared him for. Among his papers were loans he had made—some still outstanding, some that he knew would never be repaid. A bank statement once found by his grandson, Oscar, showed a balance of $140,000. There were others who had risen out of poverty or otherwise succeeded in gaining wealth, property, or businesses of their own. By the time Regino Nuñez died in 1933, he basically owned the town of Ruidosa. Demetrio Vasquez, who ranched at Ojo Carrizo several miles north of Pinto Canyon, had been another. But no other Hispanic residents in the lower Sierra Vieja had ranched as much land with as much success for as long as José Prieto.[15]

If José's accomplishments made him a role model for other Hispanic ranchers in the borderlands, perhaps his greater legacy stemmed less from his success than his humanity. Graciousness had always been his code, one he had labored to instill in his own children—with mixed success. He'd always counseled them to be honest and fair in their dealings with others, to be tolerant of fools and kind to the weak and downtrodden, to be gentle with animals and attentive to their needs. He disdained cruelty of any kind. Animals should not be killed except for food. Horses were not to be run unless there was a good reason. Goats should not be roped for sport. And the land itself—the grasses, the mountains, the rocks—were all things to be respected. He once admonished his grandchildren for rolling boulders off a cliff. But anger had never been a part of his being. The worst word he could muster was *caramba*. Despite having lived in and ranched one of the most rugged, most difficult, and harshest landscapes anywhere, and having also succeeded in an industry that requires fortitude and rewards grit, somehow it had not hardened the man. José's innate sensitivities never wavered.[16]

Only three years later, the family was to be visited by tragedy yet again. In 1966, Juan Morales broke his hip when he was thrown off a horse. As he recovered, he had to rely on Sue to take up the slack. The problem was she didn't know how to drive. She asked her sister, Lupe, who did, to come and stay with them while Juan got back on his feet. One afternoon, the two sisters were headed back to the Morales ranch house from Marfa, the truck loaded with supplies. For the first time, the Pinto Canyon road was being paved, but afternoon thundershowers had left the oiled dirt road muddy. There should have been a flagger when they entered the single-lane detour, but there was not. By the time they saw the oncoming car careening toward them, it was too late. As Lupe hit the brakes, the truck began to skid. Neither vehicle was able to slow down before colliding head on.[17]

The driver of the other car—one of the construction workers—was the only one to emerge unscathed. The others were rushed to the Brewster County Hospital in Alpine. The *Big Bend Sentinel* reported that the passenger in the worker's car had broken his leg but that the two women had received "only minor cuts and bruises." In fact, their injuries were much more extensive. Sue would be left with a permanent limp. Lupe, in far more critical condition, was transferred to Southwestern General Hospital in El Paso, where three weeks later, at the age of fifty-five, she died of internal bleeding. The funeral in Marfa must have started to seem routine, the dwindling family in agony as they mourned yet another loss. The family was being whittled away even as internal divisions were widening—and would widen further in the years to come.[18]

Fred and the Super Cub

Meanwhile, Fred Shely had become increasingly involved in civic affairs in Marfa—with his children's sports, his position on the school board, and a lively social life, all of which left him with less time to attend to the ranch. Weekends the family packed up and drove down to the canyon, but daily commutes during the week were growing tiresome and too time consuming. Fred wasn't long in finding a solution that addressed his need and suited his style. He purchased a Piper 150 Super Cub fixed-wing airplane—one of the best-known light aircraft of all time. Introduced in 1949, the planes were favored for banners, glider towing, and bush flying—landing and taking off in rough country that lacked prepared landing strips. Simple in design and relatively affordable, the "Model T of airplanes" was also favored by ranchers. Often equipped with especially large tires for rocky runways, it could take off and land within a distance of only five times the length of the airplane. The high initial cost was offset over time. Marfa rancher Russell White, who in 1945 bought one of the first airplanes in the region, claimed, "Airplanes pay for themselves in ways that are hard to measure—in time savings and easy mobility, for instance." For people who, like Fred, ranched below the Rimrock, their value was even more evident.[19]

With his airplane, Fred could now reach the ranch in a fraction of the time it took him to drive. And in being able to fly over rather than drive through the ranch, he could do inside of an hour what once took a couple of days to accomplish. "I can come down and check my water and look at the livestock in about a tenth of the time that I can in a pickup," he claimed. As a pilot, Fred was a natural. The many hours he spent flying between Marfa and the ranch, navigating the mountains and canyons below the Rimrock with their unusual air currents, allowed him to hone his skills, which was a good thing considering he was fearless. Rain or snow, wind or dust, hardly anything discouraged him from flying. "He would jump in that plane like it was a car and say, 'I'll be right back' and just take off," recalled Adam Miller. "If it was windy or snowing, he didn't care." One time, after hitting a mesquite tree and bending the propeller, Adam found Fred beating it back into shape with a hammer. "Yeah, it shakes like hell. I think it's out of balance," he said. Reconnecting the straightened propeller to the shaft, he was airborne once again.[20]

Gregorio and Benavidez

Gregorio and Benavidez destroyed what my grandfather built.

—Manny Perez

Having positioned himself as manager of the Prieto Ranch, Gregorio commuted from El Paso when something needed to be done. But mostly he just collected the rent money and cash from the hunting leases. Some of it he invested back into the ranch—repairing fences, replacing sucker rods, or patching the roof of the ranch house. Or at least that's what he claimed. Where the rest of it went was anybody's guess. What was certain was that it wasn't being distributed to his sisters. Regardless, Gregorio never hesitated to ask them to pay the property taxes. In 1968 he told Petra it was her turn. But having received none of the ranch income, she didn't have the money. "Gregorio was milking my aunt and pocketing the money," said Manny Perez. "The work was not getting done." Still, Gregorio insisted, and recklessly so, since the Benavidez family knew what was getting done at the ranch and what wasn't.[21]

One evening Ida overheard her mother complaining to her father, Juan Benavidez, that Gregorio was demanding the tax money; it made her blood boil. To withhold the income and then have the audacity to demand the tax payment from her poor mother was too much. Ida spoke to her father about it, but Juan wanted nothing to do with the Prietos. Even so, he agreed to support her if she decided to take action. That was all she needed to hear. Ida scheduled a meeting with Lucius D. Bunton, former district attorney for the Eighty-Third Judicial District of Texas. Bunton's own roots ran deep in the region, and his history had intersected the Prietos' in subtle ways. Raised on a ranch south of Marfa, Lucius was ten years old when his half uncle, Presidio County sheriff Joe Bunton, testified at Walter Hale Jr.'s murder trial. Although Ida admitted to having no money with which to pay his fee, Bunton agreed to take the case anyway. The game was on.[22]

Gregorio's demands were certainly the catalyst, but the truth was that Juan had never been satisfied with Petra's inheritance. While Petra's sisters were content to own the ranch in common, Juan had far greater use for land than the meager income it drew. It was no secret that his mind was set on the Ojo Acebuche section of the Prieto Ranch. Acquiring it would not only provide a reliable source of water but also connect his back sections with the Pinto Canyon road, allowing him better access. The problem was that José had wanted them to keep the ranch intact—shares, not sections, having been the inheritance he bequeathed.

Among the first few sections Juan purchased had been one located just south of the Prieto Ranch on the northern flanks of Sierra Parda. He might have sold that section to his father-in-law, José Prieto, or he could have swapped it out for a section closer to his headquarters. But it was good grazing country, and Juan wasn't about to let it go. Instead, he had long wanted to add onto it, and in 1947 he finally got his chance. That year José sold Juan two noncontiguous sections—one located just west of Juan's headquarters and the second catty-corner from his Parda section. But the second conveyance had been a mistake. José had only intended to lease the mountain section to Juan, not sell it. Already losing his eyesight and unable to read the papers he signed, he later claimed Juan had deceived him.[23]

The episode did not sit well with José's other children and served to further divide the Prieto and Benavidez families. But now, with the lawsuit, it was all-out war, and José's heirs were forced to take sides. As it turned out, the Benavidezes were not the only ones feeling shortchanged. Thanks to Gregorio, Victoriano's children had been left in poverty following their father's death. And following Lupe's death, the Millers felt similarly distanced. A rift that had existed —either obviously as in the case of the Benavidezes and Victoriano's family, or quietly, as in the case of the Millers—began to widen. The three families agreed that Gregorio and, by extension, Jane and Frances, must be keeping all the ranch income for themselves. And they rightfully wanted their part.[24]

The Lawsuit

They believed what the lawyers told them: "Either you sell this ranch now or the judge is going to take it away from you."

—Oscar Perez

When Gregorio, Jane, and Frances Prieto were served the papers, they assumed Juan Benavidez was behind the lawsuit and that it was just another of his efforts to gain control of the Acebuche section. Just as the Benavidezes had done, the Prieto contingent lawyered up, retaining Presidio County attorney Norman C. Davis and El Paso attorney Frank Owen III, a Texas legislator who had served several terms in both houses of the Texas legislature. The original petition alleged that Jane and Gregorio "kept and retained all monies, rents, and revenues from the estate . . . [for] their own use and benefit" and had failed to provide any accounting to the remaining heirs. Bunton subpoenaed Gregorio to produce receipts as evidence of the repairs he claimed to have made. But Gregorio had no receipts. The defense responded with a blanket denial that funds had been mismanaged and countered that Juan had grazed livestock "upon the property

[of José Prieto], destroyed fences, altered the fences so that their livestock would be able to pass freely from their ranch which adjoins the
ranch in question and grazed very freely and liberally" without ever
paying rent.[25]

The court date was finally set, but through multiple motions for
continuance the trial was delayed and rescheduled and canceled
and rescheduled again. Meanwhile, the attorneys met privately
and with the families but could not get them to come to an agreement. The defense claimed that Petra "refused to consider any
offer of settlement by way of division or otherwise, having made
demands upon the Defendants which have been unreasonable and
unyielding . . . in that said Plaintiff desires to secure certain designated sections which in effect would destroy, as an economic unit,
the balance of the ranch." The Benavidez family either wanted the
Acebuche section or they wanted the ranch sold so they could get
their money out of it. Nothing else would do. Closing the door to
a room in the basement of the courthouse, perhaps weary of the
impasse, the attorneys told Gregorio and his sisters they must either
agree to sell the ranch and divide the proceeds or the judge would
order it sold on the courthouse steps to the highest bidder. Anxious,
exhausted, and confused, they felt they had no choice. "None of
them had enough education to know better," recalled Manny Perez.
"They were victimized because of their ignorance."[26]

Chinati Peak Ranch

In 1970, a Midland real estate agent named Flake Tompkins listed
the nineteen-section "Chinati Peak Ranch" for twenty-five dollars
an acre. "This ranch . . . is in excellent condition as far as vegetation is concerned [and] the hunting is the finest in Presidio County
with abundant deer, quail, and dove," Tompkins wrote in the listing packet. That was at least partially true. His other claims were
less defensible, including the "thirteen to fifteen inches a year" of
rain he said was average. But his assertion that the ranch received
thirty-five inches in 1970 and caused poor road conditions went
beyond mere embellishment, considering Candelaria had received
less than half that amount. Although the ranch was advertised as a

NO. 4570

PETRA BENAVIDEZ, et al   )   IN THE DISTRICT COURT OF
)
vs.   )   PRESIDIO COUNTY, TEXAS
)
GREGORIO PRIETO, et al   )   83RD JUDICIAL DISTRICT

JUDGMENT

ON the _23_ day of October, A.D., 1975, the above entitled

and numbered cause came on to be heard. The Plaintiffs appeared in

person and by and through their Attorney of Record, and the Defendants

appeared in person and by and through their Attorney of Record. Where-

upon the attorneys advised the Court, and stated, that, subject to the

approval of the Court, they had agreed to a compromise of all matters

in controversy between them as follows:

"It is agreed that Judgment may be entered in this cause decree-

ing that all properties belonging to the Estates of JOSE PRIETO and

wife, JUANITA BARRERAS PRIETO, have heretofore been sold, and all of

the proceeds divided equitably and lawfully by and between the heirs

and in the proper and proportionate fractions to which each of the parties

hereto was entitled, SAVE AND EXCEPT the minerals under eight sections

of land located in Presidio County, Texas.

Court judgment of *Benavidez v. Prieto*, 1975. Courtesy of the Presidio County District Court, Marfa, Texas.

whole, Tompkins enthused that it could be operated as three separate ranches. Or, alternatively, it could be divided up. "[There is] the potential of subdividing into smaller ranchettes that could be resold," he wrote.[27]

Maybe because there had been no serious offers, the asking price was lowered a year later. But the lack of offers was likely a result of the ranch not having been advertised at all. Sale notices of ranches Tompkins represented had appeared regularly in the *El Paso Herald*, the *Odessa American*, the *Abilene Reporter-News*, and the *Lub-*

bock Avalanche-Journal, but there was no such advertisement for the Prieto Ranch. In fact, the first ad would not appear until August 1971, when he offered the ranch at "less than $22 an acre." Probably trying to appeal to hunters, the advertisement read, "Joint venture have 5 men need 5!" If each of ten partners put $5,000 down, he enthused, the rest could be paid over ten years at 7 percent interest. "Good tax advantage—No personal liability," he added.[28]

About that same time, Tompkins told the family he'd received an offer of $20.56 an acre—a good price considering there had been no better ones. But for Frances and Jane, the offer likely felt insulting. It was as if some dollar figure could be placed upon their precious family kingdom, one that could serve the family as a geographical anchor and provide a memorial to their father's success. But, if so, they were also weary of the lawsuit, which had now dragged on for three years. With increasing pressure from their attorneys, they also felt as if they had no other option. Finally, tearfully, they agreed to the price. When it came time to sign, however, there was no buyer other than Tompkins himself. Either the buyer had backed out or the prospective purchase had been no more than a ruse. Regardless, time had run out.[29]

With aching hearts, each of José's remaining daughters in turn signed the deed. What had taken a lifetime to build took only minutes to lose—this small family empire their father had built from nothing slipped quietly from the family's grasp. Tompkins agreed to pay 20 percent down with the remainder payable over ten years at 7 percent interest. But the interest payments never came, and within a year he had sold the ranch to Richard H. Johnson, a Dallas-based aeronautical engineer and glider pilot, for $26 an acre, netting a cool $66,000 profit—far more than any of the Prieto heirs received. But Tompkins's sleight of hand had not gone unnoticed. "The ranch was stolen out from under them and the guy that bought it turned around and sold it," Oscar Perez recalled. Recognition of the shady land deal has meant a half century of mourning for the Prietos.[30]

Some certainly breathed a sigh of relief, but many in the family felt cheated. They had all been operating at a marked disadvantage. Lacking fluency in English and absent an understanding of the US judicial system, they had been left at the mercy of their attorneys.

And although both Bunton and Owen were well-respected litigators, some even suspected the attorneys had been colluding with Tompkins. What is certain is that the details surrounding the sale of the Prieto Ranch left much to question and, in the end, Tompkins was the only one who seemed to have come out ahead.[31]

Leaving Alamitos

Throughout the lawsuit and sale of the Prieto Ranch, Juan and Sue Morales had been caught in the middle and—being the only ones still ranching—were most deeply affected. After breaking his hip, Juan had already been sidelined. Now they had to leave the only life they had known for the last quarter century. In 1971, using what money they'd received from selling the livestock, as well as their share of the ranch money, they built a small apartment behind Jane's house, in the backyard where José Prieto had tapped away his final years.[32]

Juan and Sue's new life in El Paso may have been more of a shock to them than it had been for José, the full effect tempered by his inability to see. After living in one of the most remote places anywhere, with a backyard as monumental as a national park, they were suddenly confined to a tiny stick-frame house in the sprawling suburbs of El Paso. Juan, being who he was, took it in stride, still full of life and his youthful proclivities. Even so, breaking his hip was also the beginning of a slow decline. Juan had known something was wrong for some time but stubbornly refused to go to the doctor. By the time he finally did, he was diagnosed with bone cancer. It had started around the fractures in his hip but had since spread throughout his body. Advanced as it was, he was told he had only about six months to live. But Juan wasn't having it. Defiantly, he said to his nephew Oscar, "What the hell do they know? If I want to live six more years, then I'll live six more years!" And he did. But in the meantime, the debilitating and painful disease slowly spread throughout his body. On 13 December 1979, Juan died at the age of eighty-one. Sue lived another six years before joining him in death on 15 September 1985, at age eighty-six. Both were buried at the Merced Cemetery in Marfa.[33]

A New Leaf

Nearing the end of his own life, Gregorio began to soften. He joined the Abundant Living Faith Center—an evangelical church in El Paso that claimed sinners could be forgiven, washed clean again, and achieve salvation. If he was growing more reflective and circumspect in his later years, perhaps he saw the church as a way to rectify all the damage he'd done. If he was so inclined, he certainly had plenty of things to repent. What seemed clear was that much of his arrogance and condescension had drained out of him. Somewhere along the way, he'd gained a measure of humility. One day while visiting Petra in a nursing home in El Paso shortly before her death, he did a thing no one had ever seen him do. He apologized. "I always thought badly of you," he said to Petra's daughter, Ida, "but I was wrong and I am sorry. I know now that you were the best daughter that Petra could have had." For a man who had never admitted any wrong, it was as if he'd been reinvented.[34]

But nothing Gregorio could say or do would diminish the destruction he'd left in his wake—the damage to his family's honor, the violence he brought to his family, causing the death of his youngest brother, the man he murdered in Mexico, the poverty in which he left his eldest brother's family, and, finally, the conflict he triggered that caused them to lose the family ranch. But if his family judged him harshly, life did not. His wife, Pilar, preceded him in death by nearly a quarter century. Gregorio's final years found him living in El Paso with his two devout daughters, who perhaps maintained a calm counterbalance to their father's episodic rages. On 24 July 2011, Gregorio died at the age of 107. The one bad seed had outlived all but one of his siblings.[35]

〰〰〰

The decade of the 1960s was a period of slow recovery from the drought of the 1950s, as rainfall failed to return to normal. In the country below the Rimrock there was even less to recover. The Shelys explored both uranium and perlite mining, but neither turned out to be the windfall they had hoped. For the Prietos, it was a time of transition and tragedy—when everything came crashing down around them, when injuries and deaths seemed to constantly whittle the family away, when the family would turn upon itself, cannibalizing the small empire that José had worked so hard to build.

In the end, the Benavidez family members were the only ones left standing. Unable to secure their share of either the land from Petra's inheritance or the income from it, they chose the only way they believed they could ever get anything at all. Gregorio's mismanagement of funds ultimately forced the showdown that took the Prieto Ranch out of the family's hands forever. To Juan Benavidez, with his long-standing resentment toward the Prietos, it might have seemed like a success. For the remaining Prieto heirs, it was nothing short of tragic. Each received a wad of cash but lost their family ranch and, with it, much of their history. For Pinto Canyon as for the Prieto family itself, it was the end of an era.

10

No Ranch for Old Men

Changing Times

I think the good Lord got mad that morning when he made all this. It ain't a ranch for old men.

—Apache Adams

APACHE ADAMS pulled the latigo snug and rebuckled the cinch on his saddle. Where the breast collar would have gone, he strapped a leather apron around the front of his horse, securing it to the cinch rings—a trick learned from Mexican cowboys accustomed to riding through catclaw and cactus. Stepping up into the saddle, he touched his horse's ribs with his spurs and turned in to the dense thicket. He heard it—a sudden commotion, a breaking of branches—before catching sight of the massive bull. Wild as a gazelle, it exploded out of the brush at a dead run and went blazing up the side of a rocky hill. The chase was on.

Apache charged after him, gaining ground as he spurred his horse forward. Uncoiling his rope as he went, he closed in on the bull as they raced down the far side of the hill. Just as the bull was crossing an old two-track road, Apache threw a wide loop that cleared the animal's horns and dropped neatly around its neck. Jerking the rope tight, he pitched the slack toward the bull and turned his horse sharply to the left. The rope popped taut, yanking the bull's head around, nearly pulling Apache's little horse down with it. The bull stood facing them, panting in a cloud of dust. Apache clicked his horse forward just enough to gain some slack in his rope. Then he pitched the slackened rope on the ground in front of the bull and waited. A standoff in the still desert heat. Apache tried to urge it forward. He threw up an arm. Nothing. He raised it again and, this time, smacked the side of his leggings. Finally, the bull took a step forward. No sooner had its front foot cleared the rope than Apache spurred his horse and the rope popped taut again. Now the bull's head was being pulled down, following the rope's path under its front leg. Apache spurred the little horse forward as it struggled against the weight, gaining a foot. Two feet. Finally, the bull began to go down, landing on his side with a great thud.

Apache turned his horse to face the bull and backed it, stretching the rope tight again as he stepped down and walked over to the bull. Puffs of dust rose just beyond its moist nostrils in time with its breathing. The whites of its upturned eye made it look the wild beast it was. Apache pulled his pigging string out of his belt, doubled it, and slung the loop end over the bull's rear leg. He ran the loose ends through the loop, making a lark's head, and drew it up and tied it to the bull's foreleg. Then he stepped away and stood catching his breath. The furious buzz of the cicadas seemed to rise in pitch as the sun baked the dry ground.

A half hour later, Apache backed his stock trailer up to the bull, ran the end of his rope through the open back of the trailer, and tied it off to the front slats. Then he walked back to the bull, untied its feet, and nudged it with his boot. But it wouldn't budge. He poked its flanks with a stick. Nothing. He yelled and waved his hands over his head, kicking dust in the air. Nothing. Finally, Apache walked around to the back of his truck, pulled out a chain, and wrapped it around the bull's horns. He unspooled the cable to the eight-thousand-pound winch welded to the front of his trailer, secured the winch hook to the chain around the bull's horns, then pressed the controller switch. The cable slowly retracted until it grew taut and the chain tightened around the bull's horns and the trailer rolled back a few inches before catching. Then, slowly, it began to drag the massive beast forward. Apache lifted the bull's head just enough to clear the back of the trailer as it was dragged inside.

The only problem was that now there was no room left for his

Apache Adams tightening the cinch of his saddle in Pinto Canyon. Photo courtesy of Don Cadden.

horse. The bull took up almost the entire trailer. Apache shrugged and wrapped his reins around his horse's neck and led it to the back of the trailer and smacked it on the rump. The horse jumped into the trailer, right on top of the bull, and Apache pulled the trailer gate shut and latched it. A half hour later, the sun riding the crest of the Sierra Quemada in Mexico, Apache topped the rim of Pinto Canyon where the road leveled out, and he motored on toward Marfa with his recumbent quarry. For Apache, it was just another day's work.[1]

Apache Adams would be the last cowboy to ranch in Pinto Canyon. Over the years he'd perfected the art of gathering remnants of cattle herds after the others had been rounded up and shipped—the wild ones no one else could catch, the ones that bolted at the first scent of a human. It was the early 1990s, and the new absentee owner of Pinto Canyon Ranch had enlisted Apache to clear off the cattle as they prepared to put the land on the market. Everything was different now. The Shelys' departure a decade earlier marked the end of the long tradition of sheep and goat ranching in the canyon. Ranches were no longer being run by families but by newcomers who'd purchased the land for reasons that had nothing to do with deriving an income from it. The passage of family ranches was part of a much larger trend across the region, but the marginal productivity of Pinto Canyon hastened its transition out of agriculture. The modern age had arrived.

The 1970s and most of the 1980s would be remarkably wet ones in the Big Bend, and the rain came when it was most needed. In September 1970, Hurricane Celia slammed into the Texas coast, killing twenty-eight people and causing nearly a billion dollars in damage before heading inland. By the time it reached the Big Bend, it had calmed, releasing a slow, soaking two inches. It accurately portended a decade filled with record-setting rains, as well as the month in which they'd occur. Nearly three inches fell on 12 September 1975 in Candelaria. Three years later, tropical storms dropped more than six and a half inches of rain in the same month. The Presidio international bridge was underwater for days. It would have been more helpful had the rains arrived a decade earlier, while the region was still struggling to recover from the drought. As it was, however, by the time the rains began falling in earnest, ranching in Pinto Canyon was at an end and there was little the rain could do to save it.[2]

It was just as well that goat and sheep ranching had ended in the canyon. The bottom had fallen out of the market for natural fibers, and the market didn't seem to be coming back. Between 1962 and 1971 the use of nylon, polyester, and acrylic more than quadrupled, causing wool prices to plummet to new lows. Although wool effected a partial recovery in the early 1970s, it ultimately couldn't keep pace with skyrocketing inflation. Mohair prices did much better initially, due to a strong export market. Most often used in blends, it was far less affected by synthetics than wool, and with a growing preference for mohair in Europe and Japan, prices rose during the 1970s before plummeting again during the 1980s. For producers still in the game, government price supports meant the difference between success and failure.[3]

Now, when the ranges were being restocked in the Presidio County lowlands, it was not sheep and goats but usually bovines, which required less care and were less susceptible to predators. As sheep and goat operations folded, predator control eased. Without a concerted effort to fight them, the wool and mohair ranchers who remained risked being forced out of business. Meanwhile, the beef market stayed stronger than that of wool and more stable than that of mohair, haltingly reaching a peak at more than sixty-six dollars per hundredweight in 1979 before declining again during the 1980s. By the 1990s, however, it had risen to nearly seventy-five dollars and would remain above sixty dollars per hundredweight for the rest of the decade.[4]

The Shelys Leave Pinto Canyon

When prices are good, you think it's the only way to live, and when prices are down and you're losing, you think you ought to be doing something else.

—Fred Shely, 1978

By 1970, the Shelys had been ranching in Pinto Canyon for more than twenty years. In that time, they had enjoyed a few years of good weather and decent market prices, but by and large it had been a tough run. Their first decade was dominated by the seven-year drought—an unwelcoming introduction to the lowlands. So they were certainly due the reprieve that the strong market and ample rainfall of the 1970s promised. But life would intervene in ways that limited their options.

On the first day of the new decade, Terry and Belle had just hosted a large New Year family gathering at the ranch. There was much work to do—readying the space, laying in supplies, cooking for and cleaning up after some thirty people. Perhaps the extra exertion was to blame. But a couple of days later Belle was folding clothes when she suddenly collapsed and died from a stroke at the age of seventy-eight. After her death, Terry became more isolated than ever. Without his outgoing and sociable partner of nearly fifty years, he grew increasingly withdrawn.[5]

Meanwhile, Fred and Kathryn had been living in Marfa for several years when she gave birth to their third child and only son, Freddie, in December 1959. If he was the apple of his father's eye, it only served as an even greater blow when his life was cut short by a tragic bicycle accident. In August 1973, as Freddie pedaled down Salarosa Street in Marfa, he failed to stop at an intersection and careened into the side of a southbound pickup. By the time he arrived at the Brewster County Memorial Hospital in Alpine, he had already died from a fractured skull.[6]

"It is the evidence of this fellowship and love that has sustained us in our grief," the Shely family wrote in a piece published in the *Sentinel*, "and with the feeling of gratefulness that can never be adequately expressed in words, we extend to you our heartfelt appreciation." A memorial fund was set up in Freddie's honor—the money earmarked to purchase lights for the new tennis courts at the Marfa high school. A crushing blow had struck at the very core of Fred Shely. "When my dad lost my brother, he kind of lost his heart," recalled Terry Jean. If his interest in the ranch had already started to flag, the loss of his son was just one less reason to care.[7]

Even so, Fred and Terry pushed past their collective sorrow and carried on. Livestock and windmills and water gaps beckoned. But if things returned to normal for a while, such routine didn't last for long. In 1976, the Chaffin family sold the old Sutherlin place on Horse Creek to a Houston man, T. Frank Bartle, causing Fred to lose the lease he'd held for almost thirty years. The property transfer effected an immediate constriction of the ranch operation and forced the Shelys to sell about half their livestock. But it also eased Fred's workload, the majority of which fell on his shoulders as Terry slowed down. Meanwhile, good workers were becoming increasingly hard to find—especially shearing crews. After Fred found his last crew passed out by the shearing pens following a bender in Mexico, he told them to pack up and leave. Weary of finding reliable shearers, Fred built a small shearing shed with drop-down shearers behind the ranch house where his men could do the work themselves. It meant a decline in speed and efficiency but at least resolved the uncertainty of itinerant shearing crews.[8]

Still other problems arose. As neighboring ranches were sold out of agriculture and others were converting from sheep and goats to cattle, predator control came to a standstill. By the late 1970s, the Shelys were among a small handful of holdouts: the Benavidez family to the northwest, the Dipper Ranch to the east, a few others up on the rim. But the sheep and goat industry in the lowlands was in freefall, and without a widespread, concerted effort to control predators, they soon grew out of hand. In 1977 alone, Fred claimed to have lost around eight hundred head to predators. They might have transitioned to cattle, but the ranch simply couldn't run enough head to make it pay. Nothing penciled out. Things were grinding down to their inevitable conclusion. Finally, Fred and Terry decided it was time to get out.[9]

Selling was mostly a financial decision. But the truth was, Fred's interest in the ranch had begun to diminish. His time was also increasingly filled with other duties. Between serving on the Marfa City Council, on the board of Rio Grande Electric, and as president of the school board, he found less time to be at the ranch. In addition, both Terry Jean and Nancy were starting lives of their own and his father had become ever more reclusive. By the end, Terry no longer left the house. Without his dad as an active partner, ranching lost much of its luster for Fred. There seemed less and less reason to try and hang on. "His heart just wasn't in it anymore," recalled Terry Jean. It was time to move on.[10]

In 1982, Fred and Terry sold their ranch in Pinto Canyon to a self-made millionaire out of Houston for about fifty-one dollars an acre. It was an incredible windfall considering that local land prices a decade earlier had been just half that. They had made some minor improvements—built a small shearing barn, added to the house, built a landing strip, and dug a new water well—but nothing that would have doubled the value of the ranch in such a short period of time. The real difference was shifting trends in land tenure, especially as well-moneyed, nonagricultural buyers entered the scene. For people who were not required to squeeze a living out of such a meager range or live off the interest that the sparse grass and browse provided, the value of the land was calculated on a completely different scale.[11]

After leaving the ranch, Terry moved to Marfa, where he lived until his death in March 1984 at the age of ninety-three. Fred and Kathryn remained in Marfa until around 1990, when they moved to Lajitas, where Fred ran the golf course and pro shop for the next ten years. Always a natural athlete, he'd gotten hooked on what he called "cow pasture pool" after hitting his first golf ball. He and Kathryn would later move to Humble, Texas, to be closer to their daughter Nancy. There, they both died within a year of each other— Kathryn in 2010 at the age of seventy-nine and Fred just shy of a year later at the age of eighty-three. Both had been considered pillars of the community. Paying tribute to Fred's life, the Texas House of Representatives later credited him with being "first and foremost a family man whose guidance, love and counsel extended far beyond the circle of his immediate family."[12]

Factors in the Demise of Pinto Canyon Ranching

The Shelys' departure did not mark the end of ranching in Pinto Canyon, but it did signify the end of family ranching. It was a significant demographic shift that seemed largely inescapable. From that point forward, if livestock was run at all, it would be run by lessees and would be cattle rather than sheep and goats. Still, it was only a matter of time before even cattle raising became a thing of the past in the canyon. There were many reasons for ranching's demise. Some were unique to the place: being remote, rugged, infested with predators, and bearing scant forage accounted for much. The decline in productivity, especially following the droughts of the 1930s and the 1950s, was likewise central. But the transition also stemmed from broader trends.

The proximity of the canyon to the national border and lax labor laws initially assured a ready supply of low-cost workers from Mexico. In response to a nationwide labor shortage following World War II, the United States had entered into agreements with Mexico that guaranteed a minimum wage and reasonable accommodations in exchange for legal but temporary contract workers. Even so, the Bracero Program was not universally embraced. Most Texas ranchers and farmers preferred to continue hiring undocumented workers, as they always had, to avoid higher costs and bureaucratic obstacles. That helped delay the participation of the state in the program, although Texas eventually signed on in 1947. When the program was revised a few years later, however, Congress included measures to crack down on illegal immigration. To do so, the US Immigration and Naturalization Service launched Operation Wetback in 1954 designed to send undocumented workers—those not cooperating with the Bracero Program—back to Mexico. Nearly four million would be repatriated over the ten-year life of the program, and many of them were sent south by rail from Presidio.[13]

The programs were only marginally successful. Ranchers and farmers who had always hired illegals continued to do so. And thousands who entered the country legally under the Bracero Program stayed illegally thereafter. Today, millions of Mexican Americans trace their roots back to the Bracero Program. Similarly, Operation Wetback was unable to slow illegal immigration; agriculturists never stopped hiring undocumented workers. In the borderlands, the greatest legacy of the program was the expansion of the US Border Patrol and a more permanent, strategic law enforcement presence. If that presence proved worrisome to borderland ranchers, the passage of the Immigration Reform and Control Act of 1986, which required workers to carry I-9 forms, proved even worse. For the first time, the federal government imposed strict penalties on employers who violated rules on hiring foreign workers. The inability to secure cheap labor along the border removed one of very few offsets in a struggling industry.[14]

Regulations continued to pile up. After constricting ranchers' workforce, the federal government began cracking down on the free-for-all predator war that had been waged for more than a century. After Congress had already prohibited the killing of golden eagles, the Environmental Protection Agency in 1972 banned the use of strychnine and Compound 1080—the primary methods of controlling coyotes. The passage of the Endangered Species Act in 1973 was seen as yet another threat to the ranching industry, although the dearth of public lands in Texas and absence of data on threatened and endangered species populations limited the local impact of the law. Still, some felt it was a further intrusion, another ply in the rope that seemed tied around the necks of ranchers. Against a broader background of declining productivity, regulations made a marginal enterprise more marginal still.[15]

Finally, with the transition of land out of agriculture, the "economy of scale" provided when neighbors pursued the same ends was removed. Whether pooling labor resources, sharing a pasture, or borrowing a pair of fencing pliers, when those just across the fence were fretting over the same drought and fighting the same mountain lion, the commonality of purpose brought benefits. And for predator control to amount to anything more than plugging a dam with a wine cork, pretty much everyone had to be doing it. As ranchlands began to pass from ranch families to absentee owners with different priorities, the ranchers who remained were fewer in number and increasingly isolated.

Playgrounds for the Prosperous

When A. J. Rod purchased the Pinto Canyon Ranch from the Shelys, it heralded a fundamental shift in land tenure in the canyon—from hardscrabble goat ranchers to absentee millionaires—one that would prove lasting. Like those who would follow, Arnold J. Rod had not made his fortune from livestock. Born in 1918 and raised on a cotton farm in Wharton County, Texas, Rod began to learn the machine tool and cutting tool business at sixteen, while attending Southern Methodist University. After serving an apprenticeship in Ohio with Warner & Swasey, a manufacturer of machine tools, instruments, and special machinery, he returned to the Houston area, where he and his wife, Earline, started the A. J. Rod Company out of their garage in 1948. Incorporating the business a few years later, they moved to a permanent location in east Houston, where the company soon became one of the most successful industrial cutting tool distributors in the state, boasting sales of around $30 million a year.[16]

By the 1970s, Rod had begun investing in land and livestock. He bought ranches in Bastrop County that he stocked with showcase cattle along with herds of deer and bison, and he constructed fish ponds equipped with automatic feeders. He purchased the Pinto Canyon Ranch for different reasons entirely. Unlike his extravagant Bastrop ranches, the Pinto Canyon Ranch was primarily an investment and secondarily a hunting retreat for his family and associates in the corporate world—a place to go once or twice a year to bag a few bucks, maybe an aoudad or a turkey or two, and lesser creatures that might pass before them. To ensure there would be enough deer to hunt, he installed an array of deer feeders and employed trappers to keep predators in check.[17]

Soon after purchasing the ranch, Rod had a sign erected over the Pinto Canyon road declaring his ownership. But the sign served to confuse travelers who mistook it for a private road and felt compelled to turn around and return to Marfa. Residents in Ruidosa and at the hot springs resented the sign and complained bitterly until the county commissioners finally took it down in 1996 at county expense. By that time it had served to deflect travelers for more than a decade. In the meantime, Rod had his attorneys hard at work trying to clear defects against the title. He brought suit against one J. J. Swofford contesting an affidavit filed in 1930 claiming he owned eight sections of land in Pinto Canyon. The suit countered that Swofford was "a mere naked trespasser" and had never, in fact, occupied or possessed the land in question. Rod secured affidavits from the Shelys stating they had possessed and operated the ranch continuously since 1948. Unable to locate Swofford, the suit was instead "cited by publication" where it ran in the *Sentinel* for a month and a half. Ultimately Swofford's heirs signed a quitclaim deed, settling the suit out of court.[18]

The ranch was secured by sign and by deed, but Rod was not content with the access. Aside from a few barely maintained ranch roads, the only way to get around was by horse or on foot. For Rod—a pear-shaped man, heavy of gut—and his well-heeled companions, that wouldn't do. To allow them to hunt from a Jeep or their high-seated customized hunting rigs, he hired a heavy machine operator to bulldoze new access roads all across the ranch—places where roads had never been and arguably were never meant to be. About a year after selling the ranch, Fred and his daughter Terry Jean flew over Pinto Canyon and were horrified by Rod's improvements. "They had made all these hunting roads. It was like they had just cut it into pieces," recalled Terry Jean.[19]

The old Wilson Ranch house was remade into a hunting lodge. Enclosing a porch, the new owner added rows of bunkbeds as ordered as an army barrack. Come hunting season, and momentarily freed from the shackles of their workaday schedules, Rod and his son, Robert, and their buddies could stalk their quarry and bond in the singular way male hunters tend to bond. But they rarely came at other times of the year, and if their arrival was heralded by engines grinding over the hills and rifle shots echoing through the canyon, they never stayed for long. The few who remember recount stories of corn-fed deer, aoudads, and hogs being hauled out of the canyon by the truckload.[20]

Ayala de Chinati: Donald Judd and Appropriateness of Place

My first and largest interest is in my relation to the natural world, all of it, all the way out.

—Donald Judd

In 1976, artist Donald Judd purchased Rancho Alamito—the old Juan Morales ranch—from Richard Johnson. He was unlike any landowner who had come before. Judd's preservation ethic coupled with his philosophical embrace of nature, art, and community brought a new vision to the canyon that would ultimately forge a more secure future. By the time Judd bought the Morales place, he'd been living in the region for the better part of five years and had been increasingly drawn to the vast and uninhabited landscape. He first rented a house in Marfa in 1971 to escape the hectic New York art world, where he had gained widespread acclaim as a pioneer in the minimalist school of art. But Judd disliked the term "minimalist" and had grown frustrated with New York, where he had lived for most of the previous twenty years.[21]

Marfa was not his first choice. He'd spent summers in his Land Rover exploring back-country dirt roads of the American Southwest in search of places to escape. Ultimately, the quiet, unpeopled expanses of the Big Bend caught his attention and convinced him to stay—that and many large, empty buildings where he could install his art. The decline of the cattle industry along with the closing of Fort D. A. Russell and the Marfa Army Air Field following World War II had sent Marfa into economic freefall. From an all-time high of five thousand people during the war, the population of the town by 1970 had declined by nearly half, leaving many downtown buildings unoccupied and the fort shuttered. As much as anything, however, Judd felt a connection to the landscape and the indigenous rock and adobe architecture that graced it. Ironically, it was not the sweeping grasslands of the Marfa Plain—which might have reminded him of his childhood in the Midwest and which seemed more consistent with his austere artistic sensibilities—but the dramatic and rugged Chinati Mountains and the lowlands of the Sierra Vieja that captivated his imagination.[22]

Born in Missouri in 1928, Judd spent summers on his grandparents' farm in Excelsior Springs, where his grandmother encouraged his fascination with the natural world. After high school, he served in the Army Corps of Engineers during the Korean War before earning a degree in philosophy from Columbia University in 1953. Turning his attention to art, he took graduate courses in art history and attended night classes at the Art Students League. Initially a painter, Judd grew disillusioned with the form in favor of three-dimensional objects and nonrepresentational designs. He continued to refine his thinking about art and architecture and gained a reputation for his terse and searing criticism while writing for major art magazines. By the 1960s, as his work gained recognition, Judd began exhibiting widely, particularly in New York and Europe.[23]

In 1964, Judd married Julie Finch, a dancer and choreographer, and after purchasing a historic building in the cast-iron district of Manhattan (later known as SoHo), they co-founded Artists against the Expressway to defeat a project that threatened the district and some six thousand artists estimated to live there. Around the same time, the couple had two children: Flavin in 1968 and Rainer in 1970. But even as Judd became more involved in civic life, he grew to disdain the frenetic energy of the big city. As an antidote, he packed up the Land Rover and took his family on months-long camping trips in Baja California and, later, in West Texas. Judd loved the peace and quiet, but Julie's career was rooted in New York and she had no desire to live in the desert. They separated in 1976 and divorced amid a contentious custody battle that Judd eventually won. In a characteristic about-face, Judd never spoke to her again. "He had a way of cutting people off," Julie recalled.[24]

In the interim, Judd purchased a full city block of Marfa, including two former airplane hangars and a building that had been offices for the US Army Quartermaster Corps. With the addition of an adobe perimeter wall that was nine feet high, this complex would serve as his home and studio. In 1977, Judd became a full-time resident of Marfa, about the time the newly formed Dia Foundation out of Houston took an interest in his ideas for siting permanently installed art. Established by one of Judd's art dealers, German-born Heiner Friedrich, along with his wife, Philippa de Menil, of the Schlumberger oilfield services dynasty, the foundation supported large art projects and the artists who produced them. Heiner may have been the driving force, but it was Philippa who had the stock required to fund Dia's large and expensive art exhibits. In 1979, the Dia Foundation purchased the former Fort D. A. Russell for permanent installations of Judd's work and those of several other artists handpicked by Judd. Having seen his own work displayed poorly and carelessly damaged during transport, he deplored museums and galleries that he felt degraded and commodified art.[25]

Encouraged by the Dia Foundation's eagerness to invest locally and their financial ability to do so, Judd introduced the art patrons to the wild and rugged lands of Pinto Canyon. They were undoubtedly impressed, for the same year that they purchased the old fort they also bought the old Prieto and Perez ranch properties from Richard Johnson, in addition to the old Mesquite Ranch that spanned the better part of the Chinati Mountains—more than forty thousand acres in all. They agreed to sell Judd the Perez property after the existing grazing lease expired, but they also had more personal reasons. By the time they purchased the ranches, both Heiner and Philippa had become devout practitioners of Sufism—the mystical branch of Islam—and the land was to serve as a place of spiritual retreat.[26]

Judd did not share Heiner and Philippa's spiritual mysticism, but his own near-religious zeal for the landscape became an increasing focus for the rest of his life. Fortunately, his success as an artist afforded him the ability to support his bourgeoning land habit. After buying the Morales Ranch, Judd purchased land with increasing frequency, acquiring the Perez Ranch in 1983, the Vizcaino Ranch in 1988, and the Ruidosa Hot Springs in 1990, along with several smaller parcels. By the early 1990s, his holdings had expanded to almost forty thousand acres, stretching from Sierra Parda northward almost to Candelaria. He called it collectively Ayala de Chinati. Much like his art, which was decidedly nonrepresentational, the words had little to do with the place. Ayala had been the surname of a New York neighbor, and Chinati was the name of the mountain range. Judd simply liked the way they sounded, befitting the beauty of the landscape. And that was justification enough.[27]

With the aid of a range management plan obtained through the local Natural Resources Conservation Service office, Judd leased the grazing to gain the agricultural tax exemption but otherwise wanted the land to be allowed to recover. He restored the Morales, Perez, and Vizcaino ranch houses to his exacting standards, derived from a sound understanding of the tenets of historic preservation. He had attained fluency in such matters before leaving New York, where he had restored the historic cast-iron building at 101 Spring Street in SoHo. Following his purchase of that landmark structure in 1968, he slowly transformed the dilapidated building, integrating art with living space—a practice he would follow with all subsequent restorations. In the process, Judd developed strong ideas about striking a balance between historic integrity, beauty, and appropriateness of place. By the time he set to restoring the rock and adobe structures in Texas, he had developed a solid philosophical foundation from which to draw.[28]

Mostly that philosophy was to leave the buildings alone. Despite his reputation for being headstrong and cantankerous, Judd had peculiar sensitivities and approached the ranches and the ranch houses with a care and respect normally reserved for sanctuaries. "I don't think anything should be done in the buildings that goes against what they were, or their original nature," he said. "I think there should be a greater respect for the past, for the buildings and of all things than there generally is." At his ranches and in his ranch houses, as with the buildings in Marfa, Judd had found the perfect opportunity to put his words into practice.[29]

Still, Judd's attempts to protect the land sometimes put him at odds with the county. After purchasing the Perez Ranch, he was outraged when a county grader widened the Pinto Canyon road far beyond what seemed necessary. In response, Judd had a barbed wire fence erected on both sides of the road. Only sixteen feet apart in places, the fence left no shoulder on which road crews could maneuver. Judd's attorney finally reached an agreement with the county in 1989, but some locals resented what they called his "spite fence," believing it was an effort to keep people out. But if Judd felt spite against anyone, and there were certainly a few he did, he held nothing against the broader community. In fact, he'd entertained plans to create a more sustainable regionally based economy, envisioning fruit and produce markets, local crafts, organically grown beef, and possibly even a sotol brewery. And when the old adobe church in Ruidosa was threatened with demolition, Judd wrote an impassioned plea to preserve it and drew up plans for doing so.[30]

In fact, most who were critical of Judd didn't really know him. His friend Jamie Dearing, who also served as his studio assistant throughout the 1970s, was one who did. For several years he would drive a truckload of art from New York to Marfa, slowly moving Judd's expansive collection to its new home. One year the drive was particularly hot and slow—the truck refused to travel faster than 40 miles per hour. Worse yet, it had no air-conditioning. Accustomed to the milder temperatures of the Northeast, by the time Jamie arrived he was shirtless and sweating profusely. "You look like a wreck," Judd told him, before suggesting he take a swim. "A swim?!" Jamie asked incredulously, as he looked out at heat waves shimmering

Donald Judd, Flavin Judd, and Jamie Dearing outside Marfa in 1975. Photo courtesy of Jamie Dearing © Judd Foundation.

over the Marfa Plain. Judd pointed to a stock tank on top of the hill near the house. Without missing a beat, Jamie threw off the rest of his clothes, ran to the tank, and jumped in, enjoying instant relief. Wiping the water from his eyes, he blinked in disbelief to see an ice cube the size of a small refrigerator floating beside him. As if on cue, Judd suddenly appeared at the side of the tank, holding out a big glass of bourbon—with a single ice cube. "Welcome to Texas," he said, a grin stretching across his face. "He'd clearly done it for me and I was so moved," Jamie recalled. "It was funny and it was typical of Don's generosity."[31]

That Judd also had an abiding affection for the landscape became increasingly evident to Jamie the more time they spent together outdoors. "There was always something about that land, the high desert, and the change of the topography and the nature of the botanical shift from the high plateau to the river valley that just enthralled him," he recalled. Despite perceptions of Judd's precisely rendered geometric art as being decidedly urban, over time Jamie came to see it as less architectonic, less inspired by urban structures than by the natural world. "Almost all of [his artworks] have a precedent in the land," he said. "He put more geography, topography, geometry, spatial interest, visual glint of light and shadow into his work than almost anybody . . . and I think it was the same thing he saw in that land."[32]

Few aside from art critics could write about Judd's art in an informed way, but even they tended to focus on the strict geometry of his work and always felt compelled to place it within the broader context of art history, of minimalism, of the work of his peers. Yet the lofty ramblings of the highbrow art literati almost always sprang from an intensely urban context. Not many people really knew Judd, and without that kind of familiarity the inspiration for much of his work remained elusive. It was true that Judd had a fondness for European architecture, but "even more important than anything was that land," recalled Jamie, who points to works like Judd's untitled round concrete piece on the grounds of the Philip Johnson Glass House in New Canaan, Connecticut, as evidence. Twenty-five feet across, and ranging from three to four feet tall, it was Judd's largest work to date, as well as his first outdoor concrete sculpture. From a distance, it appears similar to open concrete tanks that dot the ranchlands of the Big Bend. But closer inspection reveals the interior edge of the "tank" to be level while the outside edge is sloped, running parallel to the ground surface. "I think it's what he saw in those cow tanks," said Jamie, "the idea that there can be both horizontality in fact as defined by a flat body of water held by gravity in a plane and something that slants around it."[33]

Slope and aspect, the geometric perfection of cacti, the almost playful verticality of the faulted landscape likely served as conceptual fodder for Judd's art. And, similar to his art, he perceived in the land a certain transcendent integrity. As such, to witness the destruction of that land was agonizing. One day as he and Jamie were driving to Shafter, they came upon a new ranch road someone had bulldozed up the side of a hill. Seeing this, Judd pulled over onto the shoulder and, coming to a stop, slowly leaned his head against the steering wheel. "I thought he was going to cry," recalled Jamie. "Look at that, look at what they've done to the land," Judd said. "What an ugly scar." More troubling yet were the roads A. J. Rod had dozed in Pinto Canyon, which infuriated him. "Here, everywhere, the destruction of new land is a brutality," he wrote.

Donald Judd and Dan Flavin at Casa Morales in 1981. Photo courtesy David Zwirner, New York/London.

"Nearby a man bought a nearly untouched ranch three or four years ago, bulldozed roads everywhere so he could shoot deer without walking, and last fall died. . . . Within a real view of the world and the universe this violence would be a sin."[34]

"He felt that the land was sacred, not to be torn up," said his daughter, Rainer. "Buying land gave him the empowered feeling that he was protecting it." Indeed, Judd had gone far to protect his own land against such violence and would have continued to expand his ranch further still. But time was running out. In 1993, his health—as well as his relationship with Marfa—began to deteriorate. Over the summer, a noisy ice machine across from his Marfa complex instigated a virtual war between him, the ice plant operator and, eventually, the city. The fight was exacting a toll, but behind the mounting stress, he was being stalked by a silent killer. Marfa mayor Jake Brisbin recalled his last meeting with Judd. "Don looked terrible that day," he said. "It looked like this ice machine thing was really taking its toll." But even as his fight against the ceaseless noise of the ice machine raged on, in quieter moments, Judd seemed to be turning inward—and outward, his focus reaching existential proportions. "My first and largest interest is in my relation to the natural world," he once wrote. "All of it, all the way out. This interest includes my existence . . . the existence of everything and the space and time that is created by the existing things." His last land purchase was four sections north of Pinto Canyon, including a tiny and picturesque rock house in a wild and narrow canyon. Already weak, he enticed his neighbor, Sherman Bales, to drive him there, where he sat down on a blanket with his notebook and a jug of water and asked to be left alone.[35]

By late fall, friends had noticed Judd was having trouble breathing. But with exhibitions, awards, and talks scheduled in Europe, he summoned the strength to meet his partner, Marianne Stockebrand, a Cologne art historian, in Germany. Once there, he finally went in for a full medical evaluation. What he'd thought was giardia (a parasitic infection) was identified as an advanced case of non-Hodgkin's lymphoma. There was little they could do. Judd and Stockebrand flew to Amsterdam, where he received the Sikkens Prize for his use of color, but he felt too ill to give the talk he had written. Back in the United States, he was able to spend Christmas at 101 Spring Street with his family, but, soon after the New Year, he checked himself into the New York Hospital. He died at nine in the morning on 12 February 1994 at the age of sixty-five. In his will, Judd had made specific provisions that if his ranch was to be sold, it should go to an entity that "will be best able to preserve this property, in its natural state for the benefit of the scientific community and the general public." The task of honoring his wishes would fall upon his children, who would soon discover it would be anything but easy.[36]

The Last Holdouts

My grandfather tried to get all of them to be big ranchers. And the only ranch left is my dad's.

—Ida Benavidez

After the forced sale of the Prieto ranch properties, Petra was the last of José's family who remained on the land, even if Juan and Petra's success had been gained partly at his expense. Following Ida's move to El Paso, Juan Benavidez and his son Ben were left to run the ranch by themselves. In 1967, after nearly thirty years of ranching the lowlands, Juan handed the operation over to Ben. But Juan's retirement was anything but idle. He continued to live and work on the ranch and would do so until the day he died.[37]

By the mid-1970s, the Benavidez and Shely families were the last holdouts running sheep and goats in Pinto Canyon. Juan had about three thousand animals in total—mostly Angoras, along with some Spanish meat goats. He employed at least one man whose only job was to check fences and trap and hunt predators. "He was a loner and that is just what he loved to do," recalled Juan's granddaughter Teresa. Policing the ranch for four-legged intruders was one thing, but Juan and Ben were soon facing predators of another sort. Around 1976, they discovered goats missing. After an extensive search, they found some of the goats, but not where they should have been. And there was one other troubling matter—among the tiny goat tracks was something else: fresh human footprints.[38]

It appeared the goats were being herded westward through the old Nuñez Ranch. Back at headquarters, Juan and Ben jumped into the truck and raced toward Ruidosa, hoping to intercept the thieves. Passing through Ruidosa, they continued driving south on Highway 170 before turning east on a dirt road headed toward Sierra Parda. That's when things started to get strange. They hadn't gone far when they were suddenly confronted with a roadblock—in the middle of nowhere. Sheriff Rick Thompson was there, and he said they could not pass. Juan earnestly explained that they were in hot pursuit of goat thieves. Instead of offering help, however, Thompson questioned them suspiciously and then told them to go home. With no other option, they did.[39]

Juan filed a report, at which point Sheriff Thompson promised to launch an investigation, maybe even pull in the Texas Rangers. But if he did anything at all, they never knew it. Frustrated by the sheriff's disinterested response, Ben told Thompson he planned to camp out and post guard in the pasture. If he saw the thieves, he was going to shoot them. Thompson responded that it was a good way to get himself killed. Ben didn't post guard and they never found their goats. "They stole them all. In the end we didn't have any more," recalled Teresa. "It was very fishy."[40]

The fishiness made better sense after Thompson was busted in 1991 with a horse trailer full of cocaine—the largest load ever confiscated in West Texas. Caught red-handed, he and his partner, a lowland rancher named Robert Chambers, both ultimately pleaded guilty. The prosecuting attorneys recommended the minimum sentence. Instead, Chambers got twenty-two years and Thompson got life. The mystery of the stolen goats was never resolved. Thompson's presence that day may have been merely coincidence, but one thing was certain—the Benavidez family was out of the hair goat business. When they restocked, they turned to cattle. Their shearing days were over.[41]

Despite his nominal retirement, Juan hardly slowed down. Unable to remain idle, his final years were filled with horses and mules and livestock, just as they had always been. And he sure as hell wasn't moving to town. Even after Ida and her husband, Paul Taulbee, helped fix up Juan and Petra's house in Marfa, he refused to move in. He had always said he would die in the mountains, on the ranch he'd built from nothing. He proved prescient. While rounding up mules in August 1985, Juan Benavidez dropped dead at the age of eighty from a heart attack. He might have protested the funeral mass held for him at St. Mary's Catholic Church in Marfa the following Sunday. But he had little reason to complain. Active until the day he died, and doing the work he knew and loved, he'd been gifted the kind of death most anyone would envy.[42]

Apache Adams: The Last Cowboy

They nearly had to roll the rocks around [to eat], but they made it.

—Apache Adams

After buying the Pinto Canyon Ranch, A. J. Rod enlisted nearby rancher Sherman Bales to serve as the on-site manager who would oversee the property and watch after his cattle. A game warden with the Texas Parks and Wildlife Department, Bales had first moved to Presidio County in 1972. When not on patrol, he ran cattle along Capote Creek and other leased land below the Rimrock. He later leased the Briscoe Ranch near the Ruidosa Hot Springs and eventually gained a section of his own where he built a modest ranch house and a set of pens. Meanwhile, he agreed to work with Rod's general manager, Mance Draper, foreman of Rod's Flying Bar Ranch, located near Bastrop.[43]

After getting the title and livestock and management all set up, however, Rod was barely able to enjoy the ranch he'd bought. In 1988—only six years after the purchase—he died at the age of sixty-nine. Toward the end, he found it increasingly difficult to control his weight. As a result, his health began to decline. By the time he died, he weighed 350 pounds. For a while his son, Robert, continued to fly out in his airplane during hunting season. But with his father gone, there seemed fewer reasons to keep it, and Robert's sister, Sandra, needed the money. In 1994, the Rod family decided to put the ranch on the market. Before they could sell, however, they needed to gather their cattle, which was no easy task considering the beasts had gone so feral no one could catch them. To remedy the situation, they called in the expert—a man nearly as wild as the cattle he pursued, and with a name befitting his reputation: Apache Adams.[44]

Born in 1937 on his parents' ranch along the Rio Grande south of Marathon, Apache was riding before he could walk and breaking colts by the time he was ten. He learned the art of tracking from Mexican cowboys when he was still a boy and became an expert roper from years of catching wild cattle and feral burros during the screwworm epidemic. By the early 1990s, Apache was running cattle across thirty leased sections on Terlingua Ranch in southern Brewster County. That wasn't enough for his herd, but he was able to graze many thousands of acres more by leaving his gates open. "Back then it was open range law," recalled Apache. "If you didn't want my cattle on you, you had to fence me out." But there were few neighbors and no other competing livestock in the area, so it was a hell of a deal for the money. Or at least it was until the county voted out the open range law, forcing him to sell off most of his cattle. In the meantime, he started looking for better land.[45]

Apache had earned a reputation for his ability to catch wild cattle, as well as just about anything else that could be caught (among other things, he once roped a mountain lion). He struck the standard agreement with the Rods—to gather the remnant cattle "on halves," meaning he got to keep half of whatever he caught. "The cattle had been there about seven or eight years, was all mavericks, and they were just pretty wild and never been penned or worked," he said. With a couple of his regular hands, Juan Villa and Dogie Delaney, Apache arrived early one morning, unloaded their horses, and got down to work. Although the cattle were scattered across nearly thirty sections of land, the men worked fast. In three days they were finished, without having penned a single one. The cattle were too wild to herd into pens, so Apache and his men chased, roped, and tied down every last one. In the process, Apache took note of the quality of the range. Coming on the heels of a run of wet years, the canyon looked lush—especially compared to his Terlingua Ranch.[46]

Apache didn't waste time in securing an agreement with the Rods to lease the ranch until it sold. Hiring several eighteen-wheel cattle trucks, he had about two hundred head from his Terlingua Ranch trailered over. Arriving at the edge of the canyon, the drivers refused to descend. No problem. Apache had them back the trucks over the cattle guard, and they turned the cattle loose, one truckload at a time. Without their horses, Apache and his helper Chris Spencer had no option but to herd the cattle down into the canyon on foot. About halfway down, Apache noticed Chris suddenly freeze in his tracks. He'd just walked into a detail of US Marines hidden under a camouflage net watching for drug smugglers, and he now had eight rifles pointing right at him. He'd startled them, and they were not happy about it. After a gruff interrogation, they finally let him go.[47]

The next day trucks brought the last of the cattle, and the animals finally fanned out across the ranch. "They nearly had to roll the rocks around [to eat], but they made it," Apache recalled. Although he stayed down in Pinto Canyon when he needed to, Apache usually commuted from Alpine, leaving town by 4:00 a.m. and arriving at the ranch by 6:00 to beat the heat. He always kept at least one man stationed down in the canyon to watch over the livestock. Two brothers, Juan and Santos Villa, from San Carlos, Chihuahua, worked in shifts. One brother would stay about a month or two and then they would trade places. When on their shift, they slept in a metal shack beside the old George Sutherlin homestead, which by then was falling into ruin. But the shack, lacking insulation or interior sheathing, was so hot in the summer they called it "the microwave."[48]

Starting early, before the broiling heat turned work into drudgery, was standard practice in the lowlands. But the ruggedness of the terrain and the sinister brush—in places as thick as a bramble—presented additional challenges. Good leggings were always mandatory. But on days when Apache knew he'd be brush-crashing, extra precautions were called for, such as leather sleeves to knock it aside. Sometimes he took additional measures, fitting his horse with a leather apron where the breast collar would usually go. "A lot of horses got to where they wouldn't hit that catclaw," said Apache. "They'd go to ducking off on you. So you had to change horses pretty regular."[49]

There being nearly thirty miles of perimeter fence on the ranch, much of it in disrepair, Apache and his men spent a lot of their time riding fence. One day Apache rode up on a small pink object on the ground. Stepping down to look, he saw it was a half set of false teeth. Apache and Juan scoured the pasture for several hours thinking they might find the rest of the man. No luck. On his way out later that evening, Apache met Sherman Bales coming down Pinto

Canyon road and showed him the teeth. Sherman laughed and said they belonged to Rod's foreman, old man Draper. Months earlier, he and Sherman had been out working cattle and the old man put them in his pocket so they wouldn't clatter in his mouth as he rode. At some point, they fell out. They tried for hours to find them but finally gave up. Sherman mailed the denture to the owner, and a couple of weeks later Apache got a thank-you letter in the mail. Draper said it was good to have his teeth back, but unfortunately they'd dried out and no longer fit.[50]

Going Native: Jeff Fort

Apache had been leasing the Pinto Canyon Ranch for a couple of years while it lingered on the market. One day the real estate agent asked if he'd show it to a prospective buyer—a man from Houston. That Jeff Fort was in the area at all was a chance event—he was a New England businessman on his first tour of the Texas Big Bend. He didn't come to buy a ranch, but here he was. His unassuming demeanor could make one mistake him as ordinary, at least until you knew his history. Born in 1941 in the Manhattan borough of New York City, John F. "Jeff" Fort III spent his childhood in Maryland and Virginia before earning a degree in aeronautical engineering from Princeton. While attending graduate school at MIT in industrial management, he got a job working as a production clerk for Simplex Wire and Cable Company. Ambitious and driven, Jeff was good at his job. So good, in fact, that by the time the company was acquired by Tyco in 1974, he had been promoted to general manager. At Tyco, he quickly rose through the ranks. Within a decade, Jeff was sitting at the helm. While he was CEO, the company grew dramatically through a dizzying run of acquisitions. Meanwhile, sales skyrocketed: from $250 million to $3.5 billion a year.[51]

Jeff's success was the stuff of storybooks. But after ten years of tense business meetings and high-rise conference calls, he realized it was killing him. At fifty-one, at the peak of his career, he retired from Tyco, quit smoking, and started climbing mountains. In earnest. After scaling the highest in New England (Mount Washington), he climbed the highest in North America—Denali (20,310 feet), and then—at almost 23,000 feet—the tallest in the entire Western Hemisphere: Aconcagua, in the Andes. Somewhere among the *krumholtz* and chirping marmots above the treeline and snow-packed alpine peaks he found renewal, a vitality absent during his corporate years. He felt alive again.[52]

Even so, the business world beckoned. In 2002, a scandal at Tyco threatened to sink the company. Dennis Kozlowski, who had succeeded Jeff as CEO, was indicted for tax fraud and faced thirty-eight felony counts for pocketing $170 million from Tyco and more than $400 million in sales of tainted stock. Desperate to avoid financial collapse and a public relations disaster, the board urged Jeff to return. They needed the kind of thrift and stability he represented. While Kozlowski had been known for jets, yachts, and mansions, Jeff's tenure as CEO saw him driving to work in a Pontiac with no air-conditioning. Despite the press's excoriation of Kozlowski and the cloud it cast over Tyco, Jeff was able to set the company back on course even as the scandal continued to dominate headlines.[53]

In the meantime, Jeff had relocated to Houston with his wife, Marion Barthelme, a prominent civic art patron and the widow of famed fiction writer Donald Barthelme. Realizing there must be more to Texas than the Gulf prairies and urban jungles of southeastern Texas, in a quintessentially American way he headed west and did what so many have done before: he fell in love with the land. He drove wide-eyed through Big Bend National Park and toured the small towns of the region: Marathon, Fort Davis, Alpine. He found himself in Marfa when he heard about Pinto Canyon. As Jeff descended the steep and rocky road into the canyon for the first time, he was stunned. The land he saw was every bit as monumental as the national park; he couldn't believe it was private property. The kicker was, he heard it was for sale.[54]

By the time Apache had dropped him off back where he'd left his car, Jeff knew he wanted to buy the Pinto Canyon Ranch if there was a way. But he also realized how little he knew about ranch real estate. There was research to be done. So Jeff told the real estate agent he was interested in seeing some other ranches for sale, anything roughly comparable. He had toured a handful before he came to realize there wasn't anything comparable. Yet in the process of being shuttled down dirt roads, seeing the country, noting prices, and comparing terms, he'd received the crash course he needed. With that,

Jeff made an offer that the Rods refused. Following several rounds of negotiations, they finally reached an agreement. In December 1996, Jeff signed the paperwork making him the new owner of the Pinto Canyon Ranch.[55]

Meanwhile, Apache continued to lease the ranch. His cattle allowed Jeff to claim the agricultural tax exemption, and Apache watched over the place in his absence, for even in retirement Jeff continued to serve on boards and even bought and sold a few companies just to maintain his chops. Visiting as often as he could, however, Jeff began exploring and learning the ways of the land. As strong a hiker as he was, he might have done it afoot. But cowboys don't hike, and, with Apache there, it made better sense to ride. That was probably because Jeff wasn't fully aware of Apache's reputation. "He liked going with me because most people would try to take care of him," recalled Apache. "Hell, I just took him like one of the cowboys. We'd go do stuff." Sometimes that stuff involved pretty sketchy situations. One time he and Jeff cornered a wild bull in a box canyon. As they approached, the bull suddenly turned and charged them. "I just barely got out of the way," Apache remembered. "Bumped me pretty good and I think he kinda bumped Jeff pretty good . . . but that was an everyday occurrence."[56]

Jeff Fort and Apache Adams on the Pinto Canyon Ranch, ca. 1996. Photo courtesy of Don Cadden.

Over the course of their long rides and near wrecks, Apache and Jeff developed an unlikely friendship. It was clear that Jeff enjoyed having Apache around, enough so that he grew lax in collecting the lease payments, but in his own quiet way. "I'd send him a check and he'd tear it up or something. I don't know. Never did go through the bank," claimed Apache. Even after getting "screwed" on a horse deal with Apache, Jeff was quick to forgive. "It was just in his blood," Jeff recalled almost wistfully. "He couldn't help himself."[57]

Although Pinto Canyon was lush when Apache first arrived, between 1994 and 1999, precipitation began to taper off, with Candelaria averaging only about 75 percent of normal. Pinto Canyon was worse yet. Tanks went dry, and the creek ceased to flow until the only wells still running were at the headquarters, the old George Sutherlin place, and one windmill on the north end of the ranch. The grass grew scarce and Apache's animals started to decline. The watered country left to him wasn't enough to support his herd. He finally told Jeff, "Your little ranch and my cattle aren't doing any good. Let's just terminate this for a while." Jeff agreed, but only if Apache would stay on as foreman. Apache continued to oversee the ranch for several more years. But after buying a ranch outside Fort Stockton, the commute got to be too much. Ultimately, he had to give it up.[58]

Preserving the Land

Meanwhile, Judd's ranches hung in the balance. His death had left his estate more than $5 million in debt, and his money was tied up in art and property. It was now largely up to his son and daughter, Flavin and Rainer, to find a way out. One of the co-executors of his estate, Marianne Stockebrand, believed the Chinati Foundation and Judd estate should merge and Judd's properties be sold to settle the debt. His children felt strongly otherwise. Ultimately, Stockebrand agreed to relinquish her executorship in exchange for certain Judd artworks and payment of legal fees but would remain director of the Chinati Foundation. That may have eased their burden, but when the state attorney general got involved, the gravity of their task became evident.[59]

One thing was clear: the ranches had to go. With no income with which to service the debt or pay the taxes, the Judd heirs had no options except to sell their father's art or his property, and the art had to stay. No sooner was the word out than they started receiving offers: from neighboring ranchers, from the Lannan Foundation in Marfa, from speculative investors. Lowball all. Their desire to retain the three ranch houses while selling the land presented insurmountable complications. Prospective buyers either wanted the houses with the land or balked at providing permanent easements to access them. And there was one further issue: Judd had wanted his ranches protected. They needed to find someone willing to place the land in a conservation easement—protecting it from development in perpetuity. So, not just any buyer would do.[60]

When Flavin and Rainer first approached Jeff Fort about buying Judd's ranches, he politely declined. Jeff didn't mind the stipulations but countered that he already had far more land than he knew what to do with. But seeing in Jeff someone already conservation oriented (he had placed the Pinto Canyon Ranch under a conservation easement) and who had the means to make things happen, they weren't ready to take no for an answer. They challenged him to simply make them an offer. He finally made one he was sure they would reject. But he was wrong. In February 2000, Jeff purchased thirty-three thousand acres from Judd's estate. Added to the eighteen-thousand-acre Pinto Canyon Ranch, the purchase made him one of the largest landowners in the region.[61]

The deed carried several provisions. First, the Judd estate would retain ownership of the three houses and about thirty-five acres surrounding each. For their part, they agreed to keep the grounds clean and the houses in good repair. No permanent residents would be allowed. For Jeff's part, he would allow ingress and egress so the houses could be maintained. More importantly, in accordance with the terms of the new conservation easement, he agreed to keep the property "forever predominantly in its natural and scenic condition; to protect any rare plants, animals, or plant communities . . . and to prevent any use of the protected property that significantly impairs or interferes with the conservation values or interests of the property." Notably, subdivision was prohibited, and there were limits placed on construction, destruction of plants, mining, dumping, and other uses that might damage the land. Most important of all, the restrictions would be carried with the deed in perpetuity. It was as close to permanent protection—a guarantee of "forever"—as any legal document could be.[62]

Judd's preservation ethic had guaranteed the protection of his ranches. But he also, indirectly, helped protect tens of thousands of acres of land he never owned. When he introduced Heiner Friedrich and Philippa de Menil to Pinto Canyon and the Chinati Mountains, he was almost certainly thinking they might help him acquire more property or finance a massive outdoor art project that he had planned. Indeed, they did facilitate his acquisition of the Perez Ranch. But after Schlumberger stock began to tumble in the early 1980s, forcing the Dia Foundation to scale back operations, Judd and the Friedrichs suffered a well-publicized and contentious separation. Although their friendship was severed for good, the couple shared Judd's preservation leanings and found their own way to guarantee the protection of the land.[63]

In 1996, the couple donated all but three thousand acres of the property to the Texas Parks and Wildlife Department "to be held as a natural area so that its scenic, biological, and aesthetic qualities are preserved for the benefit of the people of Texas in perpetuity." At more than thirty-eight thousand acres, it was the largest land gift in state history. Presciently, the couple also established a private endowment to maintain the property and offset the loss of property taxes to the county. For years, access issues and a lack of funds hamstrung the ability of the department to open the Chinati State Natural Area to the public. In 2013, they came one step closer after obtaining necessary easements, and in 2017 they presented a draft public use plan in a series of public meetings. Although access may still be years away, pending natural and cultural resource surveys and visitor infrastructure, it will continue to be held in trust for the people of Texas to be managed as an "outdoor laboratory for scientific exploration and education."[64]

Indeed, Judd's legacy looms larger than even he could have predicted. Since his death, his artistic vision and the philosophy surrounding it have become enshrined. In Marfa, Judd's name is inextricably linked to the town's self-image, not that either are necessarily what Judd would have wanted. In life, his peace and privacy were paramount. He may have envisioned Marfa as one day having "the greatest visible concentration of contemporary art in the world," but throwing open the gates to throngs of art seekers seemed of less interest to him than his relationship to place. "I'm not keen on people," he said during an interview with the *Dallas Morning News* in 1991. The only reason he was not a full-fledged recluse, he added, was simply because he "hadn't worked up to it yet." Perhaps growing frustrated with his cynical responses, the journalist challenged him to name just one thing he liked. For a moment, Judd appeared stumped. Finally, he answered, "The land."[65]

A Partnership Is Born

As it had been with the Pinto Canyon Ranch, the Judd ranches were under a grazing lease when Jeff Fort bought them. Jason Sullivan, the lessee, had been running cattle below the Rimrock since graduating from high school in Fort Davis in the early 1990s. Although the country was notably poorer than the Wild Rose Pass Ranch in the Davis Mountains where he was raised, it was his first "big project" on his own. By buying stocker cattle out of Mexico, he'd been able to profit from the sparse grasses with a minimum of infrastructure. Jason also gave Jeff his first tour of the ranch. Despite his initial concerns that Jeff was a "bunny hugger," he's remained as local manager and has been Jeff's right-hand-man ever since.[66]

Not long after Fort purchased the land, the two were riding in a remote part of the ranch when Jason pointed out a particular boulder he thought might be of interest. When Jeff saw it, he was floored. "It was this fifteen-foot-tall petroglyph with all these carvings," Jeff recalled. He sent photographs of the panel to Bob Mallouf, then director of the Center for Big Bend Studies at Sul Ross State University in Alpine. "The phone started ringing off the hook," Jeff said. When Bob first visited the ranch, he was clearly impressed, and contagiously so, for it would be the beginning of an enduring collaboration between Jeff and the center. In addition to allowing access to his ranch and funding nearly all of the work, Jeff has become an avocational archeologist extraordinaire. Jeff and Jason have since discovered hundreds of prehistoric and historic sites across the ranch. And they're still at it.[67]

Jeff Fort with a boulder petroglyph he discovered on the Pinto Canyon Ranch in 2001. Photo courtesy of Bob Mallouf.

For several years, archeologist John Seebach served as the lead Center for Big Bend Studies researcher on the Pinto Canyon Ranch, and during that time he directed excavations at three rock shelters and reported the findings in a monograph published in 2007. After he left to accept a professorship, the project was adopted by Sam Cason, who has since conducted systematic surveys and excavations in both rock shelters and open campsites. In addition to the site database (which has grown to include more than eight hundred sites) and an increasing array of radiocarbon dates, the project has produced session posters, conference presentations, and a comprehensive field guide to prehistoric sites on the ranch. More recently Bryon Schroeder has joined the project, focusing his research on a remote rock shelter containing deep cultural deposits and a complex history of excavations—some scientific and some not.[68]

Inspired by the results of the research on the prehistory of the area, in 2008 Jeff approached me about investigating the historic sites on the ranch and writing its history. I should have been ecstatic at the privilege of receiving such an offer. And yet I found the crumbling adobe remains within that crazed geography mystifying and—for reasons difficult to explain—somehow unsettling. But over the years

of interviews, archival research, and fieldwork, a story emerged
that was far more nuanced, complex, and compelling than I ever
could have imagined. In the process, the wild desert lowlands edged
into my subconscious as the land's recent human past was slowly
revealed. In peering beneath the shadow, what was once to me only
a bewildering chaos of mountains and canyons has become something much greater.

⌃⌃⌃

By the time Apache Adams turned his cattle down into Pinto Canyon, its final passage out of agriculture was almost complete, awaiting only the next dry spell. A. J. Rod was the first of a new breed
whose wealth made the ranch a disposable luxury, not a lifeline. So
it was for Donald Judd, although he approached the land from a
radically different perspective—one whose purchases were less an
investment than a method for protecting a place he'd come to love.
In many ways, Jeff Fort expanded on and completed the work that
Judd had started. It may not have been what Jeff originally set out
to do, but his own evolving land ethic aligned with that of Judd in
fundamental ways. In the end, his own preservation ethic made for
a conservation success story unlike any other in the region.

11

Only the Land Remains

The West is less a place than a process. And the western landscape has now become our most valuable natural resource.

—Wallace Stegner, *Where the Bluebird Sings to the Lemonade Springs*

CHINATI PEAK casts a broad shadow over Pinto Canyon today, just as it has for eons, bearing witness to the epic play of human drama that has unfolded beneath its summit. Long a passageway between two worlds, for a brief moment the canyon hosted an unlikely array of homesteaders who sought to overcome obstacles the land imposed upon them. Aridity, meager forage, and far greater ruggedness placed limitations on ranchers' ability to succeed. Periods of favorable market and climate conditions allowed such limitations to be ignored for a time, but there was eventually a reckoning that could not be avoided. Pinto Canyon is owned today for different reasons—ones that do not require the land to turn a profit, but allow profiting from things the land passively and abundantly provides.

The lowlands were among the last places in the Big Bend to be settled, awaiting laws that made such rough, marginal country worth claiming. The first run of settlers succeeded in bringing the canyon a road and a school, suddenly connecting it to the larger world and making it more of a community than it would ever be again. José Prieto was the first to arrive and one of the last to leave, having prospered against great odds and despite a string of tragedies that clouded over his accomplishments. Although the canyon was largely unaffected by the bloody Mexican Revolution, it triggered a minor outmigration of people like Albin O'Dell, who felt the risk was too great. Others would follow—including the Sutherlins and the Wilsons—both families braving the revolutionary years only to find that success in the canyon remained elusive.

The drought and Great Depression of the 1930s served as a stark backdrop to ethnic tensions as the stress of the times added fuel to a simmering feud. The murder of Pablo Prieto was only the beginning of a string of tragedies the Prieto family would endure, often because of José's one unruly son, who sowed trouble wherever he went. The seven-year drought of the 1950s left the canyon impoverished, forcing those who remained to find other sources of income. For the Prietos, it was only borrowed time. Family strife led to the forced sale of the ranch and the end of a legacy. The Shelys, who had arrived at the cusp of the drought, proved remarkably resilient until mounting challenges on the ranch, along with their own tragedies, forced the only logical outcome. Their departure signaled the end of family ranching in Pinto Canyon.

The frailty of the land hastened its transition out of agriculture, accelerating its historical evolution. Although plants and animals have been in continuous flux for millennia, the changes in the twentieth century were probably among the most sweeping since the Chinati volcano obliterated all life some thirty-three million years ago. Recurrent drought amid continuous grazing reduced grasses and forbs, causing soil to erode and shrubs to proliferate. Those least palatable to livestock have increased at least fourfold over the last hundred years. But it was never a pastoral paradise. More than a century ago, land agent William Baines emphasized the land's failings. It has always been so.

That environmental factors were largely determinant of the historical trajectory in Pinto Canyon is perhaps most easily supported by the fact that the only surviving family ranches aside from the Benavidez Ranch are above the Rimrock: those of the Loves, the Brites, and the Millers. And of them all, only the ranches above still operate at a profit. The lowlands, by contrast, have reverted to wildlands once again. To be sure, historical contingency—the accidents of history—account for much. But ultimately the land itself was reason enough.

By the time the Prieto Ranch was appraised for Judd and Heiner Friedrich and Philippa de Menil, there was little left to recommend it. After José Prieto moved to El Paso, the care of the ranch had been entrusted to a string of lessees whose short-term interests were not the long-term health of the land. An appraiser found the range "extremely overgrazed," with more than 95 percent of the vegetation unpalatable for livestock. Although he estimated the carrying capacity for the entire twelve-thousand-acre ranch to be 50 head of cattle, he counted more than 285 head—rawboned and gaunt—roaming the hills, with spines sticking out of their lips and noses from browsing cactus and lechuguilla.[1]

If the decline of the land was somehow inevitable, protecting it has been anything but. When the Prieto family members had to sell their ranch, it narrowly escaped being subdivided. Had the perlite or uranium deposits been more robust, or the financial incentives stronger, much more of the land could have been irreversibly scarred. Had ownership fallen to individuals lacking the means or the vision of Donald Judd or Heiner Friedrich or Jeff Fort, the land could have suffered devastating insults. As it is, Pinto Canyon and most of the Chinatis have been protected from the kind of development that would diminish their spirit of place.

Much of the land's value now stems from the very things that were once its greatest liabilities. The volcanism, faulting, and erosion that tortured and exposed the bones of this ragged land made it hell for ranching but prime for wildlife and a paradise for renewal and scientific inquiry. Recognition of these less traditional values—ones that do not depend on rainfall or the price of mohair—has been a growing trend. *Texas Monthly* counted Pinto Canyon as one of "the most visually stunning and remote" places remaining in the state. And images of the canyon have graced magazine covers and been showcased by photographers since at least the 1920s.[2]

A drive down the Pinto Canyon road today offers only the faintest of hints the place was once a community of ranches bound by a schoolhouse and a common plight. Gone are the sounds of thousands of bleating sheep and goats, the cries of children at play, the creak of wagons struggling up the steep grade of the road. Most of the adobe homes that dotted the canyon are slowly melting back into the earth, the corrals long since fallen into disrepair. Only a few walls of the old Horse Creek schoolhouse still stand. Nobody lives here anymore, and, aside from a few ranch workers, no one stays for long. The canyon feels empty and timeless. Other than the crumbling houses, there is little to provide any continuity with its human past.

Yet behind these fading ruins are epic stories of human triumph and tragedy, of trying to survive in a hostile and difficult land. Throughout its history, Pinto Canyon has served as a perfect repository for such stories, containing them as intimately as a mescal bean within its pod. In their telling, they provide context, a more meaningful connection to the past, a deeper sense of place. But stories must be told to be remembered. And remembered to be told. Although a few stories endure, most are forgotten. In the end, only the land remains.

Notes

Chapter 2. The Terrible Mountains

1. Joseph C. Cepeda, "The Chinati Mountains Caldera, Presidio County, Texas," in *Cenozoic Geology of the Trans-Pecos Volcanic Field of Texas*, ed. Anthony W. Walton and Christopher D. Henry, Bureau of Economic Geology Guidebook 19 (Austin: University of Texas, 1979), 106–25; Kevin Urbanczyk, "Geologic History of Pinto Canyon Area," public lecture, Casa Perez open house, Casa Perez, Pinto Canyon, Tex., 6 September 2014; Christopher D. Henry, telephone interview by author, 26 February 2015; Timothy W. Duex and Christopher D. Henry, *Calderas and Mineralization: Volcanic Geology and Mineralization in the Chinati Caldera Complex, Trans-Pecos Texas*, Geological Circular 81–2, Bureau of Economic Geology (Austin: University of Texas at Austin, 1981), 10; William MacLeod, *Big Bend Vistas: A Geological Exploration of the Big Bend* (Alpine: Texas Geological Press, 2002), 178; Stephen Self, "The Effects and Consequences of Very Large Explosive Volcanic Eruptions," *Philosophical Transactions of the Royal Society* 364, no. 1845 (2006): 2081–83.

2. Eyewitness quoted in Clive Oppenheimer, "Climatic, Environmental and Human Consequences of the Largest Known Historic Eruption: Tambora Volcano (Indonesia) 1815," *Progress in Physical Geography* 27, no. 2 (2003): 236; Self, "Volcanic Eruptions," 2080–90.

3. David L. Amsbury, "Geology of Pinto Canyon Area, Presidio County, Texas" (PhD diss., University of Texas, 1957), 96–101; David L. Amsbury, Pinto Canyon Area, Presidio County, Texas, Geological Quadrangle Map No. 22 (Austin: Bureau of Economic Geology, University of Texas, 1958); Kevin Urbanczyk, David Rohr, and John C. White, "Geologic History of West Texas," in *Aquifers of West Texas*, ed. Robert E. Mace, William F. Mullican III, and Edward S. Angle, Texas Water Development Board Report No. 356 (Austin: Texas Water Development Board, 2001), 22; Duex and Henry, *Calderas and Mineralization*, 3–11; Cepeda, "Chinati Mountains Caldera," 106; John Andrew Wilson, "Geochronology of the Trans-Pecos Texas Volcanic Field," *New Mexico Geological Society Guidebook, 31st Field Conference, Trans-Pecos Region* (Socorro: New Mexico Geological Society, 1980), 206–7.

4. Here, as throughout this book, distances and other spatial data were derived from land satellite imagery using Google Earth Pro as well as Environmental Systems Research Institute's (ESRI) ArcMap—the industry standard for geographic information systems (GIS) as used by land managers, planners, geographers, and archeologists worldwide. The many layers I utilized were created from digitized US Geological Survey (USGS) maps (for elevation, topography, and place-names), from the Texas General Land Office land survey data, personal fieldwork, and a range of historical sources. The equivalent printed versions of the primary maps used include USGS, Cuesta del Burro West [map], 1:24,000, 7.5 Minute Series (Reston, Va.: US Department of the Interior, USGS, 1983); USGS, Ruidosa Hot Springs [map], 1:24,000, 7.5 Minute Series (Reston, Va.: US Department of the Interior, USGS, 1979); USGS, Pueblo Nuevo [map], 1:24,000, 7.5 Minute Series (Reston, Va.: US Department of the Interior, USGS, 1979); USGS, Las Conchas [map], 1:24,000, 7.5 Minute Series (Reston, Va.: US Department of the Interior, USGS, 1980); USGS, Sierra Parda [map], 1:24,000, 7.5 Minute Series (Reston, Va.: US Department of the Interior, USGS, 1978); and USGS, Chinati Peak [map], 1:24,000, 7.5 Minute Series (Reston, Va.: US Department of the Interior, USGS, 1979) (hereafter referred to as USGS, Maps); ESRI, ArcMap Version 10.1 (Redlands, Calif., 1999–2012).

5. Because much of the lowlands area has similar geology, ecology, and culture, this discussion applies beyond the strict confines of the canyon to encompass the southern Sierra Vieja and the lowlands around the flanks of the Chinatis. This wider area, alternatively called the Sierra Vieja, the Viejas, or simply the lowlands, serves as the broader region treated here.

6. Tucker F. Hentz, "Geology," *Handbook of Texas Online*, accessed 13 December 2016, http://www.tshaonline.org/handbook/online/articles/swgqz; Ronald K. De Ford, "Some Keys to the Geology of Northern Chihuahua," in *Guidebook of the Border Region*, ed. Diego A. Cordoba, Sherman A. Wengerd, and John Shomaker (Socorro: New Mexico Geological Society, 1969), 61–65; Patricia Wood Dickerson, "Structural Zones Transecting the Southern Rio Grande Rift—Preliminary Observations," *New Mexico Geological Society Fall Field Conference Guidebook – 31, Trans-Pecos Region*, ed. Patricia W. Dickerson, Jerry M. Hoffer, and Jonathan F. Callender (Socorro: New Mexico Geological Society, 1980), 69; William R. Muehlenberger, "Texas Lineament Revisited," in *New Mexico Geological Society Fall Field Conference Guidebook – 31, Trans-Pecos Region*, ed. Patricia W. Dickerson, Jerry M. Hoffer, and Jonathan F. Callender (Socorro: New Mexico Geological Society, 1980), 113–16; Julius E. Dasch, Richard L. Armstrong, and Stephen E. Clabaugh, "Age of Rim Rock Dike Swarm, Trans-Pecos, Texas," *Geological Society of America Bulletin* 80 (September 1969): 1819–23.

7. Ronald K. De Ford, "Tertiary Formations of Rim Rock Country, Presidio County, Trans-Pecos Texas," *Texas Journal of Science* 20, no. 1 (1958): 1–37; Amsbury, "Geology of Pinto Canyon," 108.

8. Christopher D. Henry and Jonathon G. Price, "Early Basin and Range Development in Trans-Pecos Texas and Adjacent Chihuahua: Magmatism and Orientation, Timing, and Style of Extension," *Journal of Geophysical Research* 91, no. 86 (1986): 6213–24. It should be noted that the name Sierra Quemada is used here to simplify, by conflating, a number of distinct mountains in Mexico, most of which have their own names. These include (from north to south) Sierra Pilares, Sierra el Pajarito, Sierra los Fresnos, Sierra la Ventana, Sierra Ojo Caliente, Sierra la Pinosa, Sierra la Chiva, Sierra la Parra, Sierra la Muelle, and Sierra la Quemada. See Instituto Nacional de Estadística, Geografía, e Informática (INEGI), Carta Topográfica, México Norte, Escala 1:250,000, Serie I de Imágenes Cartográficas Digitales (Mexico City: INEGI, 2000). The admittedly weak precedent for applying a single name for this chain of mountains was set out in INEGI, Mapa General del Estado Chihuahua, Escala 1:1,192,180 (Mexico City: INEGI, n.d.).

9. J. R. Mraz and G. R. Keller, *Structure of the Presidio Bolson Area, Texas, Interpreted from Gravity Data*, Geological Circular 80–13, Bureau of Economic Geology (Austin: University of Texas, 1980), 4–8; Eddie Joe Dickerson, "Bolson Fill, Pediment, and Terrace Deposits of Hot Springs Area, Presidio County, Trans-Pecos Texas" (MS thesis University of Texas, 1966), 18–46.

10. E. Dickerson, "Bolson Fill," 36, 85–87.

11. Wayne D. Pennington and Scott D. Davis, "Notable Earthquakes Shake Texas on Occasion," *Texas Almanac.com*, accessed 19 February 2015, https://texasalmanac.com/topics/media/notable-earthquakes-shake-texas-occasion; Cliff Frohlich and Scott D. Davis, *Texas Earthquakes* (Austin: University of Texas Press, 2002), 152–54 (quote, 153).

12. Amsbury, "Geology of Pinto Canyon," 6, 9, 28, 67; Amsbury, Pinto Canyon Area map. The steepest gradient in the state is the western scarp of the Guadalupe Mountains. USGS, Maps; Wikipedia contributors, "List of Mountain Peaks of Texas," *Wikipedia*, accessed 11 February 2015, http://en.wikipedia.org/w/index.php?title=List_of_mountain_peaks_of_Texas&oldid=629332097.

13. Charles Laurence Baker, "Exploratory Geology of a Part of Southwestern Trans-Pecos, Texas," *University of Texas Bulletin No. 2745* (Austin: Bureau of Economic Geology, University of Texas, 1927), 53.

14. USGS, Maps; W. R. Osterkamp, *Annotated Definitions of Selected Geomorphic Terms and Related Terms of Hydrology, Sedimentology, Soil Science and Ecology*, Open-File Report 2008–1217 (Reston, Va.: USGS, 2008), 27.

15. USGS, Maps; Franklin T. Heitmuller and Brian D. Reece, Database of Historically Documented Springs and Spring Flow Measurements in Texas, US Geological Survey Open-File Report 03–315, Geographical Information Systems (GIS) database (2003), accessed 12 May 2014, https://databasin.org/datasets/2400de0b78284e0fa44083e78824ff24.

16. Shirley C. Wade, William R. Hutchison, Ali H. Chowdhury, and Doug Coker, *A Conceptual Model of Groundwater Flow in the Presidio and Redford Bolsons*

Aquifers, Texas Water Development Board Report (Austin: Texas Water Development Board, 2011), 1–2.

17. Osterkamp, *Annotated Definitions of Selected Geomorphic Terms,* 45.

18. Thomas J. Larkin and George W. Bomar, *Climatic Atlas of Texas,* Texas Department of Water Resources Report LP-192 (Austin: Texas Department of Water Resources, 1983), 2–3.

19. Although some climate records exist for Ruidosa, which is closer, the period of record is short and the data suspect. The period of record for Candelaria spans 1940 to 2011, whereas Viejo Pass (Valentine 10WNW) spans a much longer period: 1897 to 2012. Climate statistics for each station are derived from data provided by the Western Regional Climate Center (WRCC), "Period of Record Monthly Climate Summaries," accessed 4 March 2015, http://www.wrcc.dri.edu/cgi-bin/cliMAIN/ (no longer available).

20. The average temperature at Candelaria in January (typically the coldest month) is 49 degrees F. Robert H. Schmidt, "The Climate of Trans-Pecos, Texas," in *The Changing Climate of Texas: Predictability and Implications for the Future,* ed. Jim Norwine, John Giardino, Gerald North, and Juan Valdez (College Station: Cartographics, Texas A&M University, 1995), 125; WRCC, "Period of Record Monthly Climate Summaries."

21. David K. Adams and Andrew C. Comrie, "The North American Monsoon," *Bulletin of the American Meteorological Society* 78, no. 10 (1997): 2197–2213; Schmidt, "Climate of Trans-Pecos, Texas," 124–29.

22. Schmidt, "Climate of Trans-Pecos, Texas," 127; WRCC, "Period of Record Monthly Climate Summaries."

23. Schmidt, "Climate of Trans-Pecos, Texas," 124, 131–32; National Oceanic and Atmospheric Administration, "Beaufort Wind Scale," NOAA, accessed 5 March 2015, http://www.spc.noaa.gov/faq/tornado/beaufort.html.

24. Schmidt, "Climate of Trans-Pecos, Texas," 125; World Health Organization, "Global Solar UV Index: A Practical Guide," WHO, accessed 6 March 2015, http://www.who.int/uv/publications/en/UVIGuide.pdf; Environmental Protection Agency, "Sun Safety," EPA, accessed 6 March 2015, http://www2 .epa.gov/sunwise/uv-index.

25. Natural Resources Conservation Service (NRCS), USDA, *Soil Survey of Big Bend National Park, Texas* (Washington, DC: GPO, 2011), 219, 235; NRCS, "Custom Soil Resource Report for Presidio County, Texas," Web Soil Survey, accessed 20 February 2015, http://websoilsurvey.sc.egov.usda.gov/App/ HomePage.htm.

26. NRCS, USDA, "Dominant Soil Orders in the United States" (1998), accessed 20 February 2015, https://www.nrcs.usda.gov/Internet/FSE_MEDIA/ stelprdb1237749.pdf; NRCS, "Aridisols Map," accessed 20 February 2015, http://www.nrcs.usda.gov/wps/portal/nrcs/detail/soils/survey/class/ maps/?cid=nrcs142p2_053595.

27. John Karges (Texas Nature Conservancy associate director of field science), email correspondence with author, 12 August 2013.

28. A. Michael Powell, email correspondence with author, 2 April 2015; A. Michael Powell, *Trees and Shrubs of the Trans-Pecos and Adjacent Areas* (Austin: University of Texas Press, 1998), 27, 29, 36, 37, 63, 65, 73, 80, 83, 91, 70, 242, 316; Emily J. Lott and Mary L. Butterwick, "Notes on the Flora of the Chinati Mountains, Presidio County, Texas," *SIDA: Contributions to Botany* 8, no. 4 (1979): 348–51.

29. Ben Buongiorno, "Handbook of the Tierra Vieja Mountains, Presidio and Jeff Davis Counties, Trans-Pecos Texas" (MA thesis, University of Texas, 1955), 44–45; Powell, *Trees and Shrubs*, 3–12.

30. L. C. Hinckley, "Contrasts in the Vegetation of Sierra Tierra Vieja in Trans-Pecos Texas," *American Midland Naturalist* 37, no. 1 (1947): 162–78; A. Michael Powell, *Grasses of the Trans-Pecos and Adjacent Areas* (Marathon, Tex.: Iron Mountain Press, 2000), 9–17.

31. W. Frank Blair and Clay E. Miller Jr., "The Mammals of the Sierra Vieja Region, Southwestern Texas, with Remarks on the Biogeographic Position of the Region," *Texas Journal of Science* 1, no. 1 (1949): 89.

32. Ibid.; Clyde Jones, Mark W. Lockwood, Tony R. Mollhagen, Franklin D. Yancey II, and Michael A. Bogan, *Mammals of the Chinati Mountains State Natural Area, Texas*, Occasional Papers, Museum of Texas Tech University, no. 300 (Lubbock: Texas Tech University, 2011).

33. "Carmen Mountain Whitetail," Western Whitetail.com, accessed 25 April 2016, http://www.westernwhitetail.com/whitetail/carmen-mountain -whitetail/; Jeff Fort, interview by author, Pinto Canyon Ranch, 10 May 2013; Texas Parks and Wildlife Department, "Desert Bighorn Sheep," Wildlife Management in Texas, TPWD website, accessed 2 June 2016, http://tpwd .texas.gov/landwater/land/habitats/trans_pecos/big_game/desertbighorn sheep/.

34. David J. Schmidly, *Texas Natural History: A Century of Change* (Lubbock: Texas Tech University Press, 2002), 436; Fort interview, 10 May 2013; Crown X Safaris Ranch: A West Texas Game Ranch website, accessed 25 April 2016, http://www.cxsafaris.com/. (As of May 2018 the website notes that the ranch no longer offers hunting and is now a breeding ranch.)

35. Jim Heffelfinger, "The Texas Black Bear," *Tracks Magazine*, July–August 2015, 121–27; James F. Scudday, "Two Recent Records of Gray Wolves in West Texas," *Journal of Mammalogy* 53, no. 3 (1972): 598; James F. Scudday, "The Vertebrate Fauna of Capote Canyon," in *Capote Falls: A Natural Area Survey*, ed. Office of Research, Lyndon B. Johnson School of Public Affairs (Austin: University of Texas, 1973), 197; Schmidly, *Texas Natural History*, 432.

36. David L. Jameson and Alvin G. Flury, "The Reptiles and Amphibians of the Sierra Vieja Ranch of Southwestern Texas," *Texas Journal of Science* 1, no. 2 (1949): 54–79; Karges email.

37. Homer W. Phillips and Wilmot A. Thornton, "The Summer Resident Birds of the Sierra Vieja Range in Southwestern Texas," *Texas Journal of Science* 1, no. 4 (1949): 101–31; Buongiorno, "Handbook," 47; Arnold F. Van

Pelt, *An Annotated Inventory of the Insects of Big Bend National Park, Texas* (Big Bend National Park, Tex.: Big Bend Natural History Association, 1999).

38. John D. Seebach III, "El Despoblado: Folsom and Late Paleoindian Occupation of Trans-Pecos, Texas" (PhD diss., Southern Methodist University, 2011), 263–76; William A. Cloud, "An Update on the Genevieve Lykes Duncan Site," *La Vista de la Frontera*, Center for Big Bend Studies, Sul Ross State University 23 (Fall 2012): 1–2, 14. For an excellent overview of the prehistoric to historic period in the Trans-Pecos, see Texas Beyond History Web Team, "Trans-Pecos Mountains and Basins," *Texas beyond History*, University of Texas at Austin, accessed 23 February 2015, http://www.texasbeyond history.net/trans-p/index.html.

39. Robert J. Mallouf, "A Synthesis of Eastern Trans-Pecos Prehistory" (MA thesis, University of Texas at Austin, 1985), 23, 101; David W. Keller et al., *A Sampling of Archeological Resources in Big Bend National Park, Texas* (Alpine, Tex.: Center for Big Bend Studies, Sul Ross State University, forthcoming).

40. Sam Cason, email correspondence with author, 15 October 2016.

41. Ibid.

42. J. Charles Kelley, *Jumano and Patarabueye: Relations at La Junta de los Rios*, Anthropological Papers No. 77, Museum of Anthropology (Ann Arbor: University of Michigan, 1986), 119–33.

43. The exact location of San Bernardino remains uncertain. The location described was the pioneer archeologist J. Charles Kelley's best guess. J. Charles Kelley, "The Historic Indian Pueblos of La Junta de los Rios [part 1]," *New Mexico Historical Review* 27, no. 4 (1952): 262; J. Charles Kelley, "The Historic Indian Pueblos of La Junta de los Rios [part 2]," *New Mexico Historical Review* 28, no. 1 (1953): 21–23; George P. Hammond and Agapito Rey, *Expedition into New Mexico Made by Antonio de Espejo, 1582–1583, As revealed in the Journal of Diego Pérez de Luxán, a Member of the Party* (Los Angeles: Quivira Society, 1929), 59–67.

44. Texas Historical Sites Atlas [restricted database], Texas Historical Commission website, accessed 4 November 2016, https://atlas.thc.state.tx .us/Account; Kelley, *Jumano and Patarabueye*, 51–55.

45. John D. Seebach, *Late Prehistory along the Rimrock*, Papers of the Trans-Pecos Archeological Program No. 3 (Alpine, Tex.: Center for Big Bend Studies, Sul Ross State University, 2011); Samuel S. Cason, "Archaeology on the Pinto Canyon Ranch: Field Guide to Prehistoric Occupation in the Sierra Vieja Breaks, Presidio County, Texas" (in-house document, Center for Big Bend Studies, Sul Ross State University, 2014), 81–86. In one of the largest projectile point collections in the region (from Big Bend National Park), arrow points (Late Prehistoric projectiles as opposed to Archaic-aged dart points) represented 37 percent of the total number of projectiles—a percentage that is close to that of other collections in the region. By contrast, arrow points collected from Pinto Canyon and the adjacent Sierra Vieja lowlands represent less than 8 percent of the total. Whether or not the "villager gath-

ering grounds" hypothesis is valid, this stark discrepancy undoubtedly signifies a localized phenomenon. See Keller et al., *Sampling of Archeological Resources in Big Bend National Park*, for projectile point statistics.

46. Enrique Rede Madrid, trans., *Expedition to La Junta de los Ríos, 1747–1748: Captain Commander Joseph de Ydoiaga's Report to the Viceroy of New Spain*, Office of the State Archeologist, Special Report 33 (Austin: Texas Historical Commission, 1992), 66–67; George P. Hammond and Agapito Rey, *The Rediscovery of New Mexico, 1580–1594: The Exploration of Chamuscado, Espejo, Castaña de Sosa, Morlete, y Leyva de Bonilla y Humaña* (Albuquerque: University of New Mexico Press, 1966), 73; James H. Gunnerson, "Southern Athapaskan Archeology," *Handbook of North American Indians*, ed. William Sturtevant (Washington, DC: Smithsonian Institution, 1979), 9:162–63.

47. Oscar Rodriguez, Amber Rodriguez, and David Gohre, "The Chinaitih Mountains: The Case for a More Plausible Narrative for How the Chinati Mountains Got Their Name," *Journal of Big Bend Studies* 26 (2014): 27–42. Providing further evidence of the Indian trail through Pinto Canyon, an expedition of fortune seekers from San Antonio exploring the mineral wealth of the Chinatis in 1879 pitched camp near the head of Pinto Creek beside a trail that descended into the canyon. "Where we camped is an old Indian trail," noted the explorer Burr G. Duval, "the Capote valley being one of the few passes from Texas to Mexico for many miles north and south." Quoted in Sam Woolford, ed., "Notes and Documents: The Burr G. Duval Diary," *Southwestern Historical Quarterly* 65 (July 1961): 501.

48. Álvar Núñez Cabeza de Vaca, *The Narrative of Cabeza de Vaca*, ed. Rolena Adorno and Patrick Charles Pautz (Lincoln: University of Nebraska Press, 2003), 148–51; W. H. Timmons, "Rodríguez-Sanchez Expedition," *Handbook of Texas Online*, accessed 12 December 2016, http://www.tshaonline.org/handbook/online/articles/upr01; Hammond and Rey, *Expedition into New Mexico*, 59–67.

49. Maria F. Wade, *The Native Americans of the Texas Edwards Plateau, 1582–1799* (Austin: University of Texas Press, 2003), 88; Jerry L. Eoff, *Just Doing the Math: Juan Dominguez de Mendoza Trail to San Clemente, 1683–1684* (Alpine, Tex.: privately published, 2015), 36–37; Rosalind Z. Rock, "San Francisco de La Junta Pueblo," *Handbook of Texas Online*, accessed 5 May 2016, http://www.tshaonline.org/handbook/online/articles/uqs45.

50. Carlos E. Castañeda, *Our Catholic Heritage in Texas, 1519–1936*, vol. 3, *The Mission Era: The Missions at Work 1731–1761* (New York: Arno Press, 1976), 229–32; Madrid, *Expedition to La Junta de los Ríos*, xv; Jefferson Morgenthaler, *La Junta de Los Rios: The Life, Death, and Resurrection of an Ancient Desert Community in the Big Bend Region of Texas* (Boerne, Tex.: Mockingbird Books, 2007), 123–24.

51. Castañeda, *Our Catholic Heritage in Texas*, 3:231–32; James Ivey, *Presidios of the Big Bend*, Southwest Cultural Resources Center Professional Papers No. 31 (Santa Fe, N.Mex.: National Park Service, US Department of the Interior,

1990), 1; Julia Cauble Smith, "Pilares, TX," *Handbook of Texas Online*, accessed 7 December 2016, http://www.tshaonline.org/handbook/online/articles/ hrp87; Carlysle Graham Raht, *The Romance of Davis Mountains and Big Bend Country*, Edition Texana (Odessa, Tex.: Rahtbooks Company, 1919), 120; Morgenthaler, *La Junta de Los Rios*, 134–35.

52. Jefferson Morgenthaler, *The River Has Never Divided Us: A Border History of La Junta de Los Rios* (Austin: University of Texas Press, 2004), 25–26.

53. This version of the story appears in multiple sources, but all seem derived from Oscar Waldo Williams, *O. W. William's Stories from the Big Bend* (El Paso: Texas Western College Press, 1965). See also Cecilia Thompson, *History of Marfa and Presidio County, Texas, 1535–1946* (Austin, Tex.: Nortex Press, 1985), 1:37; and Paul Horgan, *Great River: The Rio Grande in North American History*, 4th ed. (Hanover, N.H.: University Press of New England), 1984, 471. A very different version has the party ambushed in a canyon near San Esteban Spring en route to the Salt Lakes. See O. L. Shipman and Jack Shipman, "The Savage Saga: A Vivid Story of the Settlement and Development of Presidio County," *Voice of the Mexican Border*, 1938, 13.

54. Victor Orozco, *Las guerras indias en la historia de Chihuahua* (Juarez: Instituto Chihuahuense de la Cultura, Universidad Autónoma de Ciudad Juárez, 1992), 205, 211, as cited in Morgenthaler, *River Has Never Divided Us*, 30; Ralph A. Smith, "Mexican and Anglo-Saxon Traffic in Scalps, Slaves, and Livestock, 1835–1841," *West Texas Historical Association Year Book* 36 (October 1960): 101.

55. Library of Congress, "Treaty of Guadalupe Hidalgo," Primary Documents in American History, Virtual Services Digital Reference Section, accessed 7 June 2016, https://www.loc.gov/rr/program/bib/ourdocs/ Guadalupe.html.

56. William Henry Chase Whiting, *Journal of William Henry Chase Whiting, 1849*, in *Exploring Southwestern Trails, 1846–1854*, ed. Ralph P. Bieber (Philadelphia: Porcupine Press, 1974), 267–91 (quote, 290).

57. Ibid., 291.

58. Moritz von Hippel's report is included in William H. Emory, Report on the United States and Mexican Boundary Survey, House Ex. Doc. no. 135, 34th Cong., 1st sess. (Washington, DC: Cornelius Wendell, 1857), 89–90.

59. Clifford B. Casey, "The Trans-Pecos in Texas History," *West Texas Historical and Scientific Society Publication no. 5* (1933): 12; Thompson, *History of Marfa and Presidio County*, 1:55. The historian Ralph Smith offers some anecdotal evidence suggesting that Leaton and Spencer were likely involved in the scalp trade in the late 1830s and 1840s. See Ralph Smith, "Poor Mexico, So Far from God and So Close to the Tejanos," *West Texas Historical Association Year Book* 44 (1986): 96.

60. Spencer Tucker, *The Encyclopedia of the Mexican-American War: A Political, Social, and Military History* (Santa Barbara, Calif.: ABC-CLIO, 2013), 204–5;

William E. Russell V, "William Edward Russell: Pioneer of the Big Bend" (2003), unpublished manuscript in possession of author.

61. Virginia Duncan Madison and Hallie Crawford Stillwell, *How Come It's Called That? Place Names in the Big Bend Country* (New York: October House, 1968), 105; Raht, *Romance of Davis Mountains*, 194; Thompson, *History of Marfa and Presidio County*, 1:128; William Russell, letter to the *El Paso Herald*, 14 May 1874, reprinted in *Voice of the Mexican Border*, December 1933, 187.

62. Thompson, *History of Marfa and Presidio County*, 1:125; W. Russell, "William Edward Russell."

63. Thompson, *History of Marfa and Presidio County*, 1:120; L. H. Carpenter, Capt. 10th Cavalry, report to Commander at Fort Davis, 1880, NMRA Roll 66–783, Fort Davis National Historic Site Archives, Fort Davis, Tex.

64. Hugo O'Conor, *The Defenses of Northern New Spain: Hugo O'Conor's Report to Teodoro de Croix, July 22, 1777*, ed. and trans. Donald C. Cutter (Dallas: DeGolyer Library and Southern Methodist University Press, 1994), 71; Robert G. Smither, Capt. 10th Cavalry, letter to Commander at Fort Davis, 11 March 1883, NMRA Roll 66–783, Fort Davis National Historic Site Archives, Fort Davis, Tex.

65. Robert J. Mallouf, "The Dancing Rocks Petroglyphs: Horse Nomads of the Sierra Vieja Breaks of the Texas Big Bend," in *Archaeological Explorations of the Eastern Trans-Pecos and Big Bend: Collected Papers, Volume 1*, ed. Pat Dasch and Robert J. Mallouf (Alpine, Tex.: Center for Big Bend Studies, 2013), 235–60; Cason, *Archaeology on the Pinto Canyon Ranch*, 41–52. It is noteworthy that, while some of these boulder glyphs were discovered during archeological surveys, many, if not most, of them were found by the intrepid landowner, Jeff Fort.

66. William H. Leckie and Shirley A. Leckie, *Unlikely Warriors: General Benjamin H. Grierson and His Family* (Norman: University of Oklahoma Press, 1998), 250–75.

67. Thomas Smith, *The Old Army in Texas: A Research Guide to the US Army in Nineteenth-Century Texas* (Austin: Texas State Historical Association, 2000), 163–64; James B. Gillet, *Six Years with the Texas Rangers, 1875–1881* (Lincoln: University of Nebraska Press, 1963), 201–2.

68. Raht, *Romance of Davis Mountains*, 255–56.

69. Ibid.; Jim N. Hammond, "Big Bend Justice," *Cenizo Journal* 5, no. 4 (2013): 20–21.

70. Tenth Census of the United States, 1880, Ruidoso [*sic*], Presidio County, Texas, Enumeration District 126, NARA microfilm publication T9, 1,454 rolls, Records of the Bureau of the Census, Record Group 29, National Archives, Washington, DC, accessed at Ancestry.com; Thompson, *History of Marfa and Presidio County*, 1:140, 152.

71. William M. Baines, *State School Land Agents to State Land Board* (Austin, Tex.: State Printing Office, 1885), 11–12.

72. Ronnie C. Tyler, *The Big Bend: A History of the Last Texas Frontier* (Washington, DC: National Park Service, 1975), 113, 120; Paul Wright, "Build It and They Will Come? Boom and Bust in Presidio," *Journal of Big Bend Studies* 20 (2008): 40.

73. John Ernest Gregg, "The History of Presidio County" (MA thesis, University of Texas, 1933), 72.

74. Baines, *State School Land Agents*, 22.

75. Quoted in *Big Bend Sentinel*, Centennial Edition, 1 September 1950.

Chapter 3. Cañon del Pinto

1. Unless otherwise indicated, the Wilson story is derived from an untitled reminiscence written in 1966 by James Wilson's daughter, Millie Wilson-Dowe, as well as a student essay by Mike Shurley. The former was likely written by Millie upon request for Shurley's project and was supplemented with oral history. Dora Amelia (Millie) Wilson-Dowe, untitled reminiscence, 1966, Dowe Family Folder, Junior Historian Collection, Marfa Public Library, Marfa, Tex.; Mike Shurley, "The Dowes of Pinto Canyon," *Junior Historian* 28, no. 5 (March 1968): 12–24.

2. *Big Bend Sentinel*, 8 May 1930; Shurley, "Dowes of Pinto Canyon," 13. In Shurley's essay, he noted that James had suffered from "infantile paralysis," another term for poliomyelitis or polio.

3. Thomas Lloyd Miller, *The Public Lands of Texas, 1519–1970* (Norman: University of Oklahoma Press, 1972), 185–204; Thompson, *History of Marfa and Presidio County*, 2:57.

4. Wilson-Dowe reminiscence; Shurley, "Dowes of Pinto Canyon," 14.

5. James Wilson likely had had spinal polio, which attacks nerve cells and leads to muscle atrophy. Wikipedia contributors, "Poliomyelitis," Wikipedia, accessed 8 July 2015, https://en.wikipedia.org/w/index.php?title=Poliomyelitis&oldid=669813888; Florence Fenley, "Cowboy and Mounted Customs Inspector Went Back to Old Love . . . Ranching," *The Cattleman*, January 1968, 96; Wilson-Dowe reminiscence; Shurley, "Dowes of Pinto Canyon," 14.

6. This fictionalized scene is based on Millie Wilson's memoir, in which she writes, "Papa met José on his first trip alone, when he was looking for land to buy." The details are conjecture. Wilson-Dowe reminiscence.

7. Declaration of Intention for José Prieto, dated 2 August 1918, in Presidio County, Book of Petitions for Naturalization, Presidio County District Court Office, Presidio County Courthouse, Marfa, Tex.; "Prieto" family tree by David Keller, profile for José Prieto (1866–1963, d. El Paso, Tex.), undocumented data updated June 2015, in "Public Member Trees" database, accessed 11 June 2015 at Ancestry.com; Oscar Perez, interview by author, El Paso, Tex., 18 February 2013; Nancy Perez, interview by author, El Paso, Tex., 18 February 2013.

8. Declaration of Intention for José Prieto; 1900 US Census, Presidio County, Texas, population schedule, Justice Precinct 6, Enumeration District 0066, p. 2b, dwelling 33, family 34, Francisco Prieto, NARA microfilm publication T623, roll 1665, digital image accessed 11 June 2015 at Ancestry.com; Marriage License for José Prieto and Juanita Varreras [*sic*], Presidio County, 10 June 1891, Marriage Record 2, Presidio County Clerk's Office, Presidio County Courthouse, Marfa, Tex.; Oscar Perez, telephone interview by author, 16 January 2013.

9. Oscar Perez telephone interview, 16 January 2013.

10. Wilson-Dowe reminiscence.

11. *Big Bend Sentinel*, 16 May 1968; *Encyclopædia Britannica Online, s.v.* "rheumatism," accessed 8 April 2015, http://www.britannica.com/EBchecked/topic/501241/rheumatism.

12. Wilson-Dowe reminiscence.

13. Oscar Perez interview, 18 February 2013; Nancy Perez interview, 18 February 2013.

14. Baines, *State School Land Agents*, 12; Vernon Sayles, *Vernon Sayles' Annotated Civil Statutes of the State of Texas, with Historical Notes, Embracing the Revised Statutes of the State of Texas Adopted at the Regular Session of the Thirty-second Legislature, 1911; Incorporating under Appropriate Headings of the Revised Statutes, 1911, the Legislation Passed at the Regular and Special Sessions of the Thirty-second and Thirty-third Legislatures, to the Close of 1913*, (Kansas City, Mo.: Vernon Law Book Company, 1914), vol. 3, chap. 9.

15. Oscar Perez telephone interview, 16 January 2013; David Keller, "Archeological Site Recording of Prieto Homestead," 20 October 2010, on file at the Center for Big Bend Studies, Sul Ross State University, Alpine, Tex.; Presidio County, Records of the Certificates of Occupancy, vol. 1, 65, no. 8272, Presidio County Clerk's Office.

16. If a trail already existed where Wilson placed the road, it would have been a primitive affair—more for goat and burro than horse and wagon. More likely the original trail remained in the canyon bottom crossing and recrossing Pinto Creek for another three miles before ascending a fairly gentle southward-facing slope. It was this latter road, with more river crossings but a more gradual incline, that became the military road during the Mexican Revolution. Presidio County, Minutes of the Presidio County Commissioners, book 5, 12 November 1908, 56, Presidio County Clerk's Office; Thompson, *History of Marfa and Presidio County*, 2:60.

17. Wilson-Dowe reminiscence; Shurley, "Dowes of Pinto Canyon," 14–15.

18. Wilson-Dowe reminiscence; Stella Traweek, *The Production and Marketing of Mohair in Texas*, Bureau of Business Research Monograph No. 12 (Austin: University of Texas, 1949), 2–10; "Worksheet for Determining Range Condition and Evaluating Forage Preference and Use," Jesse Vizcaino Ranch Files, Natural Resources Conservation Service, Marfa, Tex.

19. Shurley, "Dowes of Pinto Canyon," 14–15; Wilson-Dowe reminiscence.

20. Wilson-Dowe reminiscence; David W. Keller, "Archeological Site Recording of Wilson Homestead," 18 November 2008, report on file at the
Center for Big Bend Studies, Sul Ross State University, Alpine, Tex.

21. Presidio County, Deed Records, vol. 40, 68, vol. 50, 33, Presidio County
Clerk's Office.

22. Elmer Compton, *Philip Sutherlin* (Sarasota, Fla.: privately published,
2002; updated 2009); Nancy Sutherlin-Gibson, telephone interview by author,
3 October 2014; Nancy A. Sutherlin-Gibson, "Alexander Sutherland 1620,"
December 2005, unpublished family genealogy in possession of author; Brett
J. Derbes, "Second Texas Cavalry," *Handbook of Texas Online*, accessed July 06,
2016, http://www.tshaonline.org/handbook/online/articles/qks08.

23. Compton, *Philip Sutherlin*; Certificate of Death for Grover Martin
Sutherlin, in Texas Death Certificates, 1903–1982 [online database], digital
image accessed 14 August 2015 at Ancestry.com.

24. Texas General Land Office, "Sutherlin" land grant search, accessed 28
April 2015, http://www.glo.texas.gov/history/archives/land-grants/index.

25. "San Esteban Lake," *Handbook of Texas Online*, accessed 30 April 2015,
http://www.tshaonline.org/handbook/online/articles/ros05.

26. Frederick Dale Sutherlin, telephone interview by author, 25 September 2014.

27. Presidio County, Deed Records, vol. 27, 302; Presidio County, Records
of the Certificates of Occupancy, vol. 1, 109, no. 10261; Texas General Land
Office, "Sutherlin" land grant search; F. Sutherlin telephone interview, 25
September 2014.

28. National Weather Service, NOAA, El Paso Monthly Precipitation Totals, El Paso, Tex., accessed 2 July 2018, https://www.weather.gov/epz/elpa
so_monthly_precip; Presidio County, Minutes of the District Court, August
1911, Cause No. 1119, vol. 6, 288, Presidio County Courthouse, Marfa, Tex.;
Thompson, *History of Marfa and Presidio County*, 2:85.

29. Wilson-Dowe reminiscence; "Sutherlin" family tree by David Keller,
profile for Martin Davis Sutherlin (1864–1918, d. Presidio County, Tex.),
undocumented data updated March 2015, in "Public Member Trees" database, accessed 23 April 2015 at Ancestry.com; "Sutherlin" family tree
by David Keller, profile for George B. Sutherlin (1878–1937, d. McKinley
County, N. Mex.), undocumented data updated March 2015, in "Public
Member Trees" database, accessed 24 April 2015 at Ancestry.com.

30. Wilson-Dowe reminiscence; *Big Bend Sentinel*, 14 September 1953.

31. Wilson-Dowe reminiscence.

32. Max Berger and Lee Wilborn, "Education," *Handbook of Texas Online*, accessed 26 October 2015, http://www.tshaonline.org/handbook/online/
articles/khe01; *Big Bend Sentinel*, 13 February 1964, 26 November 1953; Julia
Cauble Smith, "Indio, TX (Presidio County)," *Handbook of Texas Online*, accessed
14 May 2015, http://www.tshaonline.org/handbook/online/articles/hri03;
Thompson, *History of Marfa and Presidio County*, 2:85.

33. John Newby, email correspondence with author, 26 September 2014; Fenley, "Cowboy and Mounted Customs Inspector," 96.

34. Presidio County, Records of the Certificates of Occupancy, vol. 1, 65, no. 8272; Presidio County, Petition for Naturalization for José Prieto, dated 2 August 1918, Book of Petitions for Naturalization, Presidio County District Court Office.

35. Charles M. Hohn, "ABC's of Making Adobe Bricks," Guide G-521, New Mexico State University, Cooperative Extension Service, 2003; Oscar Perez interview, 18 February 2013.

36. Keller, "Archeological Site Recording of Prieto Homestead."

37. Presidio County, Tax Rolls, 1890–1920, payments by José Prieto, Tax Office, Presidio County Courthouse, Marfa, Tex.

38. *Levelland Daily Sun News*, 18 May 1958; Frank O'Dell, written correspondence to author, undated (ca. 2011); O'Dell Family Tree, unpublished document in possession of author.

39. *Levelland Daily Sun News*, 18 May 1958; O'Dell Family Tree, unpublished document in possession of author; O'Dell correspondence, undated (ca. 2011).

40. Presidio County, Records of the Certificates of Occupancy, vol. 1, 125, no. 10709; David Keller, "Archeological Site Recording of O'Dell Homestead," 19 November 2008, on file at the Center for Big Bend Studies, Sul Ross State University, Alpine, Tex.

41. The scene that follows is derived from a story passed down through the O'Dell family. The version I received identified the lead man as Pancho Villa. Based on the timing, however, it is unlikely that he was the culprit. Like the local bandit Chico Cano, Villa was invoked far more often than is supportable. O'Dell, written correspondence, undated (ca. 2011).

42. O'Dell, written correspondence, undated (ca. 2011); Thompson, *History of Marfa and Presidio County*, 2:103–4; Presidio County, Deed Records, vol. 43, 196, vol. 42, 161; *Levelland Daily Sun News*, 18 May 1958.

43. Presidio County, Records of the Certificates of Occupancy, vol. 1, 106, no. 10070; Weather Bureau, USDA, Cooperative Observer's Meteorological Record, 02 Ranch, Tex., February 1914–September 1928. *El Paso Herald*, 1 September 1914. The newspaper entry did not specify how big the "lamb crop" was.

44. John Newby, email correspondence with author, 23 September 2014; John Newby, email correspondence with author, 24 September 2014; El Paso, Texas, City Directory, 1917, US City Directories, 1821–1989 [online database], accessed 19 February 2016 at Ancestry.com.

45. "Sutherlin" family tree by David Keller, profile for Francis Marion Sutherlin; *El Paso Herald*, 21 September 1916; Weather Bureau, 02 Ranch, February 1914–September 1928.

Chapter 4. Signal Fires in the Chinatis

1. This story is based on the official report of Captain Kloepfer to commanding officer Colonel Langhorne. The conclusion that James Wilson had been issued a field telephone is deduced considering there is no other way he could have made the call at that point in time. The telephone would have been the 1913 or 1917 model widely used by the Signal Corps during World War I. See US Army Signal Corps, *Technical Equipment of the Signal Corps: 1916* (Washington, DC: Government Printing Office, 1917); and H. E. Kloepfer to G. T. Langhorne, Engagement with Bandits in Mexico, 23 March 1919 in RG 407 [Adjutant General's Office] Central Decimal Files, Project Files 1917–25, Mexican Border, File 319.1 (5–24–19) to 319.1 (3–23–19), Box 1370, Earl H. Elam Collection, Archives of the Big Bend, Bryan Wildenthal Memorial Library, Sul Ross State University, Alpine, Tex. (hereafter, ABB).

2. Stuart Easterling, *The Mexican Revolution: A Short History 1910–1920* (Chicago: Haymarket Books, 2012), 15–28, 42–43; B. T. Davenport, "The Watch along the Rio Grande," *Journal of Big Bend Studies* 7 (1995): 149; Thompson, *History of Marfa and Presidio County*, 2:80; *Marfa New Era*, 24 December 1910; Thomas T. Smith, "Chronological Matrix, Army Unit Locations, Big Bend, Texas 1911–1921," 30 September 2015, manuscript in possession of author.

3. Tyler, *Big Bend*, 161; Thompson, *History of Marfa and Presidio County*, 2:82–89; T. Smith, "Chronological Matrix"; "Post Returns, Fort Sam Houston, Texas, January 1912–December 1916," M617, roll 1082, Microfilm Series, National Archives Records Administration (NARA),Washington, DC.

4. Easterling, *Mexican Revolution*, 61–68; Glenn Justice, *Revolution on the Rio Grande: Mexican Raids and Army Pursuits, 1916–1919* (El Paso: Texas Western Press, 1992), 1–9.

5. Justice, *Revolution on the Rio Grande*, 1–9; Clarence C. Clendenen, *Blood on the Border: The United States Army and Mexican Irregulars* (New York: Macmillan, 1969), 116–37; Thompson, *History of Marfa and Presidio County*, 2:101–5. According to Maj. Michael M. McNamee's report of the refugee march, there were 3,352 officers and men, 1,607 women and children, and 1,762 horses and mules. Commanding Officer, Big Bend District [Maj. Michael M. McNamee, 15th Cavalry], Marfa, Tex., 10 February 1914, to Commanding General, Southern Department, in US Army Continental Commands, 1821–1920, Entry 393 Camp Marfa, Misc. Files, Orders, "Memorandum," Daily Diary of Events, NARA RG 393, Elam Collection, ABB.

6. W. D. Smithers, *Chronicles of the Big Bend: A Photographic Memoir of Life on the Border* (Austin, Tex.: Madrona Press, 1976), 13, 22–25.

7. Ibid., 19, 26.

8. Thompson, *History of Marfa and Presidio County*, 2:82–89; Thomas T. Smith, "The Old Army in the Big Bend, 1911–1921," 5 October 2015, manuscript in possession of author; Thompson, *History of Marfa and Presidio County*, 2:101; Justice, *Revolution on the Rio Grande*, 9.

9. Tony Cano and Ann Sochat, *Bandido: The True Story of Chico Cano, the Last Western Bandit* (Canutillo, Tex.: Reata Publishing, 1997), 15, 28, 44.

10. Ibid., 54–56; Thompson, *History of Marfa and Presidio County*, 2:96–97.

11. Glenn Justice, telephone interview by author, 1 February 2016; Bob Alexander, *Riding Lucifer's Line: Ranger Deaths along the Texas-Mexico Border* (Denton: University of North Texas Press, 2013), 237–45. Other sources tell a different story and provide a different date, but the official record indicates Howard's death was on 12 February, which conforms to the version related here. See also Thompson, *History of Marfa and Presidio County*, 2:97; and Cano and Sochat, *Bandido*, 54–60.

12. Cano and Sochat, *Bandido*, 68, 75–79; Alexander, *Riding Lucifer's Line*, 245; Testimony of R. M. Wadsworth, in *Investigation of Mexican Affairs: Hearing before a Subcommittee of the Committee on Foreign Relations United States Senate, Part 10*, Senate Ex. Doc. no. 285, 66th Cong., 2nd sess., Serial 7665 (Washington, DC: Government Printing Office, 1920), 1532–35. While Chico Cano may or may not have been present at the ambush, he is invariably credited with masterminding it.

13. Maj. W. H. Hay, Chief of Staff, Fort Sam Houston, Tex., telegram to District Commander, Big Bend District, Marfa, Tex., 27 May 1915, RG 393, E4440, Box 9, in Folder "1915," Box 3, Elam Collection, ABB.

14. Name Index to Correspondence of the Military Intelligence Division of the War Department Staff, 1917–41, microfilm M 1194, roll 39, RG 165, Records of the War Department General and Special Staffs, 1860–1952, National Archives, College Park, Md.

15. Justice, *Revolution on the Rio Grande*, 9–10; Thompson, *History of Marfa and Presidio County*, 2:123; Clendenen, *Blood on the Border*, 172; T. Smith, "Chronological Matrix"; T. Smith, "Old Army in the Big Bend, 1911–1921."

16. T. Smith, "Chronological Matrix"; Ground view of Ruidosa camp of 1st Squadron, Texas Cavalry, Col. Blucher S. Tharp Collection 770–2361, 2362, courtesy Texas Military Museum, Camp Mabry, Austin, Tex.

17. G. W. Fuller quoted in "First Texas Cavalry Guards State during Pershing Expedition," *Big Bend Sentinel*, Centennial Edition, 1 September 1950.

18. Name Index to Correspondence of the Military Intelligence Division, 1917–41; George Dillman, 1st Lt., 6th Cavalry, to Commanding General, Southern Department, "Weekly Report," 21 July 1916; L. M. Koehler, Commanding Officer, to Commanding General, Southern Department, "Weekly Report," 26 January 1917; J. A. Gaston, Commanding Officer, to Commanding General, Southern Department, "Weekly Report," 2 February 1917, all in RG 393, E62, Folder "1916–1917," Box 4, Elam Collection, ABB.

19. Justice, *Revolution on the Rio Grande*, 14–19; T. Smith, "Chronological Matrix"; T. Smith, *Old Army in the Big Bend*, 3.

20. J. A. Gaston, Commanding Officer, Big Bend District, to the Commanding General, Southern Department, "Weekly Report," 11 May 1917, RG 393, E 62, Folder "1916–1917," Box 4, Elam Collection, ABB; Joe M. Daniel, "Fight

between Carrancistas and Chico Cano at Santa Barbara, Chihuahua, on February 21, 1917," Ruidosa, Tex., 24 February 1917 [telegram], RG 393, E 59B, Folder "Notes and Lists," Box 3, Elam Collection, ABB.

21. J. A. Gaston, Commanding Officer, Big Bend District, to the Commanding General, Southern Department, "Weekly Report," 18 May 1917, RG 393, E 62, Folder "1916–1917," Box 4, Elam Collection, ABB; Wikipedia contributors, "José Inés Salazar," Wikipedia, accessed 22 December 2015, https://en.wikipedia.org/w/index.php?title=Jos%C3%A9_In%C3%A9s_Salazar&oldid=675460198; Friedrich Katz, *The Life and Times of Pancho Villa* (Stanford: Stanford University Press, 1998), 627.

22. J. A. Gaston, Commanding Officer, to Commanding General, Southern Department, "Weekly Report," 9 February 1917, RG 393, E 62, Folder "1916–1917," Box 4, Elam Collection, ABB; H. L. Kinnison, Sierra Blanca, telegram to Commander of the Big Bend District, 27 February 1917, RG 393, E 59B, Folder "Lists and Notes," Box 3, Elam Collection, ABB; J. A. Gaston, Commanding Officer, Big Bend District, to the Commanding General, Southern Department, "Weekly Report," 2 March 1917, RG 393, E 62, Folder "1916–1917," Box 4, Elam Collection, ABB.

23. J. A. Gaston, Commanding Officer, Big Bend District, to the Commanding General, Southern Department, "Weekly Report," 25 May 1917, RG 393, E 62, Folder "1917," Box 4, Elam Collection, ABB; Cano and Sochat, *Bandido*, 134.

24. Justice, *Revolution on the Rio Grande*, 14; Name Index to Correspondence of the Military Intelligence Division, 1917–41, Pancho Villa, 14 November 1917; Testimony of Col. Geo. T. Langhorne, in *Investigation of Mexican Affairs*, 1631; Testimony of Capt. Leonard L. Matlack, in *Investigation of Mexican Affairs*, 1648.

25. Testimony of Capt. Leonard L. Matlack, in *Investigation of Mexican Affairs*, 1648–49; Testimony of T. F. Tigner, in *Investigation of Mexican Affairs*, 1538–40. In detailing the several bandit raids and subsequent pursuits by the US Cavalry, I rely primarily on official military reports as well as the *Investigation of Mexican Affairs* Senate document. However, based on internal inconsistencies and known cover-ups (such as the one that led to the Yancey court-martial), it is evident that military accounts often inflated the number of bandits killed and booty recovered and downplayed or neglected to mention the collateral damage that resulted from military action. It is beyond the scope of this chapter to try to untangle the web of conflicting evidence. However, two other sources are important to consider: that of J. J. Kilpatrick, a farmer and merchant from Candelaria, and Harry Warren, a schoolteacher in Porvenir and Candelaria. Both of these men were scathing in their indictments against the cavalry's actions on the border. This ongoing feud is addressed in depth in Glenn Justice, *Little Known History of the Texas Big Bend* (Odessa, Tex.: Rimrock Press, 2001), 109–29.

26. Testimony of Capt. Leonard L. Matlack, in *Investigation of Mexican Affairs*, 1649; Justice, *Revolution on the Rio Grande*, 17; Return of Machine Gun Troop, 8th Cavalry, December 1917, 8th Cavalry Monthly Returns and Field Returns, 1917–19, RG 391, Records of US Regular Army Mobile Units, 1866–1918, NM-93 Entry 892, Box 1, from Proceedings of the Joint Committee of the Senate and the House in the Investigation of the Texas State Ranger Force, vol. 3, Texas State Library and Archives Commission, accessed 20 December 2015, https://legacy.lib.utexas.edu/taro/tslac/50062/tsl-50062.html.

27. Testimony of Sam H. Neill, in *Investigation of Mexican Affairs*, 1541–48; Testimony of George T. Langhorne, in *Investigation of Mexican Affairs*, 1633; Justice, *Revolution on the Rio Grande*, 31. The story of the Brite Ranch raid has been dealt with exhaustively in many other sources and is only summarily addressed here. See especially Justice, *Revolution on the Rio Grande*, 21–33; Justice, *Little Known History*, 130–46; Cano and Sochat, *Bandido*, 137–60; and Tyler, *Big Bend*, 177–80.

28. Justice, *Revolution on the Rio Grande*, 32.

29. Ibid., 36–39; Robert Keil, *Bosque Bonito: Violent Times along the Borderland during the Mexican Revolution*, ed. Elizabeth McBride (Alpine, Tex.: Center for Big Bend Studies, Sul Ross State University, 2002), 27–34. It is worth noting that Keil's version differs significantly from other versions, and it contradicts some details of the prevailing narrative. More recently, in an archeological project that I led in the fall of 2015, material evidence surfaced that strongly suggests military involvement in the massacre. David W. Keller, "Archeological Investigations at the Porvenir Massacre Site, Presidio County, Texas," 2017, report on file at the Center for Big Bend Studies, Sul Ross State University, Alpine, Tex. also in possession of author.

30. Justice, *Revolution on the Rio Grande*, 50–55; Testimony of Mr. E. W. Nevill [*sic*], in *Investigation of Mexican Affairs*, 1511–13.

31. Justice, *Revolution on the Rio Grande*, 51–52. See also Keil, *Bosque Bonito*, 47–58, for details about the pursuit. Again, some aspects of Keil's version differ from the military report that formed the basis of Justice's account. However, because this same military report constituted the only account of the battle, it easily could have exaggerated their success.

32. Signal Corps telegram, 4 August 1918, RG 393, E 58, Folder 1918, Box 4, Elam Collection, ABB; Langhorne, Commanding Officer, to Commanding General, Southern Department, 4 August 1918, RG 393, E 62, Folder "1916–1917," Box 4, Elam Collection, ABB; Testimony of Mr. E. W. Neville, in *Investigation of Mexican Affairs*, 1511–13; Thompson, *History of Marfa and Presidio County*, 2:157.

33. Testimony of Capt. Leonard L. Matlack, in *Investigation of Mexican Affairs*, 1654; Thompson, *History of Marfa and Presidio County*, 2:157–58.

34. H. E. Kloepfer to G. T. Langhorne, Engagement with Bandits in Mexico, 23 March 1919, in RG 407 [Adjutant General's Office] Central Decimal

Files, Project Files 1917–25, Mexican Border, File 319.1 (5–24–19) to 319.1 (3–23–19), Box 1370, Elam Collection, ABB; Charles H. Harris III and Louis R. Sadler, *The Texas Rangers and the Mexican Revolution: The Bloodiest Decade, 1910–1920* (Albuquerque: University of New Mexico Press, 2004), 489. It is worth noting that J. J. Kilpatrick Jr., a storeowner and farmer in Candelaria, claimed that Kloepfer's report was falsified and that, in fact, no bandits had been killed at all. He also claimed, however, that no credit had been given the civilian scouts, which was untrue. Kilpatrick harbored an intense hatred for, and maintained an ongoing feud with, the US Cavalry and especially Captain Matlack, who was in charge of Troop K of the 8th Cavalry at Candelaria. While it is unlikely that all of his many accusations against the military were fabricated, his testimony must be viewed within the context of both his hostility toward the military and his well-documented alcoholism. J. J. Kilpatrick Jr., "Nuñez Ranch Raid and the Killing of Bandits on Telephone Wires," Circular No. 3, Briscoe Center for American History, University of Texas at Austin.

35. Kloepfer to Langhorne, Engagement with Bandits in Mexico, 23 March 1919.

36. Ibid.

37. K. B. Edmunds, Marfa, to Commanding Officer, Camp Albert, Marfa, 16 September 1919, RG 393, E 62, Folder "1919 Set 5," Box 2, Elam Collection, ABB; Testimony of Capt. Leonard L. Matlack, in *Investigation of Mexican Affairs*, 1655. This event prompted backlash from the Mexican government, which took offense at the American incursion. That fact, as well as inconsistencies in the official military reports, brings the veracity of this account into question. J. J. Kilpatrick claimed the supposed thefts were pretext for crossing the river and that Matlack and his men had in effect simply massacred the sleeping Mexicans in a vendetta against Chico Cano. He may have been right, but Kilpatrick's alcohol-fueled rants typically verge on the hysterical, which also tends to place them in question. See J. J. Kilpatrick, "Value of the Evidence Collected by the Fall Senate Committee: An Examination of the Exaggerated and Fabricated Testimony of Some of the Witnesses," 45–53, unpublished manuscript, Briscoe Center for American History, University of Texas at Austin.

38. Justice, *Revolution on the Rio Grande*, 58–59.

39. Ibid. This story has been recounted in several sources in addition to Justice's *Revolution on the Rio Grande* and is only briefly summarized here.

40. Justice, *Revolution on the Rio Grande*, 70–79; Testimony of Capt. Leonard L. Matlack, in *Investigation of Mexican Affairs*, 1655–58.

41. Justice, *Revolution on the Rio Grande*, 70–79; Testimony of Capt. Leonard L. Matlack, in *Investigation of Mexican Affairs*, 1655–58.

42. T. Smith, *Old Army in the Big Bend*, 3–4; War Department, Report of Chief of Construction Division to the Secretary of War, Annual Reports, vol. 1, 1920 (Washington, DC: Government Printing Office, 1921), 1604–6.

43. J. H. Conlin, Captain, Quartermaster Corps, "List of Government Improvements at Outpost at Ruidosa, Texas," August 1921, in Mexican Border Conflicts Files, Fort Davis National Historic Site, Fort Davis, Tex.

44. War Department, "Report of the Adjutant General of the Army to the Secretary of War, October 5, 1921," in Annual Reports, Fiscal Year Ended 30 June 1921 (Washington, DC: Government Printing Office, 1921), 59; US Army, "Fifth Cavalry-Marfa, Texas," *Cavalry Journal* 30, no. 125 (October 1921): 446.

45. Colwell, Candelaria, telegram to Signal Corps, United States Army, Marfa, 26 December 1919, RG 393, E59, Folder "1919," Box 3, Elam Collection, ABB; *Marfa New Era*, 24 January 1920; Testimony of Mr. A. J. King, in *Investigation of Mexican Affairs*, 1536–37.

46. Joe Nichol, Ruidosa, to Commanding Officer at Ruidosa, April 1918, RG 394, File 143, Folder 3, Box 4, Elam Collection, ABB.

Chapter 5. El Camino Afuera

1. Oscar Perez interview, 18 February 2013. This as well as other semifictionalized scenes are derived primarily from interviews with José Prieto's descendants in addition to other sources, as indicated.

2. Gary Walton and Hugh Rockoff, *History of the American Economy*, 12th ed. (Mason, Ohio: South-Western, 2014), 408; J. R. Vernon, "The 1920–21 Deflation: The Role of Aggregate Supply," *Economic Inquiry* 29, no. 3 (1991): 572–80.

3. Shurley, "Dowes of Pinto Canyon," 16. Several sources erroneously indicate that Dowe was a Texas Ranger at this time. However, his service with the Rangers was restricted to an eight-month period from December 1917 to August 1918. See Harris and Sadler, *Texas Rangers and the Mexican Revolution*, 523.

4. Fenley, "Cowboy and Mounted Customs Inspector," 96; Shurley, "Dowes of Pinto Canyon," 16.

5. Presidio County, Deed Records, vol. 43, 597, vol. 49, 591, vol. 27, 302.

6. Fenley, "Cowboy and Mounted Customs Inspector," 94; Shurley, "Dowes of Pinto Canyon," 17.

7. Fenley, "Cowboy and Mounted Customs Inspector," 94.

8. Ibid.; Ginny Watts, interview by author, Alpine, Tex., 24 July 2009; Charles H. Harris, Frances E. Harris, and Louis R. Sadler, *Texas Ranger Biographies: Those Who Served, 1910–1921* (Albuquerque: University of New Mexico Press, 2009), 91.

9. Presidio County, Deed Records, vol. 52, 270, 428, vol. 26, 631, vol. 42, 374, vol. 62, 407.

10. Presidio County, Book of Petitions for Naturalization, José Prieto; *Big Bend Sentinel*, 28 February 1963.

11. Prieto family tree by David Keller, profile for Victoriano Prieto (1892–

1946, d. Aldama, Chihuahua), profile for Gregorio Prieto (1904–2011, d. El Paso, Tex.), profile for José Prieto Jr. (1902–61, d. Odessa, Tex.), undocumented data updated June 2015, all in "Public Member Trees" database, accessed 11 June 2015 at Ancestry.com; Presidio County, Deed Records, vol. 79, 538; Oscar Perez interview, 18 February 2013.

12. Traweek, *Mohair in Texas*, 121; Alan L. Olmstead and Paul W. Rhode, "Cotton, Cottonseed, Shorn Wool, and Tobacco – Acreage, Production, Price, and Cotton Stocks: 1790–1999 [Annual]," in *Historical Statistics of the United States, Earliest Times to the Present: Millennial Edition*, ed. Susan B. Carter, Scott Sigmund Gartner, Michael R. Haines, Alan L. Olmstead, Richard Sutch, and Gavin Wright (New York: Cambridge University Press, 2006), table Da755–765, accessed 1 September 2015, https://hsus.cambridge.org/HSUSWeb/toc/treeTablePath.do?id=Da755–765; Presidio County, Deed Records, vol. 79, 538.

13. David W. Keller, geographical information systems (GIS) map of Prieto Ranch based on deed record data (2015); Flake Tompkins, "Chinati Peak Ranch, Presidio County, Texas," real estate listing (Midland, Tex., 1970), unpublished manuscript in possession of author; Emilio F. Moran, *Human Adaptability: An Introduction to Ecological Anthropology*, 3rd ed. (Boulder, Colo.: Westview Press, 2008), 240.

14. *El Paso Herald*, 4 September 1917.

15. Thirteenth Census of the United States, 1910, Justice Precinct 1, Dimmit, Tex., roll T624_1547, p. 6A, Enumeration District 0036, FHL microfilm 1375560, in Records of the Bureau of the Census, RG 29, National Archives, accessed 23 April 2015 at Ancestry.com; "Sutherlin" family tree by David Keller, profile for Grover Martin Sutherlin (1885–1963, d. Gaines County, Tex.), undocumented data updated March 2015, in "Public Member Trees" database, accessed 23 April 2015, Ancestry.com; Presidio County, Deed Records, vol. 42, 161, 174; Presidio County, Records of the Certificates of Occupancy, vol. 1, 150.

16. Weather Bureau, 02 Ranch, February 1914–September 1928; *El Paso Herald*, 13 June 1918, 23 October 1918.

17. J. K. Taubenberger and D. M. Morens, "1918 Influenza: The Mother of All Pandemics," *Emerging Infectious Diseases* 12, no. 1 (2006), accessed 26 May 2015, http://dx.doi.org/10.3201/eid1209.050979; Presidio County, Register of Deaths, vol. 3, 1918, Presidio County Clerk's Office; Karen Green, "The Influenza Epidemic of 1918 in Far West Texas," *Journal of Big Bend Studies* 5 (1993): 125–36.

18. David Brown, "1918 Flu Virus Limited the Immune System," *Washington Post*, 18 January 2007, http://www.washingtonpost.com/wp-dyn/content/article/2007/01/17/AR2007011701113.html.

19. Last Will and Testament of M. D. Sutherlin, filed 12 November 1918, in Presidio County, Probate Records, vol. 5, 236, Presidio County Clerk's Office.

20. Ibid.

21. Inventory and Appraisement of the Estate of M. D. Sutherlin, Deceased, in Presidio County, Probate Records, February 1919, vol. 5, 238.

22. S. T. Wood v. Grover M. Sutherlin et al., Cause No. 2528, in Presidio County, Minutes of the District Court, July term, AD 1919, vol. 8, 141.

23. Martin Sutherlin, telephone interview by author, 16 October 2013; Certificate of Commissioners Court for Payment of Scalp Bounty Warrants, Presidio County, Minutes of the Commissioners Court, 13 January 1916, 7; Schmidly, *Texas Natural History*, 400.

24. Presidio County, Probate Records, vol. 5, 295; Presidio County, Register of Deaths, vol. 3, 18; Fourteenth Census of the United States, 1920 Justice Precinct 6, Presidio, Tex., roll T625_1840, p. 6A, Enumeration District 167, image 825, in Records of the Bureau of the Census, RG 29, National Archives, accessed at Ancestry.com. There is no death certificate in the Presidio County records for May Sutherlin. Her death date was derived from her Woodmen of the World tree stone grave marker in the Sutherlin family cemetery. It reads as follows: "MAY SUTHERLIN – MAR. 15, 1891 – FEB. 15, 1914 – WE LOVED HER – In Memoriam - Woodmen Circle."

25. Weather Bureau, 02 Ranch, February 1914–September 1928; *Marfa New Era*, 1 July 1922; Olmstead and Rhode, "Cotton, Cottonseed, Shorn Wool, and Tobacco"; Paul H. Carlson, *Texas Woollybacks: The Range Sheep and Goat Industry* (College Station: Texas A&M University Press, 1982), 194.

26. Presidio County, Records of the Certificates of Occupancy, vol. 1, 152; Presidio County, Deed Records, vol. 42, 477, vol. 45, 607; National Agricultural Statistics Service (NASS) Historic Annual Average Beef Prices, 1909–2001, USDA, Washington, DC, spreadsheet in possession of author; Olmstead and Rhode, "Cotton, Cottonseed, Shorn Wool, and Tobacco"; Fenley, "Cowboy and Mounted Customs Inspector," 96; O. C. Dowe to the Wool Growers' Central Storage Co., San Angelo, Tex. 21 November 1921, in File D, Dowe Collection, ABB.

27. *Marfa New Era*, 24 May 1919; Charles M. Madigan, "A Libel Case with a 6-Cent Verdict," *Chicago Tribune*, 8 June 1997, http://articles.chicagotribune.com/1997–06–08/news/9706300080_1_henry-ford-chicago-tribune-anarchist; Mead and Metcalf [attorneys at law] to Mr. Harry E. Andrews, Mr. Jas S. Smithwick, Joseph Boyer, 28 December 1921, in File D, Dowe Collection, ABB.

28. Dowe to Harry Andrews and Jim Smithwick, 17 January 1922, in File D, Dowe Collection, ABB.

29. Dowe to Harry Andrews and Jim Smithwick, 7 January 1922, in File D, Dowe Collection, ABB.

30. Dowe to Harry Andrews and Jim Smithwick, 20 March 1922, in File D, Dowe Collection, ABB.

31. Dowe to Lela McKinley, 4 April 1922, in File D, Dowe Collection, ABB.

32. Presidio County, Deed Records, vol. 62, 213; *Marfa New Era*, 2 September 1922.

33. *Marfa New Era*, 31 July 1926; Thompson, *History of Marfa and Presidio County*, 2:281.

34. Presidio County, Deed Records, vol. 64, 70, 320, 1408, vol. 64, 526, vol. 68, 65, vol. 64, 70, vol. 68, 202, vol. 62, 486, vol. 68, 537, vol. 74, 144; Watts interview, 24 July 2009.

35. Watts interview, 24 July 2009; Presidio County, Minutes of the District Court, Cause No. 2894, Dora E. Wilson v. J. E. Wilson. Levi James "Jim" Watts, who married Mamie, was the son of James "Jim" R. Watts, who had served as a scout for the 8th Cavalry in Ruidosa. Court records did not indicate what Dora's allegations were but, based on past infidelity (supposedly with his sister-in-law, Josie), it might have been adultery. Janell Stratton, telephone interview by author, 14 July 2017.

36. It is curious to note that the article in the *Sanderson Times* indicated Wilson had moved there with his wife. It is possible he married again, but, if so, no record has been found and the obituary in the *Sentinel* referred only to Dora as his widow. Like many aspects of history, this remains a mystery. Certificate of Death for James Wilson, in Texas Death Certificates, 1903–1982 [online database], digital image accessed 14 August 2015 at Ancestry.com; *Sanderson Times*, 8 May 1930; *Big Bend Sentinel*, 8 May 1930.

37. Application to Set Aside Portion of Will and for Sale of Real Estate, Estate of M. D. Sutherlin, Deceased, No. 166, County Court of Presidio County, Tex., March term, 1922, Presidio County, Probate Records, vol. 6, 11; Order of Court Setting Aside Portion of Will and Ordering Real Estate Sold, Case No. 166, County Court of Presidio County, Tex., March term, 1922, Presidio County, Probate Records, vol. 6, 18.

38. Weather Bureau, 02 Ranch, February 1914–September 1928; Olmstead and Rhode, "Cotton, Cottonseed, Shorn Wool, and Tobacco"; *Marfa New Era*, 3 March 1923.

39. Weather Bureau, 02 Ranch, February 1914–September 1928; Olmstead and Rhode, "Cotton, Cottonseed, Shorn Wool, and Tobacco; Marfa New Era, 3 March 1923. Presidio County, Deed Records, vol. 71, 168; "Sutherlin" family tree by Keller, profile for Grover Martin Sutherlin (1885–1963, d. Gaines County, Texas), undocumented data updated March 2015.

40. Martin Sutherlin telephone interview, 16 October 2013; Presidio County, Deed Records, vol. 71, 168, vol. 45, 607, vol. 62, 213.

41. Martin Sutherlin telephone interview, 16 October 2013.

42. Elsie's comment recalled in Michael Sutherlin, email correspondence with author, 5 September 2012; Martin Sutherlin telephone interview, 16 October 2013; 1930 US Census, Precinct 1, Hudspeth County, Texas, population schedule, Enumeration District 1, p. 2B, Grover Sutherlin, National Archives microfilm publication T626, roll 2359, digital image accessed 1 May 2015 at Ancestry.com; 1940 US Census, El Paso, Tex., population schedule, Enumeration District 71–10, p. 3B, Grover Sutherlin, National Archives microfilm publication T627, roll 4028, digital image accessed 1 May 2015 at Ancestry.com; Certificate of Death for Grover Martin Sutherlin, in Texas

Death Certificates, 1903–1982 [online database], digital image accessed 14 August 2015 at Ancestry.com.

Chapter 6. Sangre del Cordero

1. This story is part of the Prieto family lore recounted years later by Gregorio and passed down by his nephew. Oscar Perez, telephone interview by author, 1 October 2013. The use of the term "Mexican" in this instance reflects the way that Hispanics in the borderlands were (and often still are) called "Mexicans" regardless of nationality. In truth, Gregorio was a citizen of the United States.

2. Presidio County, Book of Petitions for Naturalization, José Prieto.

3. El Quinto Censo General de Población y Vivienda 1930, Dirección General de Estadística, México [Mexico Census, 1930], Rancho de El Alamos Fosesigua, Victoriano Prieto, digital image, accessed 4 January 2016 at Ancestry.com FamilySearch, compiler (Salt Lake City, Utah: FamilySearch, 2009); "Prieto" family tree by David Keller, profile for Victoriano Prieto (1892–1946, d. Aldama, Chihuahua, Mexico) undocumented data updated July 2015, in "Public Member Trees" database, accessed 11 June 2015 at Ancestry.com; José "Joe" Manuel Prieto, email correspondence with author, 3 October–1 November 2013; "Prieto" family tree by David Keller, profile for José Prieto Jr. (1902–1961, d. Odessa, Tex.) undocumented data updated October 2015, in "Public Member Trees" database, accessed 12 October 2015 at Ancestry.com.

4. *Marfa New Era*, 17 June 1922; Presidio County, Deed Records, vol. 62, 407; Presidio County, Book of Petitions for Naturalization, José Prieto.

5. Oscar Perez interview, 18 February 2013; Manny Perez, interview by author, Pinto Canyon Ranch, 19 April 2013.

6. The story, as told to me, was that Gregorio killed a total of seven deer, which seems improbable and likely an unnecessary embellishment. Oscar Perez interview, 18 February 2013; Manny Perez interview, 19 April 2013.

7. Presidio County, Minutes of the County Court, Presidio County Courthouse, vol. 3, 338; Manny Perez interview, 19 April 2013; Oscar Perez, interview by author, Pinto Canyon Ranch, 19 April 2013.

8. "Prieto" family tree by David Keller, profile for Gregorio Prieto (1904–2011, d. El Paso, Tex.).

9. Manny Perez interview, 19 April 2013; "El Paso, TX [Union Depot]," The Great American Stations, accessed 12 July 2016, http://www.great americanstations.com/Stations/ELP; 1930 US Census, El Paso, Texas, population schedule, Enumeration District 13, sheet 24A, p. 198 (stamped), dwelling 261, family 488, Pablo Prieto, National Archives microfilm publication T626, roll 2327, digital image accessed 24 July 2015 at Ancestry.com; Wikipedia contributors, "Lydia Patterson Institute," Wikipedia, accessed 24 July 2015, https://en.wikipedia.org/w/index.php?title=Lydia_Patterson_Institute&oldid=601667704; Walter N. Vernon, "Lydia Patterson Institute," *Handbook of Texas Online*, accessed 24 July 2015, https://www.tshaonline.

org/handbook/online/articles/kb121; "History of Blackwell School," Blackwell School Alliance, accessed 24 July 2015, https://www.theblackwellschool.org/history.

10. The photographs referred to are from the Prieto family's collection, shared with the author by the Perez family.

11. "Weatherby" family tree by David Keller, profile for James Sim Weatherby (1886–1941, d. San Angelo, Tex.), undocumented data updated August 2015, in "Public Member Trees" database, accessed 18 August 2015 at Ancestry.com.

12. Presidio County, Deed Records, vol. 71, 610.

13. Ted Gray, interview by author, Alpine, Tex., 9 October 2008; *Abilene Reporter-News*, 4 April 1934; Weatherby v. Guerrero, Cause No. 8078, Court of Civil Appeals of Texas, Austin, accessed 14 August 2015, https://casetext.com/case/weatherby-v-guerrero; Manny Perez interview, 19 April 2013.

14. Marilyn Miller and Marian Faux, eds., *The New York Public Library American History Desk Reference* (New York: Stonesong Press and the New York Public Library, 1997), 289–95; *Big Bend Sentinel*, 31 October 1929, 28 November 1929, 12 December 1929, 2 January 1930.

15. Traweek, *Mohair in Texas*, 121; Olmstead and Rhode, "Cotton, Cottonseed, Shorn Wool, and Tobacco."

16. Miller and Faux, *American History Desk Reference*, 284–85; *Big Bend Sentinel*, 17 August 1933.

17. Western Regional Climate Center (WRCC), General Climate Summary Tables (Presidio, Tex. [417262]), Summary of the Day (SOD), US Climate Archive, accessed 21 May 2015, http://www.wrcc.dri.edu/cgi-bin/cliMAIN.pl?tx7262; *Big Bend Sentinel*, 15 September 1932, 8 September 1932.

18. WRCC, General Climate Summary (Presidio, Tex.); *Big Bend Sentinel*, 15 June 1933, 13 July 1933, 10 August 1933.

19. *Big Bend Sentinel*, 8 March 1934, 14 June 1934, 21 June 1934.

20. National Centers for Environmental Information, NOAA, Historical Palmer Drought Indices, accessed 23 September 2015, http://www.ncdc.noaa.gov/temp-and-precip/drought/historical-palmers/psi/193001–201507; Col. Philip G. Murphy, "The Drought of 1934: A Report of the Federal Government's Assistance to Agriculture," report presented to the President's Drought Committee, 15 July 1935, https://fraser.stlouisfed.org/docs/publications/books/drought_1934_aaa.pdf; *Big Bend Sentinel*, 28 June 1934, 12 July 1934, 26 July 1934.

21. NASS, Historic Annual Average Beef Prices; Justice, *Little Known History*, 40; *Big Bend Sentinel*, 5 July 1934, 23 August 1934; Murphy, "Drought of 1934," 67.

22. In 1930 there were sixteen thousand sheep and twenty-one thousand goats in Presidio County; by 1935 there were eighty-five hundred sheep and seventeen thousand goats. US Department of Commerce, Sixteenth Census of the United States, 1940, Agriculture, vol. 1, part 5 (Washington DC: Gov-

ernment Printing Office, 1942), 424, 439, 556; Murphy, "Drought of 1934," 90, 195–96; *Big Bend Sentinel*, 29 November 1934.

23. The population of Presidio County in 1930 was 10,154. Elizabeth Cruce Alvarez, ed., *Texas Almanac, 2006–2007: Sesquicentennial Edition 1857–2007* (Dallas: Dallas Morning News, 2005), 369; *Big Bend Sentinel*, 6 September 1934, 26 July 1934, 30 August 1934, 23 August 1934, 22 March 1935.

24. Presidio County, Tax Rolls, unrendered property, 1934, Tax Office, Presidio County Courthouse.

25. "Walter Fleetwood Hale Jr.," unpublished manuscript, W. F. Hale Jr. File, Texas Ranger Hall of Fame Museum, Waco, Tex. Although untitled and unattributed, a newspaper article included in the Hale file at the Ranger museum strongly suggests that the author is Sgt. Lloyd Polk, who briefly owned the gun and conducted a significant amount of research on it. See Mike Cox, "Ex-Detective Tracks Gun's History," *Austin American-Statesman*, 17 June 1983.

26. "Hale" family tree by David Keller, profile for Walter Fleetwood Hale (1900–1965, d. Austin, Tex.), undocumented data updated August 2015, in "Public Member Trees" database, accessed 25 August 2015 at Ancestry.com; "Walter Fleetwood Hale Jr." manuscript.

27. *Valley Morning Star* (Harlingen, Tex.), 5 May 1931; "Walter Fleetwood Hale Jr." manuscript.

28. "Walter Fleetwood Hale Jr." manuscript; *Valley Morning Star*, 5 May 1931; *Austin American*, 5 May 1931; Consumer Price Index Inflation Calculator, Bureau of Labor Statistics, US Department of Labor, accessed 25 August 2015, http://data.bls.gov/cgi-bin/cpicalc.pl. It is noteworthy that suspicions leading to Hale's resignation were only indicated in the unpublished manuscript, which was unsourced and could not be corroborated. However, because of the level of detail provided, and general agreement with other aspects of the story, I believe it to be accurate.

29. Robert M. Utley, *Lone Star Lawmen* (New York: Penguin Group, 2007), 152–53; Thompson, *History of Marfa and Presidio County*, 2:192, 387; Harris and Sadler, *Texas Rangers and the Mexican Revolution*, 176–77.

30. The story of Pablo's murder that follows was gleaned from many sources, primarily a firsthand account by Ranger Walter Fleetwood Hale Jr. (who killed him), several newspaper articles, a sheriff's affidavit, and oral history from descendants of the Prieto family. Gregorio relayed the story many times over the years, the drama becoming part of the family lore. Because the story is a composite—the details derived from numerous sources in a mixed fashion—they are all listed below rather than as individual footnotes. While some versions are contradictory, most of the details could be cross-verified. Unless otherwise indicated, the sources are Walter Fleetwood Hale, "Rustler's Ambuscade," Folklore Project, Life Histories 1936–39, Federal Writers' Project, US Work Projects Administration, Manuscript Division, Library of Congress (1941), accessed 10 January 2013, http://www.loc.gov/

item/wpalh002443/; Sheriff J. D. Bunton, Sworn Testimony, 8 June 1934, examining trial before W. G. Young, Justice of the Peace, Precinct No. 1, Presidio County, Tex., Sheriff's Office Files, Presidio County Sheriff's Office, Marfa, Tex.; Oscar Perez telephone interview, 16 January 2013; Nancy Perez interview, 18 February 2013; Oscar Perez interview, 18 February 2013; Manny Perez interview, 19 April 2013; Oscar Perez interview, 19 April 2013; Death Certificate of Pablo Prieto, dated 7 June 1934, Presidio County, Register of Deaths, Presidio County Clerk's Office; *Big Bend Sentinel*, 7 June 1934, 2 August 1934, 9 August 1934, 16 August 1934, 27 January 1950; Thompson, *History of Marfa and Presidio County*, 2:480.

31. "Moon Phases 1934," Calendar-12.com, accessed 18 February 2013, http://www.calendar-12.com/moon_phases/1934.

32. "Rinche" was commonly used in the borderlands as a pejorative term for a Texas Ranger—and, by extension, any law enforcement officer. It is a hybrid word, a kind of portmanteau, that combines the phonetic qualities of "Ranger" and the Spanish *pinche*, which translates to "damned" or "goddamned." Jerry Javier Lujan, interview by author, Pinto Canyon Ranch, 19 October 2012.

33. Although Gregorio claimed that Hale killed Pablo with the spray of a Thompson submachine gun (and Manny and Oscar Perez remember bullet holes in the door frame of the house), Hale's account, as well as the court records and death record, suggest only a single shot killed Pablo. These two accounts are difficult to reconcile.

34. *Big Bend Sentinel*, 7 June 1934; Oscar Perez interview, 16 January 2013.

35. This scene is based on a story passed down through the Prieto family and is shamelessly embellished here. Oscar Perez interview, 18 February 2013.

36. *Big Bend Sentinel*, 7 June 1934.

37. *Big Bend Sentinel*, 9 August 1934.

38. *Big Bend Sentinel*, 9 August 1934.

39. Oscar Perez interview, 18 February 2013; Mary K. Earney, "Farther Than Nearer: Judges, Lawyers, and Cases of the 83rd State Judicial District of Texas, 1917–1983," *Big Bend Sentinel*, 29 December 1994; *Big Bend Sentinel*, 16 August 1934.

40. *Big Bend Sentinel*, 16 August 1934; Earney, "Judges, Lawyers, and Cases."

41. Presidio County, Minutes of the District Court, vol. 6, Cause No. 1296, 1297, 166; *Big Bend Sentinel*, 16 August 1934; Earney, "Judges, Lawyers, and Cases"; Hale, "Rustler's Ambuscade."

42. *Big Bend Sentinel*, 9 August 1934, 2 August 1934, 23 August 1934.

43. Francisco E. Balderrama and Raymond Rodriguez, *Decade of Betrayal: Mexican Repatriation in the 1930s* (Albuquerque: University of New Mexico Press, 1995), 116.

44. Manny Perez interview, 20 April 2013; José Adam Prieto Miller (Adam Miller), telephone interview by author, 15 February 2016.

45. James Randolph Ward, "The Texas Rangers, 1919–1935: A Study in Law Enforcement" (PhD diss., Texas Christian University, 1972), 220; Utley, *Lone Star Lawmen*, 154–55.

46. Maria Flora Vizcaino, interview by author, Marfa, Tex., 28 August 2013; Oscar Perez interview, 18 February 2013.

47. Presidio County, Index to District Court Minutes, vol. 2, Presidio County District Court, Presidio County Courthouse, Marfa, Tex., 170; Presidio County, Minutes of the District Court, vol. 9, Cause No. 1296, 1297, 159–70, 183.

48. Sim's son Harper claimed in his self-published book that $60,000 was spent on the fence. That figure seems unlikely considering that Sim was already in debt and the sum would have been more than three times the purchase price of the ranch. Willis Harper Weatherby, "Chinati (Conquistadores)," December 1989, unpublished manuscript in possession of author.

49. Weatherby, "Chinati."

50. Presidio County, Deed Records, vol. 77, 630, vol. 102, 256–58, vol. 111, 170, vol. 109, 401.

51. Presidio County, Deed Records, vol. 109, 424; Weatherby, "Chinati"; Certificate of Death for James Sim Weatherby (5 June 1941), San Angelo, Tex., in Texas Death Certificates, 1903–1982 [online database], digital image accessed 14 August 2015 at Ancestry.com; James Sim Weatherby, Find a Grave, accessed 9 October 2015, www.findagrave.com/cgi-bin/fg.cgi?page=gr&GRid=63771638&ref=acom.

52. WRCC, General Climate Summary Tables (Presidio, Tex.); *Big Bend Sentinel*, 3 May 1935, 26 April 1935.

53. *Big Bend Sentinel*, 14 June 1935, 19 July 1935, 23 August 1935, 6 September 1935; NOAA, Historical Palmer Drought Indices, accessed 1 October 2015, http://www.ncdc.noaa.gov/temp-and-precip/drought/historical-palmers/psi/193001–201507.

Chapter 7. Madre Patria

1. Manny Perez interview, 20 April 2013.

2. WRCC, General Climate Summary Tables (Presidio, Tex.); Thompson, *History of Marfa and Presidio County*, 2:524.

3. Michael Lind, *Land of Promise: An Economic History of the United States* (New York: Harper, 2013), 303–5; Traweek, *Mohair in Texas*, 121; Harold Stults, Edward H. Glade Jr., Scott Sanford, John V. Lawler, and Robert Skinner, "Fibers: Background for 1990 Farm Legislation, Agriculture Information Bulletin No. AIB-591, March 1990, 73, https://www.ers.usda.gov/publications/pub-details/?pubid=42014; Olmstead and Rhode, "Cotton, Cottonseed, Shorn Wool, and Tobacco"; Nebraska Agricultural Statistics Service, USDA, Lamb Price Dollars per CWT (hundredweight), accessed 4 February 2016, http://www.nass.usda.gov/Statistics_by_State/Nebraska/NE_Historical_Data/nebhist/historic%20prices%20-%201ambs.pdf (no longer available).

4. Carlson, *Woollybacks*, 207; Mona Lange McCroskey, "The Most Efficient Fiber Producers on Earth: Angora Goat Ranching in Yavapai County, Arizona, 1880–1945," Prescott Corral of Westerners International, accessed 9 February 2016, http://www.prescottcorral.org/TT3/V2AngoraGoatRanching. htm; Carol Canada, "Wool and Mohair Price Support," 30 October 2008, Congressional Research Service, Library of Congress, Prepared for Members and Committees of Congress, accessed 9 February 2016, nationalaglaw center.org/wp-content/uploads/assets/crs/RS20896.pdf.

5. Milton Friedman and Anna Jacobson Schwartz, *From New Deal Banking Reform to World War II Inflation* (Princeton, N.J.: Princeton University Press, 1980): 129–76; Robert A. Calvert, "Texas since World War II," *Handbook of Texas Online*, accessed 10 February 2016, http://www.tshaonline.org/handbook/online/articles/npt02; Elmer Kelton, "Ranching in a Changing Land," *Texas Almanac 2006–2007: Sesquicentennial Edition 1857–2007* (Dallas: Dallas Morning News, 2005), 26–31; Carlson, *Woollybacks*, 207–8.

6. US Department of Commerce, Sixteenth Census of the United States, 1940, Agriculture, vol. 1; US Department of Commerce, Census of Agriculture: 1950, vol. 1, part 26: Texas (Washington DC: Government Printing Office, 1952).

7. Use of the term "highland" was derived from the Highland Hereford Breeders Association, which used the word to distinguish the better quality Hereford cattle of the higher-elevation Alpine and Marfa grasslands from the lower quality mixed-breed cattle of the desert lowlands. When sheep and goat raisers followed suit by creating their own association, they utilized the same term. Although less applicable (since many sheep and goats were raised in the lowlands), the name recognition factor justified the stretch. Thompson, *History of Marfa and Presidio County*, 2:497, 509, 543; *Big Bend Sentinel*, 13 February 1942.

8. *Big Bend Sentinel*, 13 February 1942; Thompson, *History of Marfa and Presidio County*, 2:497–98; Clifford B. Casey, *Mirages, Mysteries, and Reality: Brewster County, Texas, the Big Bend of the Rio Grande* (Seagraves, Tex.: Pioneer Book Publishers), 1972, 268; Roy Bedichek, *Adventures with a Texas Naturalist* (Austin: University of Texas Press, 1947), 141; Jeff Watson, *The Golden Eagle* (London: T&AD Poyser, 2010), 314.

9. El Quinto Censo General de Población y Vivienda 1930, Dirección General de Estadística, México [Mexico Census, 1930], Rancho de El Alamos Fosesigua, Victoriano Prieto, digital image.

10. Guadalupe Prieto, telephone interview by Oscar Perez, 1 August 2015. Although the boys would have driven to Aldama, the train would have provided another option—especially for short visits. The construction of the Kansas City, Mexico and Orient Railway between Chihuahua City and Ojinaga was completed in late 1928 after its purchase by the Santa Fe Railroad. However, the US section to Presidio would not be completed until

1930. *El Paso Evening Post*, 26 December 1928; Thompson, *History of Marfa and Presidio County*, 2:310. See also John Leeds Kerr, *Destination Topolobampo: The Kansas City, Mexico and Orient Railway* (San Marino, Calif.: Golden West Books, 1968), 155–68.

11. Oscar Perez, telephone interview by author, 6 February 2013; Manny Perez, interview by Mattie Matthaei, Pinto Canyon Ranch, 20 April 2013.

12. *Big Bend Sentinel*, 1 August 1985; "Prieto" family tree by David Keller, profile for Juan M. Benavidez (1905–85, d. Presidio County, Tex.), undocumented data updated January 2016, in "Public Member Trees" database, accessed 5 January 2016 at Ancestry.com; Ida Benavidez-Taulbee, interview by author and Mattie Matthaei, El Paso, Tex., 19 February 2013.

13. Benavidez-Taulbee interview, 19 February 2013; Oscar Perez interview, 18 February 2013.

14. Benavidez-Taulbee interview, 19 February 2013; Benny Benavidez, interview by Mattie Matthaei, Chinati Hot Springs, 11 September 2013.

15. Benavidez-Taulbee interview, 19 February 2013; Benny Benavidez interview; Presidio County Abstracts of Title, D&P Railroad Block 2, Survey Section 51, Big Bend Title Company, Marfa, Tex.; Ida Benavidez-Taulbee, telephone interview by author, 16 April 2016. Although the number of livestock Juan is reported to have owned seems improbable, the information was provided by his daughter, Ida, whom I believe is a reliable source.

16. Benny Benavidez interview.

17. Benavidez-Taulbee interview, 19 February 2013; Ramona Vega Holder, *Life on My Side of the Tracks* (Bloomington, Ind.: Authorhouse, 2006), 40–41.

18. Teresa Benavidez, interview by author and Mattie Matthaei, Chinati Hot Springs, 15 June 2013; Benavidez-Taulbee interview, 19 February 2013.

19. Benavidez-Taulbee interview, 19 February 2013; Teresa Benavidez interview.

20. Oscar Perez interview, 18 February 2013; Nancy Perez interview, 18 February 2013. Most of these stories were passed down from Frances Perez to her children.

21. Teresa Benavidez interview.

22. As family lore goes, this tale doesn't seem to comport well with history. Villa was supposedly in the Juarez area during this period, and no evidence has surfaced that he made such a southward push this early. *Big Bend Sentinel*, 26 June 1936; Oscar Perez interview, 6 February 2013; Oscar Perez interview, 18 February 2013; Nancy Perez interview, 18 February 2013.

23. Oscar Perez interview, 6 February 2013; Ricardo Romo, "Responses to Mexican Immigration, 1910–1930," in *Beyond 1848: Readings in the Modern Chicano Historical Experience*, ed. Michael R. Ornelas (Dubuque, Iowa: Kendall/Hunt Publishing, 1993), 115–35; Julia Cauble Smith, "Ochoa, TX," *Handbook of Texas Online*, accessed 15 January 2016, https://tshaonline.org/handbook/online/articles/ht003.

24. J. C. Smith, "Ochoa, TX"; Oscar Perez, interview by author, Chinati Hot Springs, 21 April 2013.

25. *Big Bend Sentinel*, 17 June 1938; Ruben Perez, telephone interview by author, 28 July 2015; *Big Bend Sentinel*, 26 June 1936; 1940 US Census, Presidio County, Tex., population schedule, Justice Precinct 1, Enumeration District 189–2A, p. 9b, dwelling 166, Manuel Perez, National Archives microfilm publication, roll T627_4125, digital image accessed 19 January 2016 at Ancestry .com.

26. Oscar Perez interview, 21 April 2013.

27. Manny Perez interview, 20 April 2013.

28. Manny Perez interview, 20 April 2013; Ruben Perez telephone interview; Gary Paul Nabhan, *Desert Terroir: Exploring the Unique Flavors and Sundry Places of the Borderlands* (Austin: University of Texas Press, 2012), 44–45.

29. 1940 US Census, Presidio County, Tex., population schedule, Justice Precinct 1, Enumeration District 189–2A, p. 9a, dwelling 157, José Prieto, National Archives microfilm publication, roll T627_4125, digital image accessed 19 January 2016 at Ancestry.com.

30. Manny Perez interview, 20 April 2013; José Adam Prieto Miller (Adam Miller), telephone interview by author, 16 November 2012; 1910 US Census, Presidio County, Tex., population schedule, Justice Precinct 6, Enumeration District 201, sheet 6a, family number 66, Eunardo Morales, National Archives microfilm publication, roll T624_1582, digital image accessed 19 January 2016 at Ancestry.com; draft registration card for Juan Morales, Ruidosa, Presidio County, Tex., in US World War I Draft Registration Cards, 1917–18, roll 1983579, digital image accessed 21 January 2016 at Ancestry. com; US Social Security Applications and Claims Index, 1936–2007 [online database], entry for Guillerma Hinojos Hernandez (1901–96), accessed 20 January 2016 at Ancestry.com; Presidio County, Minutes of the District Court, Cause No. 3210, Juan Morales v. Guerma Hinojos Morales, July term AD 1935, District Clerk's Office, Presidio County Courthouse, Marfa, Tex.; Oliver Smith Jr., telephone interview by author, 4 September 2013.

31. *Big Bend Sentinel*, 30 September 1940; Manny Perez interview, 20 April 2013; José "Joe" Prieto email correspondence, 3 October–1 November 2013.

32. Manny Perez interview, 20 April 2013; *Big Bend Sentinel*, 19 September 1985.

33. Oliver Smith Jr. interview, 4 September 2013; Manny Perez interview, 20 April 2013.

34. Oscar Perez interview, 18 February 2013; Manny Perez interview, 20 April 2013; Jerry Javier Lujan, "Chinati (Chanate [sic]) Mountain Moonshine," unpublished memoir in possession of author, updated 19 September 2013.

35. Manny Perez interview, 20 April 2013.

36. Ibid.; Lujan, "Chinati (Chanate [sic]) Mountain Moonshine."

37. Manny Perez interview, 20 April 2013.

38. Ibid.

39. Oscar Perez interview, 18 February 2013; Manny Perez interview, 20 April 2013.

40. Manny Perez interview, 20 April 2013.

41. Ibid.

42. Guadalupe Prieto and Bessie Miller, interview by author, Fort Stockton, Tex., 22 May 2016.

43. Ibid.; *Big Bend Sentinel*, 15 October 1943; Ruben Perez telephone interview.

44. Guadalupe Prieto and Bessie Miller interview.

45. Ibid.

46. Janice Maupin, "Sheep and Goat Ranching—Texas Style," Oral Business History Project, Graduate School of Business, University of Texas, 1972, 358, accessed 12 February 2016, http://www.thebhc.org/sites/default/files/beh/BEHprint/v02/maupin.pdf; C. P. Bailey, *Practical Angora Goat Raising* (San Jose, Calif.: C. P. Bailey and Sons, 1905); Adam Miller interview, 19 February 2013; Carlson, *Woollybacks*, 58; Oscar Perez interview, 18 February 2013.

47. Manny Perez interview, 20 April 2013; Premier One Supplies, Shearing Instructions, accessed 12 February 2016, https://www.premier1supplies.com/img/instruction/41.pdf; Adam Miller interview, 19 February 2013.

48. Manny Perez interview, 20 April 2013; Carlson, *Woollybacks*, 59.

49. Ibid.

50. James R. Gillespie and Frank Flanders, *Modern Livestock and Poultry Production*, 8th ed. (New York: Delmar Cengage Learning, 2010), 675–77; Maupin, "Sheep and Goat Ranching," 361; Roy D. Holt, "Pioneering Sheepmen," *Sheep and Goat Raiser* 26 (December 1945): 18; Carlson, *Woollybacks*, 46.

51. David W. Keller, "Archeological Site Recording of Casa Morales," 23 May 2009, report on file at the Center for Big Bend Studies, Sul Ross State University, Alpine, Tex.; Manny Perez interview, 20 April 2013.

52. Carlson, *Woollybacks*, 114–15; Oscar Perez interview, 18 February 2013.

53. Carlson, *Woollybacks*, 59; Animal Health Publications, Infovets.com, accessed 12 February 2016, http://www.infovets.com/books/smrm/c/c104.htm; Presidio County, Marks and Brands, Book 5, Presidio County Clerk's Office; Manny Perez interview, 19 April 2013; Oscar Perez interview, 19 April 2013; Charles W. Edwards Jr., *Up to My Armpits: Adventures of a West Texas Veterinarian* (Houston: Iron Mountain Press, 2002), 66.

54. Steve Byrns, "Angora Goats: A 'Shear' Delight!," Texas Department of Agriculture and Texas Agricultural Experiment Station, November 2011, http://sanangelo.tamu.edu/files/2011/11/AngoraGoatsAShearDelight_1.pdf; Susan Schoenian, "Sheep 201: A Beginners Guide to Raising Sheep," updated October 2012, http://www.sheep101.info/201/ewerepro.html.

55. Guadalupe Prieto telephone interview, 1 August 2015; Cano and Sochat, *Bandido*, 221.

56. Nacimiento de Francisco F. Cano, Ojinaga, record number 53, and Nacimiento de Nicolas Cano, Ojinaga, record number 65, both in Chihuahua,

Mexico, Civil Registration of Births, 1861–1947, citing Ojinaga, 1918–1947 Nacimientos, digital images accessed 18 February 2016 at Ancestry.com; Cano and Sochat, *Bandido*, 224–39.

57. Oscar Perez interview, 18 February 2013; Manny Perez, interview by author, Pinto Canyon Ranch, 21 April 2013.

58. Guadalupe Prieto telephone interview, 1 August 2015.

59. A year after Victoriano was killed, the police knocked on Martina's door and told her that Chico Jr. was due to be released unless she had money to keep him in longer. But because she had nothing to offer, Chico was released. Despite what would amount to a lifelong history of crime—including rape and murder—Chico Jr. would enjoy a full if unproductive life, dying at the age of seventy-four in El Paso a free man. Guadalupe Prieto telephone interview, 1 August 2015; Guadalupe Prieto and Bessie Miller interview; Tony Cano and Ann Sochat, telephone interview by author, 18 November 2015; US Social Security Applications and Claims Index, 1936–2007 [online database], entry for Francisco Fierro Cano, accessed 18 February 2016 at Ancestry.com.

60. Immigration Card for Guadalupe Prieto, 7 March 1949, US Department of Justice, Immigration and Naturalization Service, Form I-448, in Border Crossings: From Mexico to US, 1895–1957, microfilm roll 2, Records of the Immigration and Naturalization Service, Record Group 85, National Archives and Records Administration, Washington, DC, digital image accessed 7 January 2016 at Ancestry.com; Presidio County, Book of Petitions for Naturalization, Petition for Naturalization, José Prieto Jr., 6 August 1957; José "Joe" Prieto email correspondence, 3 October–1 November 2013; Oscar Perez interview, 18 February 2013; Adam Miller interview, 19 February 2013.

Chapter 8. Tierra Seca

1. This scene is based on stories Manny Perez told the author. Manny Perez interview, 20 April 2013.

2. Presidio County, Deed Records, vol. 126, 177; draft registration card for Terry W. Shely, No. 50, 5 June 1917, Precinct 4, Brewster County, Tex., in US World War I Draft Registration Cards, 1917–18, roll 1927370, database and images accessed 29 February 2016 at Ancestry.com 2005; 1910 US Census, Kinney County, Tex., population schedule, Meed Creek, Enumeration District 0073, p. 5a, dwelling 78, family 80, Terry M. Shely, National Archives microfilm publication T624, roll 1567, digital image accessed 29 February 2016 at Ancestry.com; Terry Jean Shely-Allen, interview by author, Alpine, Tex., 24 September 2010; Casey, *Mirages, Mysteries, and Reality*, 145–47, 315.

3. *Big Bend Sentinel*, 8 August 1970; birth record for Fred Clark Shely, 5 October 1927, p. 1341, in Texas Birth Index, 1903–97 [database], Texas Department of Health, Bureau of Vital Statistics, image accessed 29 February 2016 at Ancestry.com; *Alpine Avalanche*, 8 March 2011; Marriage record for

Fred Shely and Kathryn Hegelund, 26 October 1947, in Texas, Select County Marriage Index, 1837–1977, Brewster County, Tex., roll 2031127, database accessed 29 February 2016 at Ancestry.com; "US Merchant Marine Casualties during World War II," American Merchant Marine at War, accessed 1 March 2016, http://www.usmm.org/casualty.html; Daniel Logan, telephone interview by author, 4 March 2016; Presidio County, Deed Records, vol. 126, 177.

4. Presidio County, Deed Records, vol. 74, 144, vol. 88, 610, vol. 100, 590, vol. 103, 17; Ginny Watts, telephone interview by author, 3 June 2011.

5. Logan telephone interview; South Oak Cliff High School Yearbook (Dallas, Tex., 1923), James Logan, p. 109, in US, School Yearbooks, 1880–2012, database and digital image accessed 4 March 2016 at Ancestry.com; 1930 US Census, Colorado City, Mitchell County, Tex., population schedule, Justice Precinct 13, Enumeration District 0001, sheet 13b, family number 312, Moody Logan, National Archives microfilm publication, roll 2377, digital image accessed 4 March 2016 at Ancestry.com; 1940 US Census, Alpine, Brewster County, Tex., population schedule, Justice Precinct 13, Enumeration District 22–1A, sheet 3b, family number 62, J. H. Logan, National Archives microfilm publication, roll T627_3993, digital image accessed 4 March 2016 at Ancestry.com. James Logan purchased the Pinto Canyon Ranch from Jim Watts for $4.82 an acre. Presidio County, Deed Records, vol. 114, 576.

6. James Logan's son claimed they ran ten thousand head of goats. If so, that would have represented 22 percent of the entire goat population in the county in 1945, because most ranchers had converted their operations to sheep. Logan telephone interview; Monroe Elms, email correspondence with author, 22 July 2016.

7. Logan telephone interview. More precisely, the 17,920-acre Pinto Canyon Ranch was sold to Terry Shely in exchange for 12,641 acres of the Tesnus Ranch and assumption of two outstanding notes totaling $31,313. Presidio County, Deed Records, vol. 126, 177.

8. Fred and Kathryn's half interest in the operation cost them just under $103,000. Presidio County, Deed Records, vol. 131, 68, vol. 109, 244; Shely-Allen interview; Inventory, Appraisement, and List of Claims, Estate of Belle Shely, Deceased, No. 964, Probate Court Records, July term 1970, vol. 16, 555–57, Presidio County Courthouse, Marfa, Tex.

9. Shely-Allen interview. The Shely estate at the time of Belle's death in 1970 included about four thousand Angora goats and around a thousand sheep. Inventory, Appraisement, and List of Claims, Estate of Belle Shely.

10. Shely-Allen interview; Jaime Espensen-Sturges, "XERF," *Handbook of Texas Online*, accessed 9 March 2016, http://www.tshaonline.org/handbook/online/articles/ebx01.

11. Shely-Allen interview.

12. Shely-Allen interview; Chon Prieto, interview by author, Ruidosa, Tex., 20 November 2010.

13. Chon Prieto, interview by Mattie Matthaei, Ruidosa, Tex., 13 July 2014.

14. Alan Le May, dir., *High Lonesome* (1950; Tulsa, Okla.: Vci Video, 2005), DVD; Wikipedia contributors, "John Drew Barrymore," Wikipedia, accessed 3 March 2016, https://en.wikipedia.org/w/index.php?title=John_Drew_Barrymore&oldid=707426768.

15. *Big Bend Sentinel*, 30 December 1949, 13 January 1950; Tom Vallance, "John Drew Barrymore," *The Independent* (UK), 1 December 2004, http://www.independent.co.uk/news/obituaries/john-drew-barrymore-728653.html; Le May, *High Lonesome* DVD; *Big Bend Sentinel*, 1 September 1950; Wikipedia contributors, "*Giant* (1956 film)," Wikipedia, accessed 4 March 2016, https://en.wikipedia.org/w/index.php?title=Giant_(1956_film)&oldid=708028466.

16. In 1945, Presidio County was stocked at ninety thousand animal unit months (AUMs)—an equivalency ratio for livestock grazing. By 1950 that figure had dropped to eighty-two thousand AUMs. US Department of Commerce, Censuses of Agriculture, various, from 1900 to 2002 (Washington DC: Government Printing Office); John Vallentine, *Grazing Management* (New York: Academic Press, 1992), 279.

17. Dr. Edward Cook, Dr. David Meko, Dr. David Stahle, and Dr. Malcolm Cleaveland, "Reconstruction of Past Drought across the Coterminous United States from a Network of Climatically Sensitive Tree-Ring Data," Climatic Data Center, Instrument Animation, 25 January 1999, accessed 23 February 2016, https://www.ncdc.noaa.gov/paleo/pdsiyear.html (no longer available).

18. WRCC, General Climate Summary Tables (Presidio, Tex.); WRCC, General Climate Summary Tables (Candelaria, Tex. [411416]), Summary of the Day (SOD) US Climate Archive, accessed 21 May 2015, http://www.wrcc.dri.edu/cgi-bin/cliMAIN.pl?tx1416; WRCC, General Climate Summary Tables (Valentine 10 WSW, Tex. [419275]), Summary of the Day (SOD) US Climate Archive, accessed 21 May 2015, http://www.wrcc.dri.edu/cgi-bin/cliMAIN.pl?tx9275.

19. WRCC, General Climate Summary Tables (Candelaria and Presidio, Tex.).

20. Adam Miller interview, 19 February 2013; Manny Perez interview, 20 April 2013. *Telempacate* is the common Spanish name for the desert baileya (*Baileya multiradiata*), an annual or weak perennial herb of the sunflower family. See Centennial Museum, University of Texas at El Paso, "Desert Marigold (*Baileya multiradiata*)," Chihuahuan Desert Plants, September 2002, http://museum2.utep.edu/chih/gardens/plants/baileya.htm; and Texas Agricultural Extension Service, *Integrated Toxic Plant Management* (College Station: Texas A&M University System, 2011), 7–8.

21. Adam Miller interview, 19 February 2013; Manny Perez interview, 20 April 2013.

22. WRCC, General Climate Summary Tables (Candelaria and Presidio, Tex.); *Big Bend Sentinel*, 7 May 1953, 9 July 1953, 16 July 1953, 17 December 1953.

23. The number of goats increased in Presidio County from sixteen thousand to twenty-one thousand between 1950 and 1959. US Department of Commerce, Censuses of Agriculture, various, from 1900 to 2002.

24. Stults et al., "Fibers"; Olmstead and Rhode, "Cotton, Cottonseed, Shorn Wool, and Tobacco"; Carlson, *Woollybacks*, 209–10; Jonathan Rauch, "The Golden Fleece: Why Programs Never Die," *National Journal*, 18 May 1991, 1168–71.

25. Oscar Perez, telephone interview by author, 7 August 2015. No records about José's eye condition survive, although he was probably suffering from cataracts, glaucoma, or macular degeneration—all of which are age-related causes of blindness. However, one descendant claimed it was glaucoma, which in fact is the leading cause of blindness among African Americans and Hispanics. See National Institutes of Health, "Leading Causes of Blindness," Medicine Plus, accessed 21 March 2016, https://www.nlm.nih.gov/medlineplus/magazine/issues/summer08/articles/summer08pg14–15 .html; and Benavidez-Taulbee interview, 19 February 2013.

26. Benavidez-Taulbee interview, 19 February 2013; *Big Bend Sentinel*, 4 September 1952; Certificate of Death for Juanita Barrera Prieto, filed 10 September 1952, State File No. 42704, in Texas Death Certificates, 1903–1982 [online database], digital image accessed 7 January 2016 at Ancestry.com; Wikipedia contributors, "Pancreatic Cancer," Wikipedia, accessed 22 March 2016, https://en.wikipedia.org/w/index.php?title=Pancreatic_cancer&ol did=709652842; David P. Ryan, Theodore S. Hong, and Nabeel Bardeesy, "Pancreatic Adenocarcinoma," *New England Journal of Medicine* 371 (2014): 1039–49.

27. *Big Bend Sentinel*, 28 February 1963; Adam Miller telephone interview, 16 November 2012.

28. Benavidez-Taulbee interview, 19 February 2013.

29. Adam Miller telephone interview, 15 February 2016.

30. WRCC, General Climate Summary Tables (Presidio, Tex.); *Big Bend Sentinel*, 20 May 1954, 15 July 1954.

31. WRCC, General Climate Summary Tables (Candelaria and Presidio, Tex.); *Big Bend Sentinel*, 19 May 1955, 23 August 1956, 4 October 1956 (quote).

32. WRCC, General Climate Summary Tables (Candelaria and Presidio, Tex.); *Big Bend Sentinel*, 22 October 1953, 25 July 1957; US Department of Commerce, Census of Agriculture 1950; US Department of Commerce, Census of Agriculture: 1954, vol. 1, part 26: Texas (Washington, DC: Government Printing Office, 1956).

33. *Big Bend Sentinel*, 13 November 1930, 19 November 1937.

34. *Big Bend Sentinel*, 6 December 1956; Shely-Allen interview; *The Eagle* (Bryan, Tex.), 21 December 1975, 28 December 1975.

35. Powell, *Trees and Shrubs*, 221–22; Curtis Tunnell, *Wax, Men, and Money: A Historical and Archeological Study of Candelilla Wax Camps along the Rio Grande Border of Texas*, Office of the State Archeologist Report 32 (Austin: Texas Historical Commission, 1981), 6–7.

36. Tunnell, *Wax, Men, and Money*, 7; JoAnn Pospisil, "Chihuahuan Desert Candelilla: Folk Gathering of a Regional Resource," *Journal of Big Bend Studies* 6 (1994): 62–70; Casey, *Mirages, Mysteries, and Reality*, 184; O. C. Dowe to

Joseph Boyer and Harry Andrews, 3 January 1922, in File D, Dowe Collection, ABB.

37. Casey, *Mirages, Mysteries, and Reality*, 181; Thompson, *History of Marfa and Presidio County*, 2:304; Tunnell, *Wax, Men, and Money*, 7.

38. Casey, *Mirages, Mysteries, and Reality*, 182–83; Texas Beyond History Web Team, "Wax Camps," Texas beyond History (University of Texas at Austin), 20 April 2004, http://www.texasbeyondhistory.net/waxcamps/index .html; Manny Perez interview, 20 April 2013; Casey, *Mirages, Mysteries, and Reality*, 183.

39. "Wax Camps," Texas beyond History; Tunnell, *Wax, Men, and Money*, 6–7.

40. Adam Miller interview, 19 February 2013.

41. Powell, *Trees and Shrubs*, 51–53; Lujan, "Chinati (Chanate [*sic*]) Mountain Moonshine." The story that follows was witnessed and conveyed to me by Jerry Lujan, who as a young boy spent several months with Juan Morales over a couple of summers.

42. Curtis Tunnell and Enrique Madrid, "Making and Taking Sotol in Chihuahua and Texas," *Proceedings of the 3rd Symposium on Resources of the Chihuahuan Desert: US and Mexico* (Alpine, Tex.: Chihuahuan Desert Research Institute, 1990), 145–62; Mari Carmen Serra and Carlos A. Lazcano, "The Drink Mescal: Its Origin and Ritual Uses," in *Pre-Columbian Foodways: Interdisciplinary Approaches to Food, Culture, and Markets in Ancient Mesoamerica*, ed. John Staller and Michael Carrasco (New York: Springer, 2010), 137–56; K. Austin Kerr, "Prohibition," *Handbook of Texas Online*, accessed 21 March 2016, http://www.tshaonline.org/handbook/online/articles/vap01; Adam Miller interview, 19 February 2013.

43. Lujan, "Chinati (Chanate [*sic*]) Mountain Moonshine."

44. Ibid.

45. Tunnell and Madrid, "Making and Taking Sotol," 150.

46. Lujan, "Chinati (Chanate [*sic*]) Mountain Moonshine"; Tunnell and Madrid, "Making and Taking Sotol," 152.

47. Lujan, "Chinati (Chanate [*sic*]) Mountain Moonshine." *La migra* is slang for the US Border Patrol.

48. Tunnell and Madrid, "Making and Taking Sotol," 154–55; Texas beyond History Web Team, "Sotol," Texas beyond History, University of Texas at Austin, accessed 29 January 2016, http://www.texasbeyondhistory.net/ethnobot/images/sotol.html.

49. Tunnell and Madrid, "Making and Taking Sotol," 155–58.

50. Lujan, "Chinati (Chanate [*sic*]) Mountain Moonshine"; Lujan interview; Adam Miller interview, 19 February 2013.

51. Adam Miller interview, 19 February 2013.

52. USGS, Maps.

53. Oscar Perez interview, 18 February 2013; Nancy Perez interview, 18 February 2013; Oscar Perez telephone interview, 16 January 2013; Washing-

ton State Department of Agriculture, "Foot and Mouth Disease," June 2007, https://agr.wa.gov/FP/Pubs/docs/183-FMDBrochure.pdf.

54. Oscar Perez interview, 18 February 2013; Nancy Perez interview, 18 February 2013; Oscar Perez telephone interview, 16 January 2013.

55. Manny Perez interview, 20 April 2013.

56. Shely-Allen interview.

57. Ibid.

58. *Big Bend Sentinel*, 18 September 1958, 16 April 1959, 7 July 1960.

59. There are countless studies that amply demonstrate the combined effects of grazing and drought. According to one report, "Heavy grazing has been almost universally cited as a cause for reduction in grass cover. Droughts, characteristic of the Southwest, also played an important role in reduction of grass cover. . . . The drought of the 1950s remains the most severe and long-lasting and caused large reductions in grass cover in the Jornada Basin." R. P. Gibbens, R. P. McNeely, K .M. Havstad, R. F. Beck, and B. Nolen, "Vegetation Changes in the Jornada Basin from 1858 to 1998," *Journal of Arid Environments* 61, no. 4 (2005): 665. For recent studies on the effects of drought on soil microbes, see Fernando T. Maestre, Manuel Delgado-Baquerizo, Thomas C. Jeffries, David J. Eldridge, Victoria Ochoa, Beatriz Gozalo, José Luis Quero, Miguel García-Gómez, Antonio Gallardo, Werner Ulrich, Matthew A. Bowker, Tulio Arredondo, Claudia Barraza-Zepeda, Donaldo Bran, Adriana Florentino, Juan Gaitán, Julio R. Gutiérrez, Elisabeth Huber-Sannwald, Mohammad Jankju, Rebecca L. Mau, Maria Miriti, Kamal Naseri, Abelardo Ospina, Ilan Stavi, Deli Wang, Natasha N. Woods, Xia Yuan, Eli Zaady, and Brajesh K. Singh, "Increasing Aridity Reduces Soil Microbial Diversity and Abundance in Global Drylands," *Proceedings of the National Academy of Sciences* 112, no. 51 (2015): 15684–89, accessed 20 September 2017, http://www.pnas.org/content/112/51/15684; and J. M. Vose, J. S. Clark, C. H. Luce, and T. Patel-Weynand, eds., *Effects of Drought on Forests and Rangelands in the United States: A Comprehensive Science Synthesis*, General Technical Report WO-93b (Washington, DC: US Department of Agriculture, 2016), 167–68.

60. An animal unit is simply a standardized unit used to calculate grazing impacts of different animals. A cow with a calf six months or younger equals 1.0; a sheep, .17; and a goat, .15. See Natural Resources Conservation Service (NRCS), Animal Unit Equivalent Guide, Montana Table 6–5, accessed 20 September 2017, https://www.nrcs.usda.gov/Internet/FSE_DOCUMENTS/nrcs144p2_051957.pdf; US Department of Commerce, Census of Agriculture: 1950; US Department of Commerce, Census of Agriculture: 1959, vol. 1, part 37: Texas (Washington, DC: Government Printing Office, 1961).

61. Manny Perez interview, 20 April 2013.

62. Presidio County, Deed Records, vol. 150, 153.

63. Guadalupe Prieto telephone interview, 1 August 2015; Manny Perez interview, 21 April 2013; El Paso, Texas, City Directory, 1959, US City Direc-

tories, 1821–1989 [online database], accessed 19 February 2016 at Ancestry .com; *El Paso Times*, 25 July 2011.

Chapter 9. Tierra Perdida

1. This story conflates several anecdotes about Fred's barnstorming flights. Adam Miller interview, 19 February 2013.

2. WRCC, General Climate Summary Tables (Candelaria, Tex.).

3. Stults et al., "Fibers"; Olmstead and Rhode, "Cotton, Cottonseed, Shorn Wool, and Tobacco"; Carlson, *Woollybacks*, 210; Canada, "Wool and Mohair Price Support."

4. Casey, *Mirages, Mysteries, and Reality*, 267–68; Louise S. O'Conner and Cecilia Thompson, *Marfa and Presidio County, Texas: A Social, Economic, and Cultural Study, 1937 to 2008*, vol. 1, *1937–1989* (Marfa, Tex.: Xlibris, 2014), 251, 312; Fred Shely, interview by May Quick's eighth-grade class in Pinto Canyon, November 1978 [student film], Marfa Public Library, Marfa, Tex.

5. Jonathan G. Price, Christopher D. Henry, and Allan R. Standen, *Annotated Bibliography of Mineral Deposits in Trans-Pecos Texas*, Mineral Resource Circular No. 73 (Austin, Tex.: Bureau of Economic Geology, 1983), 37–38, 41, 46; Don G. Bilbrey, "Economic Geology of Rim Rock Country, Presidio County, Trans-Pecos, Texas" (MS thesis, University of Texas, 1957), 37–44.

6. Amsbury, Pinto Canyon Area map; *Midland Reporter-Telegram*, 15 April 1960; Presidio County, Deed Records, vol. 196, 52, vol. 200, 85.

7. Amsbury, Pinto Canyon Area map; Wikipedia contributors, "Perlite," Wikipedia, accessed 11 April 2016, https://en.wikipedia.org/w/index. php?title=Perlite&oldid=703068988; Incon Corporation, "Early History," Perlite.info, 2011, accessed 11 April 2016, http://www.perlite.info/hbk/ 0031443.html; *Big Bend Sentinel*, 7 December 1961, 19 July 1962; Shely-Allen interview.

8. *Big Bend Sentinel*, 14 October 1965; Shely-Allen interview.

9. Presidio County, Deed Records, vol. 156, 471, vol. 170, 77; Vizcaino interview.

10. Manny Perez interview by Matthaei, 20 April 2013.

11. *Big Bend Sentinel*, 20 July 1961; Certificate of Death for José Barrera Prieto [Jr.], filed 11 August 1961, State File No. 39024, in Texas Death Certificates, 1903–82 [online database], digital image accessed 13 April 2016 at Ancestry.com.

12. There are discrepancies regarding José Prieto's age at death. Of six records, four indicate he was born in 1866 and two indicate 1868 (including his headstone). Complicating the matter, his daughter Frances claimed he was 104 at his death, although no records support that date (1859). The date accepted here, as the one most referenced, is 1866. Certificate of Death for José Prieto, filed 4 March 1963, State File No. 8623, in Texas Death Certificates, 1903–82 [online database], digital image accessed 13 April 2016 at Ancestry.

com; *Big Bend Sentinel*, 28 February 1963; Wikipedia contributors, "Athero-sclerosis," Wikipedia, accessed March 14, 2016, https://en.wikipedia.org/w/index.php?title=Atherosclerosis&oldid=709897206; Epourstory410, "Our Cousin Armando," posted 25 August 2014, epourstory [Wordpress website], https://epourstory.wordpress.com/2014/08/25/our-cousin-armando/.

13. Oscar Perez, interview by author, El Paso, Tex, 20 February 2013.

14. *Big Bend Sentinel*, 28 February 1963; Wikipedia contributors, "Catholic Funeral," Wikipedia, accessed 15 April 2016, https://en.wikipedia.org/w/index.php?title=Roman_Catholic_funeral&oldid=713224710.

15. The historic standard of living equivalent of $140,000 would today amount to more than a million dollars. See Samuel H. Williamson, "Seven Ways to Compute the Relative Value of a US Dollar Amount, 1774 to present," MeasuringWorth, 2017, accessed 26 July 2017, www.measuringworth.com/uscompare/; Oscar Perez interview, 20 February 2013; Estate of Rejino [sic] Nuñez, Deceased, No. 318, p. 440, Probate Court Records, January term AD 1933, Presidio County Clerk's Office.

16. Ruben Perez telephone interview.

17. Oscar Perez interview, 18 February 2013; *Big Bend Sentinel*, 17 May 1962, 3 September 1964, 23 June 1966, 6 October 1966.

18. *Big Bend Sentinel*, 6 October 1966, 3 November 1966; Certificate of Death for Guadalupe P. Miller, filed 9 January 1967, File No. 79990, in Texas Death Certificates, 1903–82 [online database], digital image accessed 13 April 2016 at Ancestry.com.

19. Shely-Allen interview; Wikipedia contributors, "Piper PA-18 Super Cub," Wikipedia, accessed January 27, 2016, https://en.wikipedia.org/w/index.php?title=Piper_PA-18_Super_Cub&oldid=701596491; Richard Collins, ed., "Tailwheel Rancher," *Flying Magazine*, June 1979, 48–49.

20. Adam Miller interview, 19 February 2013.

21. Oscar Perez interview, 18 February 2013; Manny Perez interview by Matthaei, 20 April 2013.

22. Benavidez-Taulbee interview, 19 February 2013; Rosie Flores and Peggy McCracken, "Judge's Work Praised, Even as Rulings Caused Controversy," *Pecos Enterprise*, 17 January 2001, http://www.pecos.net/news/arch2001/011701p.htm; *Odessa American*, 18 January 2001; 1940 US Census, Presidio County, Tex., population schedule, Justice Precinct 1, Enumeration District 189–1c, p. 1b, dwelling 508, Lucius Desha Bunton III, National Archives microfilm publication T627, roll 4125, digital image accessed 22 April 2016 at Ancestry.com.

23. Presidio County, Deed Records, vol. 109, 426, vol. 150, 153; Presidio County Abstracts of Title, D&P Railroad Block 2, Survey Section 51, Big Bend Title Company; Manny Perez interview, 19 April 2013; Oscar Perez interviews, 19 April 2013, 18 February 2013.

24. Benavidez-Taulbee interview, 19 February 2013.

25. Texas Obituary and Death Notice Archive, "Frank Owen III," Gen-

Lookups, 11 September 2012, p. 1132, http://www.genlookups.com/tx/webbbs_config.pl/read/1132; Petra Benavidez et al. v. Gregorio Prieto et al., District Court of Presidio County, Texas Eighty-Third Judicial District, Cause No. 4570, Plaintiff's First Original Amended Petition, 5, and Defendants First Amended Answer, 4, in District Clerk's Office, Presidio County Courthouse.

26. Benavidez v. Prieto, Defendant's Second Amended Answer, 6; Manny Perez interview, 20 April 2013; Travis County Law Library, "Steps in the Texas Civil Litigation Process," November 2008, http://www.collincountytx.gov/law_library/documents/tx_civil_litigation_steps.pdf.

27. Flake Tompkins, "Chinati Peak Ranch, Presidio County Texas," real estate listing by Flake Tompkins, Realtor, Midland, Tex., 1970, from the personal files of Ida Benavidez; WRCC, General Climate Summary Tables (Candelaria, Tex.).

28. Newspapers.com, date and subject search for [Tompkins and Chinati /1969–1971], 21 April 2016, www.newspapers.com; Odessa American, 10 August 1971.

29. Presidio County, Deed Records, vol. 183, 458, vol. 187, 376.

30. Presidio County, Deed Records, vol. 183, 458, vol. 187, 376; Dallas Morning News, 27 July 2008; Oscar Perez telephone interview, 16 January 2013.

31. Manny Perez interview, 20 April 2013.

32. Oscar Perez interview, 18 February 2013; Big Bend Sentinel, 21 March 1963, 19 September 1985.

33. Oscar Perez interview, 18 February 2013; Big Bend Sentinel, 20 December 1979, 19 September 1985; Certificate of Death for Juan Morales (d. 13 December 1979), in Texas Death Index, 1903–2000, Texas Department of Health database accessed 20 April 2016 at Ancestry.com; Certificate of Death for Sotera Morales (d. 15 September 1985), in Texas Death Index, 1903–2000, Texas Department of Health database accessed 20 April 2016 at Ancestry.com.

34. Benavidez-Taulbee interview, 19 February 2013. Gregorio did not extend his newfound humility to everyone; neither Victoriano's daughter Lupe nor Bessie Miller (Adam Miller and Guadalupe Prieto's daughter) recalled any similar apologies, but they did vividly recall numerous incidents of his profane insults. Guadalupe Prieto and Bessie Miller interview.

35. El Paso County, El Paso County Death Records, 1956–2009 [online database], El Paso County, Tex., Clerk, accessed 28 November 2016, http://www.epcounty.com/publicrecords/deathrecords/DeathRecordSearch.aspx.

Chapter 10. No Ranch for Old Men

1. Don Cadden, Tied Hard and Fast: Apache Adams, Big Bend Cowboy (Denver, Colo.: Outskirts Press, 2011), 91–92; Don Cadden, email correspondence with author, 3 May 2016. This scene is based on one from Cadden's book.

Most of the stories Apache likes to tell end with horse wrecks that left him in the hospital. This story is somewhat more representative of his day-to-day activities. In the interest of clarity, this version was simplified by omitting mention of his helper, whom he usually had with him and always needed.

2. Precipitation during this period ranged from 14 to 20 percent above average across the region. For a seventeen-year stretch, only one year, 1977, was notably dry (about 65 percent of average) and only five dipped below average at all. WRCC, General Climate Summary Tables (Candelaria, Tex.); Wikipedia contributors, "Hurricane Celia," Wikipedia, accessed 25 April 2016, https://en.wikipedia.org/w/index.php?title=Hurricane_Celia&oldid=687172432; *Big Bend Sentinel*, 28 September 1978, 6 August 1970; NRCS, "Climate Narrative for Presidio County, Texas," in *Soil Survey of Presidio County, Texas* (NRCS/USDA, 2013), accessed 26 April 2016, http://www.nrcs.usda.gov/Internet/FSE_MANUSCRIPTS/texas/presidioTX2013/presidioTX2013.pdf.

3. Whereas in 1962 synthetic fibers accounted for only 15 percent of total fibers used in the United States, a decade later they accounted for nearly half. Stults et al., "Fibers"; Olmstead and Rhode, "Cotton, Cottonseed, Shorn Wool, and Tobacco"; USDA, Economic Research Service, *Wool and Mohair: Background for 1985 Farm Legislation*, Agriculture Information Bulletin No. 466 (Washington, DC: USDA, 1984), 15–19; *Big Bend Sentinel*, 20 April 1972; Herbert Koshetz, "Wool Industry Is Losing Its Battle in Fiber Market," *Milwaukee Journal*, 29 August 1972.

4. NASS, Historic Annual Average Beef Prices.

5. Shely-Allen interview; *Big Bend Sentinel*, 8 August 1970; Certificate of Death, Belle Shely, filed 7 January 1970, Presidio County Clerk's Office.

6. Shely-Allen interview; *Big Bend Sentinel*, 23 August 1973; Certificate of Death, Fred Clark Shely Jr., filed 14 September 1973, Presidio County Clerk's Office.

7. *Big Bend Sentinel*, 23 August 1973; Shely-Allen interview.

8. Presidio County, Deed Records, vol. 203, 130; Shely-Allen interview.

9. Shely-Allen interview; Fred Shely interview by May Quick's eighth-grade class.

10. *Alpine Avalanche*, 8 March 2011; *Big Bend Sentinel*, 6 May 1982, 11 April 1963; Shely-Allen interview.

11. Presidio County, Deed Records, vol. 240, 774; Presidio County, Minutes of the District Court, vol. 20, 605.

12. *Big Bend Sentinel*, 5 April 1984; *Alpine Avalanche*, 23 March 2010, 8 March 2011.

13. Fred L. Koestler, "Bracero Program," *Handbook of Texas Online*, modified on 5 October 2015, accessed 26 April 2016, http://www.tshaonline.org/handbook/online/articles/omb01; Kelly Lytle Hernández, "The Crimes and Consequences of Illegal Immigration: A Cross-Border Examination of Operation Wetback, 1943 to 1954," *Western Historical Quarterly* 37, no. 4 (2006):

421–44; Fred L. Koestler, "Operation Wetback," *Handbook of Texas Online*, modified 25 March 2016, accessed 26 April 2016, http://www.tshaonline .org/handbook/online/articles/pq001.

14. Kelly Lytle Hernández, *Migra! A History of the US Border Patrol* (Berkeley: University of California Press, 2010), 171–200; William Odencrantz, Steven T. Nutter, and Josie M. Gonzalez, "Immigration Reform and Control Act of 1986: Obligations of Employers and Unions," *Berkeley Journal of Employment and Labor Law* 10, no. 1 (1988): 92–115, accessed 29 April 2016, http://scholarship.law.berkeley.edu/bjell/v0110/iss1/7.

15. Dale A. Wade, Donald W. Hawthorne, Gary L. Nunley, and Milton Caroline, "History and Status of Predator Control in Texas," *Proceedings of the Eleventh Vertebrate Pest Conference*, paper 42 (1984), 123–24, accessed 11 March 2016, http://digitalcommons.unl.edu/vpc11/42; Odencrantz, Nutter, and Gonzalez, "Immigration Reform and Control Act."

16. Birth Certificate for A. J. Rod, 6 September 1918, in Texas Birth Certificates, 1903–32, Register No. 498173, digital image accessed 2 March 2016 at Ancestry.com; Wikipedia contributors, "Warner & Swasey Company," Wikipedia, accessed 3 March 2016, https://en.wikipedia.org/w/index.php?title =Warner_%26_Swasey_Company&oldid=676497587; A. J. Rod Company, "About Us," A. J. Rod Company website, accessed 2 March 2016, http:// www.ajrodco.com/about-us/.

17. *Bastrop County Times*, 13 November 1975; Apache Adams, telephone interview by author, 4 December 2015; Robert John Rod, telephone interview by author, 14 September 2016; Jason Sullivan, interview by author, Alpine, Tex., 29 August 2013.

18. *Big Bend Sentinel*, 12 May 1994, 18 July 1996; Presidio County, Minutes of the District Court (Eighty-Third Judicial District), vol. 20, 567–76, 595–620, Presidio County Courthouse.

19. Fort interview, 10 May 2013; Shely-Allen interview; *Bastrop County Times*, 3 March 1977.

20. Jeff Fort, interview by author, Alpine, Texas, 10 February 2012. The critical tone is strictly my own.

21. Judd Foundation, "Donald Judd Biography," Judd Foundation.org, accessed 30 April 2016, https://juddfoundation.org/artist/biography/.

22. Donald Judd, interview by May Quick's eighth-grade class [student film], 1978, Marfa Public Library, Marfa, Tex.; Lee Bennett, "Marfa, TX," *Handbook of Texas Online*, accessed 29 November 2016, http://www.tsha online.org/handbook/online/articles/hjm04; Alvarez, *Texas Almanac 2006–2007*, 377. The population of Marfa in 1970 was 2,682; it has continued to decline, despite substantial economic renewal.

23. Judd Foundation, "Donald Judd Biography"; Andrea Walsh, email correspondence with author, 2 August 2017.

24. Aaron Shkuda, *The Lofts of SoHo: Gentrification, Art, and Industry in New York, 1950–1980* (Chicago: University of Chicago Press, 2016), 83–85; Julie

Finch, telephone interview by author, 20 June 2016; Rainer Judd, telephone interview by author, 11 June 2016.

25. Donald Judd, "Marfa, Texas 1985," in Flavin Judd and Caitlin Murray, eds., *Donald Judd Writings* (New York: Judd Foundation and David Zwirner Books, 2016), 425–32; Judd Foundation, "Marfa, Texas," Judd Foundation .org, accessed 30 April 2016, http://www.juddfoundation.org/marfa; Michael Ennis, "The Marfa Art War," *Texas Monthly*, August 1984, 138–42, 186–92; Anne Goodwin Sides, "Donald Judd Found Perfect Canvas in Texas Town," *Weekend Edition/NPR*, 31 January 2009, http://www.npr.org/2009/01/31/ 99130809/donald-judd-found-perfect-canvas-in-texas-town.

26. Jones, et al., "Mammals of the Chinati Mountains State Natural Area, Texas," Presidio County, Deed Records, vol. 223, 64.

27. "Judd Ranch Properties," Judd Foundation Archives, Judd Foundation, Marfa, Tex.; Flavin Judd, interview by author, Marfa, Tex., 19 May 2016. The surname Ayala most likely derives from the town of the same name in the Basque Country of northern Spain. It comes from the Basque words *ai* (slope) and *al(h)a* (pasture). See Wikipedia contributors, "Ayala (surname)," Wikipedia, accessed 3 August 2017, https://en.wikipedia.org/w/index.php? title=Ayala_(surname)&oldid=774386030; Wikipedia contributors, "Ayala-Aiara," Wikipedia, accessed 3 August 2017, http://fr.wikipedia.org/w/index .php?title=Ayala-Aiara&oldid=137259096; Patrick Hanks, ed., "Ayala," in *Dictionary of American Family Names* (New York: Oxford University Press, 2003; online version, 2006), accessed 3 August 2017, http://www.oxfordreference .com/view/10.1093/acref/9780195081374.001.0001/acref-9780195081374-e-2224?rskey=80kG72&result=2201.

28. Judd Foundation, "Marfa, Texas"; Rainer Judd interview.

29. *Big Bend Sentinel*, 17 February 1994; Ennis, "Marfa Art War"; Chris Felver, dir., *Donald Judd: Marfa, Texas* [documentary] (MVD Entertainment Group, 2011).

30. "La Junta de los Rios" [folder, dated January 1989, containing information related to the idea of forming a corporation under that name to market and distribute regionally grown food and merchandise], Judd Foundation Archives; *Big Bend Sentinel*, 12 January 1989, 24 October 1991, 21 November 1991, 18 July 1996.

31. Jamie Dearing, telephone interview by author, 16 September 2016.

32. Ibid.

33. Ibid.; Donald Judd, *Untitled, 1971* [artwork], Glass House, accessed 18 July 2017, http://theglasshouse.org/learn/donald-judd-untitled-1971/. Although Judd's 1971 concrete piece at the Philip Johnson Glass House predated his time in Marfa, the idea appears to have originated with the land in Baja. Judd wrote, "In 1970, in relation to a slight slope to the arroyo, I worked out a large piece for the land of Joseph Pulitzer in St. Louis. It's a rectangle of two concentric walls of stainless steel, the outer one level and the inner one parallel to the slope of the land." Judd, "Marfa, Texas 1985," in Judd and Murray, *Donald Judd Writings*, 426.

34. Dearing interview; Donald Judd, "Ayala De Chinati 1989," in Judd and Murray, *Donald Judd Writings*, 593.

35. Sterry Butcher, "Judd's Ranches: Sanctuaries for the Artist's Life and Work," *Big Bend Sentinel*, 6 October 2011; Anthony Haden-Guest, *True Colors: The Real Life of the Art World* (New York: Atlantic Monthly Press, 1996), 265; Donald Judd, "Art and Architecture 1983," in Judd and Murray, *Donald Judd Writings*, 347; Presidio County, Deed Records, vol. 298, 68; Clementine Bales, interview by Mattie Matthaei, Marfa, Tex., 12 June 2013.

36. Haden-Guest, *True Colors*, 267–68; Laura Wilson, "Donald Judd in Marfa," *Chinati Foundation Newsletter* 21 (October 2016): 12; Flavin Judd interview; Donald C. Judd, Certificate of Death, Presidio County, Probate Records, vol. 32, 344; Donald C. Judd, Last Will and Testament, 10 December 1991, Presidio County, Probate Records, vol. 32, 314.

37. Benavidez-Taulbee interview, 19 February 2013; Benny Benavidez interview.

38. Teresa Benavidez interview.

39. Benny Benavidez interview.

40. Teresa Benavidez interview.

41. Roberto Suro, "Drug Traffickers Are Reopening Old Routes in Texas Badlands," *New York Times*, 7 February 1992; Jack D. McNamera, Plaintiff, v. United States Department of Justice, Defendant, P-96-CA-050, United States District Court for the Western District of Texas, Pecos Division, 974 F. Supp. 946; 1997 US Dist. LEXIS 12059 August 12, 1997, Decided, August, 1997, Opinion Filed.

42. Benavidez-Taulbee interview, 19 February 2013; Certificate of Death, Juan Benavidez, filed 5 August 1985, in Presidio County, Register of Deaths, vol. 6, 67, Presidio County Clerk's Office.

43. Bales interview; *Bastrop County Times*, 3 March 1977.

44. Rod telephone interview; Adams telephone interview.

45. Cadden, *Tied Hard and Fast*, 4; Adams telephone interview.

46. Adams telephone interview; Cadden, *Tied Hard and Fast*, 86.

47. Cadden, *Tied Hard and Fast*, 87; Adams telephone interview. Defying the Posse Comitatus Act, which forbade the military to enforce domestic policy within the borders of the United States, Marines were sent to the US-Mexican border early in the late 1980s to conduct surveillance in support of the Border Patrol as part of a broader effort to counteract drug-smuggling operations. Although their mission was solely to observe, on 20 May 1997 an American high school student named Esequiel Hernandez Jr. was killed by Marines in ghillie suits in the hamlet of Redford, Texas, just downstream from Presidio, leading to a temporary end to military patrols in the border area. See Robert Draper, "Soldiers of Misfortune," *Texas Monthly*, August 1997, http://www.texasmonthly.com/articles/soldiers-of-misfortune-2/; and Sasha von Oldershausen, "The Tragic Story of a Texas Teen and the Marines Who Killed Him for No Reason," Fusion, accessed 19 July 2017,

http://fusion.kinja.com/the-tragic-story-of-a-texas-teen-and-the-marines-who-ki-1795441102.

48. Adams telephone interview.

49. Ibid.

50. Cadden, *Tied Hard and Fast*, 87; Adams telephone interview.

51. US Public Records Index, 1950–93, vol. 1, accessed 30 November 2016 at Ancestry.com; Landon School, "John F. 'Jeff' Fort III '59 Business Leader," Landon, accessed 2 March 2015, http://www.landon.net/page .cfm?p=4145; Tyco International Ltd., "John Fort Assumes Primary Executive Responsibility for Tyco," press release, 3 June 2002, Investors.tyco .com, http://investors.tyco.com/phoenix.zhtml?c=112348&p=irol-news Article&ID=301812 (no longer available); *Big Bend Sentinel*, 2 May 2013.

52. Fort interview, 10 May 2013; Ruben Martinez, *Desert America: A Journey through Our Most Divided Landscape* (New York: Henry Holt, 2012), 275–92; Wikipedia contributors, "Mount Washington (New Hampshire)," Wikipedia, accessed 22 September 2017, https://en.wikipedia.org/w/index.php?title =Mount_Washington_(New_Hampshire)&oldid=799042716; Wikipedia contributors, "Denali," Wikipedia, accessed 22 September 2017, https://en .wikipedia.org/w/index.php?title=Denali&oldid=800368323; Wikipedia contributors, "Aconcagua," Wikipedia, accessed 22 September 2017, https:// en.wikipedia.org/w/index.php?title=Aconcagua&oldid=801736089.

53. Martinez, *Desert America*, 278; Mark Maremont and Laurie P. Cohen, "Tyco Spent Millions for Benefit of Kozlowski, Its Former CEO," *Wall Street Journal*, 7 August 2002, http://www.wsj.com/articles/SB102867480871784 5320; Wikipedia contributors, "Dennis Kozlowski," Wikipedia, accessed 25 May 2016, https://en.wikipedia.org/w/index.php?title=Dennis_Kozlowski &oldid=710367587.

54. Fort interview, 10 May 2013.

55. Ibid; Presidio County, Deed Records, vol. 300, 232.

56. Apache Adams, interview by author, Pinto Canyon, 5 November 2009.

57. Adams interview, 5 November 2009; Fort interview, 10 May 2013.

58. WRCC, General Climate Summary Tables (Candelaria, Tex.); Adams telephone interview, 4 December 2015.

59. Joe Nick Patoski, "What Would Donald Judd Do?," *Texas Monthly*, July 2001, http://www.joenickp.com/texas/donaldjudd.html; Flavin Judd interview.

60. Flavin Judd interview; "Post 1994 Ranch Property, Legal: Morales, Casas, Shannon, and Vizcaino," Real Estate Files, Judd Foundation Archives.

61. Fort interview, 10 May 2013.

62. Presidio County, Deed Records, vol. 315, 706–12.

63. Donald Judd, Landownership Maps, Ranches Folder, Judd Foundation Archives.

64. Texas Parks and Wildlife Department, "Chinati Mountains State Natural Area Public Use Plan," Draft 4, June 2004; *New York Times*, 18 February

1996; *Big Bend Sentinel*, 1 February 1996, 10 April 2014; Texas Parks and Wildlife, "Chinati Mountains State Natural Area," accessed 15 August 2017, https://tpwd.texas.gov/state-parks/chinati-mountains; *Odessa American*, 14 March 2017.

65. Shelley Gilbert-Allison, "Reclusive Artist Hopes Colonists Stay Away," *San Angelo Standard-Times*, 2 March 1981; Diane Jennings, "Donald Judd: What's This Artist-Anarchist-Philosopher Doing in Marfa, Texas?," *Dallas Morning News*, 13 January 1991.

66. Sullivan interview.

67. Fort interview, 10 May 2013.

68. Seebach, *Late Prehistory along the Rimrock*; Cason, *Archaeology on the Pinto Canyon Ranch*.

Chapter 11. Only the Land Remains

1. Gary Rogers, Prieto Ranch Appraisal Report, 1980, manuscript in possession of author.

2. Kate Murphy, "Let's Go Wild: No. 6 Pinto Canyon Chinati Mountains," *Texas Monthly*, October 2014, http://www.texasmonthly.com/list/lets-go-wild/no-6-pinto-canyon/; West Texas Council of Governments, "Presidio County Land Use Program" (1972); *Big Bend Sentinel*, 30 June 1994.

Bibliography

Primary Sources

Archives and Collections

Dowe Collection. Archives of the Big Bend, Bryan Wildenthal Memorial Library, Sul Ross State University, Alpine, Tex.

Elam, Earl H. Collection. Archives of the Big Bend, Bryan Wildenthal Memorial Library, Sul Ross State University, Alpine, Tex.

Fort Davis National Historic Site Archives. Fort Davis, Tex.

Judd Foundation Archives. Marfa, Tex.

Junior Historian Collection. Marfa Public Library, Marfa, Tex.

National Archives. College Park, Md.

National Archives and Records Administration. Washington, DC.

Texas State Library and Archives Commission. Austin, Tex.

Tharp, Col. Blucher S. Collection. Texas Military Museum, Camp Mabry, Austin, Tex.

Vizcaino, Jesse. Files. Natural Resources Conservation Service, Marfa, Tex.

Unpublished Material

Hale, Walter Fleetwood. "Rustler's Ambuscade." Folklore Project, Life Histories 1936–39. Federal Writers' Project. US Work Projects Administration, Manuscript Division. Library of Congress, 1941. Accessed 10 January 2013. http://www.loc.gov/item/wpalh002443/.

Keller, David W. "Archeological Investigations at the Porvenir Massacre Site, Presidio County, Texas." 2017. Report on file at the Center for Big Bend Studies, Sul Ross State University, Alpine, Tex.

——. "Archeological Site Recording of Casa Morales." 23 May 2009. Report on file at the Center for Big Bend Studies, Sul Ross State University, Alpine, Tex.

——. "Archeological Site Recording of O'Dell Homestead." 19 November 2008. Report on file at the Center for Big Bend Studies, Sul Ross State University, Alpine, Tex.

⸻. "Archeological Site Recording of Prieto Homestead." 20 October 2010. Report on file at the Center for Big Bend Studies, Sul Ross State University, Alpine, Tex.

⸻. "Archeological Site Recording of Wilson Homestead." 18 November 2008. Report on file at the Center for Big Bend Studies, Sul Ross State University, Alpine, Tex.

Kilpatrick, J. J., Jr. "Nuñez Ranch Raid and the Killing of Bandits on Telephone Wires." Circular No. 3. Unpublished manuscript. Briscoe Center for American History, the University of Texas at Austin.

⸻. "Value of the Evidence Collected by the Fall Senate Committee: An Examination of the Exaggerated and Fabricated Testimony of Some of the Witnesses." Unpublished manuscript. Briscoe Center for American History, theUniversity of Texas at Austin.

Lujan, Jerry Javier. "Chinati (Chanate [sic]) Mountain Moonshine." Unpublished manuscript in possession of author, updated 19 September 2013.

"Public Member Trees" [database]. Family trees and individual profiles compiled by David Keller. Undocumented data updated 2015–16 and accessed at Ancestry.com.

O'Dell Family Tree. Unpublished document in possession of author.

Rogers, Gary. Prieto Ranch Appraisal Report. 1980. Unpublished manuscript in possession of author.

Russell, William E., V. "William Edward Russell: Pioneer of the Big Bend." 2003. Unpublished manuscript in possession of author.

Smith, Thomas T. "Chronological Matrix, Army Unit Locations, Big Bend, Texas 1911–1921." 30 September 2015. Unpublished manuscript in possession of author.

⸻. "The Old Army in the Big Bend, 1911–1921." 5 October 2015. Unpublished manuscript in possession of author.

Sutherlin-Gibson, Nancy A. "Alexander Sutherland 1620." December 2005. Unpublished family genealogy in possession of author.

Tompkins, Flake. "Chinati Peak Ranch, Presidio County Texas." Real estate listing, Midland, Tex., 1970. Unpublished manuscript in possession of author.

"Walter Fleetwood Hale Jr." W. F. Hale Jr. File. Texas Ranger Hall of Fame and Museum. Waco, Tex. Copy of unpublished manuscript in possession of author.

Weatherby, Willis Harper. "Chinati (Conquistadores)." December 1989. Unpublished manuscript in possession of author.

Interviews and Correspondence

Adams, Apache. Interview by author, Pinto Canyon, 5 November 2009.

⸻. Telephone interview by author, 4 December 2015.

Bales, Clementine. Interview by Mattie Matthaei, Marfa, Tex., 12 June 2013.

Benavidez, Benny. Interview by Mattie Matthaei, Chinati Hot Springs, 11 September 2013.

Benavidez, Teresa. Interview by author and Mattie Matthaei, Chinati Hot Springs, 15 June 2013.

Benavidez-Taulbee, Ida. Interview by author and Mattie Matthaei, El Paso, Tex., 19 February 2013.

———. Telephone interview by author, 16 April 2016.

Cadden, Don. Email correspondence with author, 3 May 2016.

Cano, Tony, and Ann Sochat. Telephone interview by author, 18 November 2015.

Cason, Samuel S. Email correspondence with author, 15 October 2016.

Dearing, Jamie. Telephone interview by author, 16 September 2016.

Elms, Monroe. Email correspondence with author, 22 July 2016.

Finch, Julie. Telephone interview by author, 20 June 2016.

Fort, Jeff. Interview by author, Alpine, Tex., 10 February 2012.

———. Interview by author, Pinto Canyon Ranch, 10 May 2013.

Gray, Ted. Interview by author, Alpine, Tex., 9 October 2008.

Henry, Christopher D. Telephone interview by author, 26 February 2015.

Judd, Donald. Interview by May Quick's eighth-grade class, Marfa, Tex. November 1978 [student film]. Marfa Public Library, Marfa, Tex.

Judd, Flavin. Interview by author, Marfa, Tex., 19 May 2016.

Judd, Rainer. Telephone interview by author, 11 June 2016.

Justice, Glenn. Telephone interview by author, 1 February 2016.

Karges, John (Texas Nature Conservancy associate director of field science). Email correspondence with author, 12 August 2013.

Logan, Daniel. Telephone interview by author, 4 March 2016.

Love, Chip. Interview by Mattie Matthaei, Marfa, Tex., 14 July 2014.

———. Telephone interview by author, 30 August 2017.

Lujan, Jerry Javier. Interview by author, Pinto Canyon Ranch, 19 October 2012.

Miller, José Adam Prieto (Adam Miller). Interview by author, El Paso, Tex., 19 February 2013.

———. Telephone interview by author, 16 November 2012.

———. Telephone interview by author, 15 February 2016.

Newby, John. Email correspondence with author, 23 September 2014.

———. Email correspondence with author, 24 September 2014.

———. Email correspondence with author, 26 September 2014.

O'Dell, Frank. Written correspondence to author, undated (ca. 2011).

Perez, Manny. Interview by author, Pinto Canyon Ranch, 19 April 2013.

———. Interview by author, Pinto Canyon Ranch, 20 April 2013.

———. Interview by author, Pinto Canyon Ranch, 21 April 2013.

———. Interview by Mattie Matthaei, Pinto Canyon Ranch, 20 April 2013.

Perez, Oscar. Interview by author, El Paso, Tex., 18 February 2013.

———. Interview by author, El Paso, Tex., 20 February 2013.

———. Interview by author, Pinto Canyon Ranch, 19 April 2013.

———. Interview by author, Chinati Hot Springs, 21 April 2013.

———. Telephone interview by author, 16 January 2013.

————. Telephone interview by author, 6 February 2013.

————. Telephone interview by author, 1 October 2013.

————. Telephone interview by author, 7 August 2015.

Perez, Nancy. Interview by author, El Paso, Tex., 18 February 2013.

Perez, Ruben. Telephone interview by author, 28 July 2015.

Powell, A. Michael. Email correspondence with author, 2 April 2015.

Prieto, Chon. Interview by author, Ruidosa, Tex., 20 November 2010.

————. Interview by Mattie Matthaei, Ruidosa, Tex., 13 July 2014.

Prieto, Guadalupe. Telephone interview by Oscar Perez, 1 August 2015.

Prieto, Guadalupe, and Bessie Miller. Interview by author, Fort Stockton, Tex., 22 May 2016.

Prieto, José "Joe" Manuel. Email correspondence with author, 3 October–1 November 2013.

————. Telephone interview by author, 1 November 2013.

Rod, Robert John. Telephone interview by author, 14 September 2016.

Shely, Fred. Interview by May Quick's eighth-grade class in Pinto Canyon. November 1978 [student film]. Marfa Public Library, Marfa, Tex.

Shely-Allen, Terry Jean. Interview by author, Alpine, Tex., 24 September 2010.

Smith, Oliver, Jr. Telephone interview by author, 4 September 2013.

Stratton, Janell. Telephone interview by author, 14 July 2017.

Sullivan, Jason. Interview by author, Alpine, Tex., 29 August 2013.

Sutherlin, Frederick Dale. Telephone interview by author, 25 September 2014.

Sutherlin, Jackie. Telephone interview by author, 6 May 2011.

Sutherlin, Martin. Telephone interview by author, 16 October 2013.

Sutherlin, Michael. Email correspondence with author, 5 September 2012.

Sutherlin-Gibson, Nancy. Telephone interview by author, 3 October 2014.

Vizcaino, Maria Flora. Interview by author, Marfa, Tex., 28 August 2013.

Walsh, Andrea. Email correspondence with author, 2 August 2017.

Watts, Ginny. Interview by author, Alpine, Tex., 24 July 2009.

————. Telephone interview by author, 3 June 2011.

Maps and Government Documents

Amsbury, David L. Pinto Canyon Area, Presidio County, Texas. Geologic Quadrangle Map No. 22. Austin: Bureau of Economic Geology, University of Texas, 1958.

Border Crossings: From Mexico to US, 1895–1957. Microfilm roll 2, Records of the Immigration and Naturalization Service, Record Group 85, National Archives and Records Administration, Washington, DC. Digital image accessed 7 January 2016 at Ancestry.com.

Bureau of Labor Statistics, US Department of Labor. Consumer Price Index Inflation Calculator. Accessed 25 August 2015. http://data.bls.gov/cgi-bin/cpicalc.pl.

Chihuahua, Mexico, Civil Registration of Births, 1861–1947. Ojinaga, 1918–1947 Nacimientos. Digital images accessed 18 February 2016 at Ancestry.com.

Canada, Carol. "Wool and Mohair Price Support." 30 October 2008. Congressional Research Service, Library of Congress. Prepared for Members and Committees of Congress. Accessed 10 February 2016. nationalaglawcen ter.org/wp-content/uploads/assets/crs/RS20896.pdf.

Cook, Edward, David Meko, David Stahle, and Malcolm Cleaveland. "Reconstruction of Past Drought across the Coterminous United States from a Network of Climatically Sensitive Tree-Ring Data." Climatic Data Center, Instrument Animation, 25 January 1999. Accessed 23 February 2016. https://www.ncdc.noaa.gov/paleo/pdsiyear.html (no longer available).

Court of Civil Appeals of Texas, Austin. Weatherby v. Guerrero, Cause No. 8078. Accessed 14 August 2015. https://casetext.com/case/weather by-v-guerrero.

Economic Research Service, USDA. *Wool and Mohair: Background for 1985 Farm Legislation*. Agriculture Information Bulletin No. 466. Washington, DC: USDA, September 1984.

El Paso County. El Paso County Death Records, 1956–2009 [database]. El Paso County, Tex., Clerk. Accessed 28 November 2016. http://www .epcounty.com/publicrecords/deathrecords/DeathRecordSearch.aspx.

El Paso, Texas, City Directory, 1917 and 1959. US City Directories, 1821–1989 [online database]. Accessed 19 February 2016 at Ancestry.com.

El Quinto Censo General de Población y Vivienda 1930, Dirección General de Estadística, México [Mexico Census, 1930]. Rancho de El Alamos Fosesigua, Victoriano Prieto. Digital image, accessed 4 January 2016 at Ancestry.com FamilySearch, compiler. Salt Lake City, Utah: Family Search, 2009.

Emory, William H. Report on the United States and Mexican Boundary Survey. House Ex. Doc. no. 135, 34th Cong., 1st sess. Washington, DC: Cornelius Wendell, 1857.

Environmental Protection Agency. "Sun Safety." EPA. Accessed 6 March 2015. http://www2.epa.gov/sunwise/uv-index.

Heitmuller, Franklin T., and Brian D. Reece. Database of Historically Documented Springs and Spring Flow Measurements in Texas. US Geological Survey Open-File Report 03–315, Geographical Information Systems (GIS) database, 2003. Accessed 12 May 2014. https://databasin.org/data sets/2400de0b78284e0fa44083e78824ff24.

Hohn, Charles M. "ABC's of Making Adobe Bricks." Guide G-521. New Mexico State University, Cooperative Extension Service, 2003.

Instituto Nacional de Estadística, Geografía, e Informática (INEGI). Carta Topográfica, México Norte. Escala 1:250,000. Serie I de Imágenes Cartográficas Digitales. Mexico City: INEGI, 2000.

———. Mapa General del Estado Chihuahua. Escala 1:1,192,180. Mexico City: INEGI, n.d.

Investigation of Mexican Affairs: Hearing before a Subcommittee of the Committee on Foreign Relations United States Senate, Part 10. Senate Ex. Doc. no. 285.

66th Cong., 2nd sess., Serial 7665. Washington, DC: Government Printing Office, 1920.

Keller, David W. Geographical Information Systems (GIS) Map of Prieto Ranch from deed record data, 2015.

Larkin, Thomas J., and George W. Bomar. *Climatic Atlas of Texas*. Texas Department of Water Resources Report LP-192. Austin: Texas Department of Water Resources, 1983.

Library of Congress. "Treaty of Guadalupe Hidalgo." Primary Documents in American History, Virtual Services Digital Reference Section. Accessed 7 June 2016. https://www.loc.gov/rr/program/bib/ourdocs/Guadalupe.html.

Public member treesName Index to Correspondence of the Military Intelligence Division of the War Department Staff, 1917–41. Microfilm M 1194, roll 39. RG 165, Records of the War Department General and Special Staffs, 1860–1952. National Archives, College Park, Md.

National Agricultural Statistics Service (NASS). Historic Annual Average Beef Prices, 1909–2001 [spreadsheet]. USDA, Washington, DC.

National Centers for Environmental Information, NOAA. Historical Palmer Drought Indices. Accessed 23 September 2015. http://www.ncdc.noaa.gov/temp-and-precip/drought/historical-palmers/psi/193001–201507.

National Institutes of Health. "Leading Causes of Blindness." Medicine Plus. Accessed 21 March 2016. https://www.nlm.nih.gov/medlineplus/magazine/issues/summer08/articles/summer08pg14–15.html.

National Oceanic and Atmospheric Administration (NOAA). "Beaufort Wind Scale." NOAA. Accessed 5 March 2015. http://www.spc.noaa.gov/faq/tornado/beaufort.html.

Natural Resources Conservation Service (NRCS), USDA. Animal Unit Equivalent Guide, Montana Table 6–5. Accessed 20 September 2017. https://www.nrcs.usda.gov/Internet/FSE_DOCUMENTS/nrcs144p2_051957.pdf.

———. "Aridisols Map." NRCS. Accessed 20 February 2015. http://www.nrcs.usda.gov/wps/portal/nrcs/detail/soils/survey/class/maps/?cid=nrcs142p2_053595.

———. "Climate Narrative for Presidio County, Texas." In *Soil Survey of Presidio County, Texas* (NRCS/USDA 2013). Accessed 26 April 2016. http://www.nrcs.usda.gov/Internet/FSE_MANUSCRIPTS/texas/presidioTX2013/presidioTX2013.pdf.

———. "Custom Soil Resource Report for Presidio County, Texas." Web Soil Survey. Accessed 20 February 2015. http://websoilsurvey.sc.egov.usda.gov/App/HomePage.htm.

———. "Dominant Soil Orders in the United States." 1998. Accessed 20 February 2015. https://www.nrcs.usda.gov/Internet/FSE_MEDIA/stelprdb1237749.pdf.

————. *Soil Survey of Big Bend National Park, Texas.* Washington, DC: Government Printing Office, 2011.

National Weather Service, NOAA. El Paso Monthly Precipitation Totals, El Paso, Tex. Accessed 2 July 2018. https://www.weather.gov/epz/elpaso _monthly_precip.

Nebraska Agricultural Statistics Service, USDA. Lamb Price Dollars per CWT (hundredweight). Accessed 4 February 2016. http://www .nass.usda.gov/Statistics_by_State/Nebraska/NE_Historical_Data/ nebhist/historic%20prices%20-%201ambs.pdf (no longer available).

Osterkamp, W. R. *Annotated Definitions of Selected Geomorphic Terms and Related Terms of Hydrology, Sedimentology, Soil Science and Ecology.* Open-File Report 2008–1217. Reston, Va.: USGS, 2008.

Presidio County. Book of Petitions for Naturalization. Presidio County District Court Office. Presidio County Courthouse, Marfa, Tex.

————. Deed Records. Presidio County Clerk's Office. Presidio County Courthouse, Marfa, Tex.

————. Index to District Court Minutes. Presidio County District Court. Presidio County Courthouse, Marfa, Tex.

————. Marks and Brands. Book 5. Presidio County Clerk's Office. Presidio County Courthouse, Marfa, Tex.

————. Marriage Records. Presidio County Clerk's Office. Presidio County Courthouse, Marfa, Tex.

————. Minutes of the County Court. Presidio County Courthouse, Marfa, Tex.

————. Minutes of the District Court. Presidio County Courthouse, Marfa, Tex.

————. Minutes of the Presidio County Commissioners Court. Presidio County Clerk's Office. Presidio County Courthouse, Marfa, Tex.

————. Probate Records. Presidio County Clerk's Office. Presidio County Courthouse, Marfa, Tex.

————. Records of the Certificates of Occupancy. Presidio County Clerk's Office. Presidio County Courthouse, Marfa, Tex.

————. Register of Deaths. Presidio County Clerk's Office. Presidio County Courthouse, Marfa, Tex.

————. Tax Rolls, 1890–1920. Tax Office, Presidio County Courthouse, Marfa, Tex.

————. Tax Rolls. Unrendered property, 1934. Tax Office, Presidio County Courthouse, Marfa, Tex.

Presidio County Abstracts of Title. Big Bend Title Company, Marfa, Tex.

Presidio County Sheriff's Office. Sheriff J. D. Bunton, Sworn Testimony, 8 June 1934. Examining trial before W. G. Young, Justice of the Peace, Precinct No. 1, Presidio County, Tex. Sheriff's Office Files. Presidio County Sheriff's Office, Marfa, Tex.

Stults, Harold, Edward H. Glade Jr., Scott Sanford, John V. Lawler, and Robert Skinner. "Fibers: Background for 1990 Farm Legislation." Agriculture Information Bulletin No. AIB-591, March 1990. https://www .ers.usda.gov/publications/pub-details/?pubid=42014.

Texas Birth Certificates, 1903–32. Register No. 498173. Digital image accessed 2 March 2016 at Ancestry.com.

Texas Birth Index, 1903–97 [database]. Texas Department of Health, Bureau of Vital Statistics. Image accessed 29 February 2016 at Ancestry.com.

Texas Death Certificates, 1903–1982 [online database]. Digital images of individual certificates accessed at Ancestry.com.

Texas Death Index, 1903–2000. Texas Department of Health database. Accessed 20 April 2016 at Ancestry.com.

Texas General Land Office. Land Grant Search. Accessed 28 April 2015. http://www.glo.texas.gov/history/archives/land-grants/index.cfm.

Texas Historical Commission. Texas Historical Sites Atlas [restricted database]. Accessed 4 November 2016. https://atlas.thc.state.tx.us/Account.

Texas Obituary and Death Notice Archive. "Frank Owen III." GenLookups, 11 September 2012, p. 1132. http://www.genlookups.com/tx/webbbs_ config.pl/read/1132.

Texas Parks and Wildlife Department (TPWD). "Desert Bighorn Sheep." Wildlife Management in West Texas, TPWD. Accessed 2 June 2016. http://tpwd.texas.gov/landwater/land/habitats/trans_pecos/big_ game/desertbighornsheep/.

Texas, Select County Marriage Index, 1837–1977. Database accessed 29 February 2016 at Ancestry.com 2014.

Travis County Law Library. "Steps in the Texas Civil Litigation Process." November 2008. https://www.collincountytx.gov/Pages/default.aspx

US Bureau of the Census. Federal census, various years. Microfilm from the National Archives, Washington, DC. Accessed via online databases provided by Ancestry.com.

US Public Records Index, 1950–93, Volume 1. Accessed 30 November 2016 at Ancestry.com.

US, School Yearbooks, 1880–2012. Digital image and database accessed 4 March 2016 at Ancestry.com.

US Social Security Applications and Claims Index, 1936–2007 [online database]. Accessed at Ancestry.com.

US World War I Draft Registration Cards, 1917–18. Digital images accessed at Ancestry.com.

Urbanczyk, Kevin, David Rohr, and John C. White. "Geologic History of West Texas." In *Aquifers of West Texas*, edited by Robert E. Mace, William F. Mullican III, and Edward S. Angle. Texas Water Development Board Report No. 356. Austin: Texas Water Development Board, 2001.

US Army. "Fifth Cavalry–Marfa, Texas." *Cavalry Journal* 30, no. 125 (October 1921): 446.

US Army Signal Corps. *Technical Equipment of the Signal Corps: 1916*. Washington, DC: Government Printing Office, 1917.

US Department of Commerce. Twelfth Census of the United States, 1900. Volume 5, Part 1: Agriculture. Washington, DC: Government Printing Office, 1902.

———. Thirteenth Census of the United States, 1910. Volume 7: Agriculture. Washington, DC: Government Printing Office, 1913.

———. Fourteenth Census of the United States, 1920. Volume 6: Agriculture. Washington, DC: Government Printing Office, 1922.

———. Census of Agriculture: 1925, part 2: Southern States. Washington, DC: Government Printing Office, 1927.

———. Fifteenth Census of the United States, 1930. Agriculture, Volumes 1 and 2. Washington, DC: Government Printing Office, 1931.

———. Census of Agriculture: 1935. Volume 1, part 2: Southern States. Washington, DC: Government Printing Office, 1936.

———. Sixteenth Census of the United States, 1940. Agriculture, Volume 1. Washington, DC: Government Printing Office, 1942.

———. Census of Agriculture: 1945. Volume 1, part 26: Texas. Washington, DC: Government Printing Office, 1945.

———. Census of Agriculture: 1950. Volume 1, part 26: Texas. Washington, DC: Government Printing Office, 1952.

———. Census of Agriculture: 1954. Volume 1, part 26: Texas. Washington, DC: Government Printing Office, 1956.

———. Census of Agriculture: 1959. Volume 1, part 37: Texas. Washington, DC: Government Printing Office, 1961.

———. Census of Agriculture: 1964. Volume 1, part 37: Texas. Washington, DC: Government Printing Office, 1966.

———. Census of Agriculture: 1969. Volume 1, part 37: Texas. Washington, DC: Government Printing Office, 1973.

———. Census of Agriculture: 1974. Volume 1, part 43: Texas. Washington, DC: Government Printing Office, 1977.

———. Census of Agriculture: 1978. Volume 1, part 43: Texas. Washington, DC: Government Printing Office, 1981.

———. Census of Agriculture: 1982. Volume 1, part 43: Texas. Washington, DC: Government Printing Office, 1984.

———. Census of Agriculture: 1987. Volume 1, part 43: Texas. Washington, DC: Government Printing Office, 1989.

———. Census of Agriculture: 1992. Volume 1, part 43a: Texas. Washington, DC: Government Printing Office, 1994.

———. Census of Agriculture: 1997. Volume 1, part 43a: Texas. Washington, DC: Government Printing Office, 1999.

———. Census of Agriculture: 2002. Volume 1, part 43a: Texas. Washington, DC: Government Printing Office, 2004.

US District Court, Western District of Texas, Pecos Division. McNamara v. US Department of Justice. No. P-96-CA-050. 949 F.Supp. 478. (December 6, 1996). Accessed 18 July 2017. https://www.leagle.com/decision/19961427949fsupp47811351.

US Geological Survey. Chinati Peak [map]. 1:24,000. 7.5 Minute Series. Reston, Va.: United States Department of the Interior, USGS, 1979.

———. Cuesta del Burro West [map]. 1:24,000. 7.5 Minute Series. Reston, Va.: US Department of the Interior, USGS, 1983.

———. Las Conchas [map]. 1:24,000. 7.5 Minute Series. Reston, Va.: United States Department of the Interior, USGS, 1980.

———. Pueblo Nuevo [map]. 1:24,000. 7.5 Minute Series. Reston, Va.: US Department of the Interior, USGS, 1979.

———. Ruidosa Hot Springs [map]. 1:24,000. 7.5 Minute Series. Reston, Va.: US Department of the Interior, USGS, 1979.

———. Sierra Parda [map]. 1:24,000. 7.5 Minute Series. Reston, Va.: US Department of the Interior, USGS, 1978.

Vose, J. M., J. S. Clark, C. H. Luce, and T. Patel-Weynand, eds. *Effects of Drought on Forests and Rangelands in the United States: A Comprehensive Science Synthesis*. General Technical Report WO-93b. Washington, DC: US Department of Agriculture, 2016.

Wade, Shirley C., William R. Hutchison, Ali H. Chowdhury, and Doug Coker. *A Conceptual Model of Groundwater Flow in the Presidio and Redford Bolsons Aquifers*. Texas Water Development Board Report. Austin: Texas Water Development Board, 2011.

War Department. "Report of Chief of Construction Division to the Secretary of War." Annual Reports 1 (1920). Washington, DC: Government Printing Office, 1921.

———. "Report of the Adjutant General of the Army to the Secretary of War, October 5, 1921." Annual Reports, Fiscal Year Ended 30 June 1921. Washington, DC: Government Printing Office, 1921.

Weather Bureau, USDA. Cooperative Observer's Meteorological Record, 02 Ranch, Tex., February 1914–September 1928.

Western Regional Climate Center (WRCC). General Climate Summary Tables (Candelaria, Tex. [411416]). Summary of the Day (SOD), US Climate Archive. Accessed 21 May 2015. http://www.wrcc.dri.edu/cgi-bin/cliMAIN.pl?tx1416.

———. General Climate Summary Tables (Presidio, Tex. [417262]). Summary of the Day (SOD), US Climate Archive. Accessed 21 May 2015. http://www.wrcc.dri.edu/cgi-bin/cliMAIN.pl?tx7262.

———. General Climate Summary Tables [Valentine 10 WSW, Texas (419275)]. Summary of the Day (SOD) US Climate Archive. Accessed 21 May 2015. http://www.wrcc.dri.edu/cgi-bin/cliMAIN.pl?tx9275.

———. "Period of Record Monthly Climate Summaries." Accessed 4 March 2015. http://www.wrcc.dri.edu/cgi-bin/cliMAIN/ (no longer available).

Books

Baines, William M. *State School Land Agents to State Land Board*. Austin, Tex.: State Printing Office, 1885.

Gillet, James B. *Six Years with the Texas Rangers, 1875–1881*. Lincoln: University of Nebraska Press, 1963.

Hammond, George P., and Agapito Rey. *Expedition into New Mexico Made by Antonio de Espejo, 1582–1583, As revealed in the Journal of Diego Pérez de Luxán, a Member of the Party*. Los Angeles: Quivira Society, 1929.

———. *The Rediscovery of New Mexico, 1580–1594: The Exploration of Chamuscado, Espejo, Castaña de Sosa, Morlete, y Leyva de Bonilla y Humaña*. Albuquerque: University of New Mexico Press, 1966.

Madrid, Enrique Rede, trans. *Expedition to La Junta de los Ríos, 1747–1748: Captain Commander Joseph de Ydoiaga's Report to the Viceroy of New Spain*. Office of the State Archeologist, Special Report 33. Austin: Texas Historical Commission, 1992.

Núñez Cabeza de Vaca, Álvar. *The Narrative of Cabeza de Vaca*. Edited by Rolena Adorno and Patrick Charles Pautz. Lincoln: University of Nebraska Press, 2003.

O'Conor, Hugo. *The Defenses of Northern New Spain: Hugo O'Conor's Report to Teodoro de Croix, July 22, 1777*. Edited and translated by Donald C. Cutter. Dallas: DeGolyer Library and Southern Methodist University Press, 1994.

Sayles, Vernon. *Vernon Sayles' Annotated Civil Statutes of the State of Texas, with Historical Notes, Embracing the Revised Statutes of the State of Texas Adopted at the Regular Session of the Thirty-Second Legislature, 1911 [. . .]*. Kansas City, Mo.: Vernon Law Book Company, 1914.

Whiting, William Henry Chase. *Journal of William Henry Chase Whiting, 1849*. In *Exploring Southwestern Trails, 1846–1854*, edited by Ralph P. Bieber. Philadelphia: Porcupine Press, 1974.

Secondary Sources

Published or Recorded Resources, Academic Reportss, and Presentations

Adams, David K., and Andrew C. Comrie. "The North American Monsoon." *Bulletin of the American Meteorological Society* 78, no. 10 (1997): 2197–2213.

Alexander, Bob. *Riding Lucifer's Line: Ranger Deaths along the Texas-Mexico Border*. Denton: University of North Texas Press, 2013.

Alvarez, Elizabeth Cruce, ed. *Texas Almanac 2006–2007: Sesquicentennial Edition 1857–2007*. Dallas: Dallas Morning News, 2005.

Amsbury, David L. "Geology of Pinto Canyon Area, Presidio County, Texas." PhD diss., University of Texas, 1957.

Bailey, C. P. *Practical Angora Goat Raising*. San Jose, Calif.: C. P. Bailey and Sons, 1905.

Baker, Charles Laurence. "Exploratory Geology of a Part of Southwestern Trans-Pecos, Texas." *University of Texas Bulletin No. 2745*. Austin: Bureau of Economic Geology, University of Texas, 1927.

Balderrama, Francisco E., and Raymond Rodriguez. *Decade of Betrayal: Mexican Repatriation in the 1930s*. Albuquerque: University of New Mexico Press, 1995.

Bedichek, Roy. *Adventures with a Texas Naturalist*. Austin: University of Texas Press, 1947.

Bilbrey, Don G. "Economic Geology of Rim Rock Country, Presidio County, Trans-Pecos, Texas." MS thesis, University of Texas, 1957.

Blair, W. Frank. "The Biotic Provinces of Texas." *Texas Journal of Science* 2 (March 1950): 93–117.

Blair, W. Frank, and Clay E. Miller, Jr. "The Mammals of the Sierra Vieja Region, Southwestern Texas, with Remarks on the Biogeographic Position of the Region." *Texas Journal of Science* 1, no. 1 (1949): 67–92.

Bridges, Luther Wadsworth II. "Revised Cenozoic History of Rim Rock Country, Trans-Pecos Texas." MA thesis, University of Texas, 1958.

Brown, David E., ed. *Biotic Communities: Southwestern United States and Northwestern Mexico*. Salt Lake City: University of Utah Press, 1994.

Buongiorno, Ben. "Handbook of the Tierra Vieja Mountains, Presidio and Jeff Davis Counties, Trans-Pecos Texas." MA thesis, University of Texas, 1955.

Butcher, Sterry. "Judd's Ranches: Sanctuaries for the Artist's Life and Work." *Big Bend Sentinel*, 6 October 2011.

Cadden, Don. *Tied Hard and Fast: Apache Adams, Big Bend Cowboy*. Denver, Colo.: Outskirts Press, 2011.

Cano, Tony, and Ann Sochat. *Bandido: The True Story of Chico Cano, the Last Western Bandit*. Canutillo, Tex.: Reata Publishing, 1997.

Carlson, Paul H. *Texas Woollybacks: The Range Sheep and Goat Industry*. College Station: Texas A&M University Press, 1982.

Casey, Clifford B. *Mirages, Mysteries, and Reality: Brewster County, Texas, the Big Bend of the Rio Grande*. Seagraves, Tex.: Pioneer Book Publishers, 1972.

———. "The Trans-Pecos in Texas History." *West Texas Historical and Scientific Society Publication no. 5* (1933): 7–18.

Cason, Samuel S. "Archaeology on the Pinto Canyon Ranch: Field Guide to Prehistoric Occupation in the Sierra Vieja Breaks, Presidio County, Texas." In-house document, Center for Big Bend Studies, Sul Ross State University, 2014.

Castañeda, Carlos E. *Our Catholic Heritage in Texas, 1519–1936*. Volume 3, *The Mission Era: The Missions at Work, 1731–1761*. New York: Arno Press, 1976.

Cather, Steven M., Nelia W. Dunbar, Fred W. McDowell, William C. McIntosh, and Peter A. Scholle. "Climate Forcing by Iron Fertilization from

Repeated Ignimbrite Eruptions: The Icehouse–Silicic Large Igneous Province (SLIP) Hypothesis." *Geosphere* 5, no. 3 (2009): 315–24.

Cepeda, Joseph C. "The Chinati Mountains Caldera, Presidio County, Texas." In *Cenozoic Geology of the Trans-Pecos Volcanic Field of Texas*, edited by Anthony W. Walton and Christopher D. Henry, 106–25. Bureau of Economic Geology Guidebook 19. Austin: University of Texas, 1979.

Clendenen, Clarence C. *Blood on the Border: The United States Army and Mexican Irregulars*. New York: Macmillan, 1969.

Cloud, William A. "An Update on the Genevieve Lykes Duncan Site." *La Vista de la Frontera* (Center for Big Bend Studies, Sul Ross State University), 23 (Fall 2012): 1–2, 14.

Collins, Richard, ed. "Tailwheel Rancher." *Flying Magazine*, June 1979, 48–49.

Compton, Elmer. *Philip Sutherlin*. Sarasota, Fla.: privately published, 2002; updated 2009.

Cox, Mike. "Ex-Detective Tracks Gun's History." *Austin American-Statesman*, 17 June 1983.

Dasch, Julius E., Richard L. Armstrong, and Stephen E. Clabaugh. "Age of Rim Rock Dike Swarm, Trans-Pecos, Texas." *Geological Society of America Bulletin* 80 (September 1969): 1819–24.

Davenport, B. T. "The Watch along the Rio Grande." *Journal of Big Bend Studies* 7 (1995): 149–56.

De Ford, Ronald K. "Some Keys to the Geology of Northern Chihuahua." In *Guidebook of the Border Region*, edited by Diego A. Cordoba, Sherman A. Wengerd, and John Shomaker, 61–65. Socorro: New Mexico Geological Society, 1969.

———. "Tertiary Formations of Rim Rock Country, Presidio County, Trans-Pecos Texas." *Texas Journal of Science* 20, no. 1 (1958): 1–37.

Dickerson, Eddie Joe. "Bolson Fill, Pediment, and Terrace Deposits of Hot Springs Area, Presidio County, Trans-Pecos Texas." MS thesis, University of Texas, 1966.

Dickerson, Patricia Wood. "Structural Zones Transecting the Southern Rio Grande Rift—Preliminary Observations." *New Mexico Geological Society Fall Field Conference Guidebook – 31, Trans-Pecos Region*, edited by Patricia W. Dickerson, Jerry M. Hoffer, and J. F. Callender, 63–70. Socorro: New Mexico Geological Society, 1980.

Duex, Timothy W., and Christopher D. Henry. *Calderas and Mineralization: Volcanic Geology and Mineralization in the Chinati Caldera Complex, Trans-Pecos Texas*. Geological Circular 81–2, Bureau of Economic Geology. Austin: University of Texas at Austin, 1981.

Earney, Mary K. "Farther Than Nearer: Judges, Lawyers, and Cases of the 83rd State Judicial District of Texas, 1917–1983." *Big Bend Sentinel*, 29 December 1994.

Easterling, Stuart. *The Mexican Revolution: A Short History, 1910–1920*. Chicago: Haymarket Books, 2012.

Edwards, Charles W., Jr. *Up to My Armpits: Adventures of a West Texas Veteri- narian*. Houston: Iron Mountain Press, 2002.

Ennis, Michael. "The Marfa Art War." *Texas Monthly*, August 1984, 138–42, 186–92.

Eoff, Jerry L. *Just Doing the Math: Juan Dominguez de Mendoza Trail to San Clemente, 1683–1684*. Alpine, Tex.: privately published, 2015.

Felver, Chris, dir. *Donald Judd: Marfa, Texas* [documentary]. MVD Entertain- ment Group, 2011.

Fenley, Florence. "Cowboy and Mounted Customs Inspector Went Back to Old Love . . . Ranching." *The Cattleman*, January 1968, 92–98.

Friedman, Milton, and Anna Jacobson Schwartz. *From New Deal Banking Reform to World War II Inflation*. Princeton, N.J.: Princeton University Press, 1980.

Frohlich, Cliff, and Scott D. Davis. *Texas Earthquakes*. Austin: University of Texas Press, 2002.

Gibbens, R. P., R. P. McNeely, K. M. Havstad, R. F. Beck, and B. Nolen. "Vegetation Changes in the Jornada Basin from 1858 to 1998." *Journal of Arid Environments* 61, no. 4 (June 2005): 651–68.

Gilbert-Allison, Shelley. "Reclusive Artist Hopes Colonists Stay Away." *San Angelo Standard-Times*, 2 March 1981.

Gillespie, James R., and Frank Flanders. *Modern Livestock and Poultry Produc- tion*. 8th ed. New York: Delmar Cengage Learning, 2010.

Green, Karen. "The Influenza Epidemic of 1918 in Far West Texas." *Journal of Big Bend Studies* 5 (1993): 125–36.

Gregg, John Ernest. "The History of Presidio County." MA thesis, Univer- sity of Texas, 1933.

Gunnerson, James H. "Southern Athapaskan Archeology." *Handbook of North American Indians*. Volume 9, *Southwest*, edited by William Sturtevant, 162–69. Washington, DC: Smithsonian Institution, 1979.

Haden-Guest, Anthony. *True Colors: The Real Life of the Art World*. New York: Atlantic Monthly Press, 1996.

Hammond, Jim N. "Big Bend Justice." *Cenizo Journal* 5, no. 4 (2013): 20–21. Online R

Harris, Charles H., Frances E. Harris, and Louis R. Sadler. *Texas Ranger Biog- raphies: Those Who Served, 1910–1921*. Albuquerque: University of New Mexico Press, 2009.

Harris, Charles H., III, and Louis R. Sadler. *The Texas Rangers and the Mexican Revolution: The Bloodiest Decade, 1910–1920*. Albuquerque: University of New Mexico Press, 2004.

Heffelfinger, Jim. "The Texas Black Bear." *Tracks Magazine*, July–August 2015, 121–27.

Henry, Christopher D., and Jonathon G. Price. "Early Basin and Range De- velopment in Trans-Pecos Texas and Adjacent Chihuahua: Magmatism and Orientation, Timing, and Style of Extension." *Journal of Geophysical Research* 91, no. 86 (1986): 6213–24.

Hernández, Kelly Lytle. "The Crimes and Consequences of Illegal Immigration: A Cross-Border Examination of Operation Wetback, 1943 to 1954." *Western Historical Quarterly* 37, no. 4 (winter 2006): 421–44.

———. *Migra! A History of the US Border Patrol*. Berkeley: University of California Press, 2010.

Hinckley, L. C. "Contrasts in the Vegetation of Sierra Tierra Vieja in Trans-Pecos Texas." *American Midland Naturalist* 37, no. 1 (1947): 162–78.

Holder, Ramona Vega. *Life on My Side of the Tracks*. Bloomington, Ind.: Authorhouse, 2006.

Holt, Roy D. "Pioneering Sheepmen." *Sheep and Goat Raiser* 26 (December 1945): 16–23.

Horgan, Paul. *Great River: The Rio Grande in North American History*. 4th ed. 2 vols. in one. Hanover, N.H.: University Press of New England, 1984.

Ivey, James. *Presidios of the Big Bend*. Southwest Cultural Resources Center Professional Papers No. 31. Santa Fe, N.Mex.: National Park Service, US Department of the Interior, 1990.

Jameson, David L., and Alvin G. Flury. "The Reptiles and Amphibians of the Sierra Vieja Ranch of Southwestern Texas." *Texas Journal of Science* 1, no. 2 (1949): 54–79.

Jennings, Diane. "Donald Judd: What's This Artist-Anarchist-Philosopher Doing in Marfa, Texas?" *Dallas Morning News*, 13 January 1991.

Jones, Clyde, Mark W. Lockwood, Tony R. Mollhagen, Franklin D. Yancey II, and Michael A. Bogan. *Mammals of the Chinati Mountains State Natural Area, Texas*. Occasional Papers, Museum of Texas Tech University, no. 300. Lubbock: Texas Tech University, 2011.

Judd, Flavin, and Caitlin Murray, eds. *Donald Judd Writings*. New York: Judd Foundation and David Zwirner Books, 2016.

Justice, Glenn. *Little Known History of the Texas Big Bend*. Odessa, Tex.: Rimrock Press, 2001.

———. *Revolution on the Rio Grande: Mexican Raids and Army Pursuits, 1916–1919*. El Paso: Texas Western Press, 1992.

Katz, Friedrich. *The Life and Times of Pancho Villa*. Stanford: Stanford University Press, 1998.

Keil, Robert. *Bosque Bonito: Violent Times along the Borderland during the Mexican Revolution*. Edited by Elizabeth McBride. Alpine, Tex.: Center for Big Bend Studies Sul Ross State University, 2002.

Keller, David W., William A. Cloud, Samuel S. Cason, Robert W. Gray, Richard W. Walter, Thomas C. Alex, Roger D. Boren, Andrea J. Ohl, and Robert J. Mallouf. *A Sampling of Archeological Resources in Big Bend National Park, Texas*. Alpine, Tex.: Center for Big Bend Studies, Sul Ross State University (forthcoming).

Kelley, J. Charles. "The Historic Indian Pueblos of La Junta de los Rios [part 1]." *New Mexico Historical Review* 27, no. 4 (1952): 257–95.

———. "The Historic Indian Pueblos of La Junta de los Rios [part 2]." *New Mexico Historical Review* 28, no. 1 (1953): 21–51.

———. *Jumano and Patarabueye: Relations at La Junta de los Rios*. Anthropological Papers No. 77, Museum of Anthropology. Ann Arbor: University of Michigan, 1986.

Kelton, Elmer. "Ranching in a Changing Land." *Texas Almanac 2006–2007: Sesquicentennial Edition 1857–2007*, 26–31. Dallas: Dallas Morning News, 2005.

———. *The Time It Never Rained*. New York: Doubleday, 1973.

Kerr, John Leeds. *Destination Topolobampo: The Kansas City, Mexico and Orient Railway*. San Marino, Calif.: Golden West Books, 1968.

Koshetz, Herbert. "Wool Industry Is Losing Its Battle in Fiber Market." *Milwaukee Journal*, 29 August 1972.

Leckie, William H., and Shirley A. *Unlikely Warriors: General Benjamin H. Grierson and His Family*. Norman: University of Oklahoma Press, 1998.

Le May, Alan, dir. *High Lonesome* [feature film]. 1950. Tulsa, Okla.: Vci Video, 2005. DVD.

Lind, Michael. *Land of Promise: An Economic History of the United States*. New York: Harper, 2013.

Lott, Emily J., and Mary L. Butterwick. "Notes on the Flora of the Chinati Mountains, Presidio County, Texas." *SIDA: Contributions to Botany* 8, no. 4 (1979): 348–51.

MacLeod, William. *Big Bend Vistas: A Geological Exploration of the Big Bend*. Alpine: Texas Geological Press, 2002.

Madison, Virginia Duncan, and Hallie Crawford Stillwell. *How Come It's Called That? Place Names in the Big Bend Country*. New York: October House, 1968.

Mallouf, Robert J. "The Dancing Rocks Petroglyphs: Horse Nomads of the Sierra Vieja Breaks of the Texas Big Bend." In *Archaeological Explorations of the Eastern Trans-Pecos and Big Bend: Collected Papers, Volume 1*, edited by Pat Dasch and Robert J. Mallouf, 235–60. Alpine, Tex.: Center for Big Bend Studies, 2013.

———. "A Synthesis of Eastern Trans-Pecos Prehistory." MA thesis, University of Texas at Austin, 1985.

Martinez, Ruben. *Desert America: A Journey through Our Most Divided Landscape*. New York: Henry Holt, 2012.

Miller, Marilyn, and Marian Faux, eds. *The New York Public Library American History Desk Reference*. New York: Stonesong Press and the New York Public Library, 1997.

Miller, Thomas Lloyd. *The Public Lands of Texas, 1519–1970*. Norman: University of Oklahoma Press, 1972.

Moran, Emilio F. *Human Adaptability: An Introduction to Ecological Anthropology*. 3rd ed. Boulder, Colo.: Westview Press, 2008.

Morgenthaler, Jefferson. *La Junta de Los Rios: The Life, Death, and Resurrection of an Ancient Desert Community in the Big Bend Region of Texas*. Boerne, Tex.: Mockingbird Books, 2007.

———. *The River Has Never Divided Us: A Border History of La Junta de Los Rios*. Austin: University of Texas Press, 2004.

Mraz, J. R., and G. R. Keller. *Structure of the Presidio Bolson Area, Texas, Interpreted from Gravity Data*. Geological Circular 80–13, Bureau of Economic Geology. Austin: University of Texas, 1980.

Muehlenberger, William R. "Texas Lineament Revisited." In *New Mexico Geological Society Fall Field Conference Guidebook – 31, Trans-Pecos Region*, edited by Patricia W. Dickerson, Jerry M. Hoffer, and J. F. Callender, 113–21. Socorro: New Mexico Geological Society, 1980.

Nabhan, Gary Paul. *Desert Terroir: Exploring the Unique Flavors and Sundry Places of the Borderlands*. Austin: University of Texas Press, 2012.

O'Conner, Louise S., and Cecilia Thompson. *Marfa and Presidio County, Texas: A Social, Economic, and Cultural Study, 1937 to 2008*. Volume 1, *1937–1989*. Marfa, Tex.: Xlibris, 2014.

Oppenheimer, Clive. "Climatic, Environmental and Human Consequences of the Largest Known Historic Eruption: Tambora Volcano (Indonesia) 1815." *Progress in Physical Geography* 27, no. 2 (2003): 230–59.

Orozco, Victor. *Las guerras indias en la historia de Chihuahua*. Juarez: Instituto Chihuahuense de la Cultura, Universidad Autónoma de Ciudad Juárez, 1992.

Phillips, Homer W., and Wilmot A. Thornton. "The Summer Resident Birds of the Sierra Vieja Range in Southwestern Texas." *Texas Journal of Science* 1, no. 4 (1949): 101–31.

Pospisil, JoAnn. "Chihuahuan Desert Candelilla: Folk Gathering of a Regional Resource." *Journal of Big Bend Studies* 6 (1994): 62–70.

Powell, A. Michael. *Grasses of the Trans-Pecos and Adjacent Areas*. Marathon, Tex.: Iron Mountain Press, 2000.

———. *Trees and Shrubs of the Trans-Pecos and Adjacent Areas*. Austin: University of Texas Press, 1998.

Price, Jonathan G., Christopher D. Henry, and Allan R. Standen. *Annotated Bibliography of Mineral Deposits in Trans-Pecos Texas*. Mineral Resource Circular No. 73. Austin, Tex.: Bureau of Economic Geology, 1983.

Raht, Carlysle Graham. *The Romance of Davis Mountains and Big Bend Country*. Edition Texana. Odessa, Tex.: Rahtbooks Company, 1919.

Rauch, Jonathan. "The Golden Fleece: Why Programs Never Die." *National Journal*, 18 May 1991, 1168–71.

Rodriguez, Oscar, Amber Rodriguez, and David Gohre. "The Chinaitih Mountains: The Case for a More Plausible Narrative for How the Chinati Mountains Got Their Name." *Journal of Big Bend Studies* 26 (2014): 27–42.

Romo, Ricardo. "Responses to Mexican Immigration, 1910–1930." In *Beyond 1848: Readings in the Modern Chicano Historical Experience*, edited by Michael R. Ornelas, 115–35. Dubuque, Iowa: Kendall/Hunt Publishing, 1993.

Russell, William. Letter to the *El Paso Herald*, 14 May 1874, reprinted in *Voice of the Mexican Border*, December 1933, 187.

Ryan, David P., Theodore S. Hong, and Nabeel Bardeesy. "Pancreatic Adenocarcinoma." *New England Journal of Medicine* 371 (2014): 1039–49.

Schmidly, David J. *Texas Natural History: A Century of Change.* Lubbock: Texas Tech University Press, 2002.

Schmidt, Robert H. "The Climate of Trans-Pecos, Texas." In *The Changing Climate of Texas: Predictability and Implications for the Future*, edited by Jim Norwine, John Giardino, Gerald North, and Juan Valdez, 122–37. College Station: Cartographics, Texas A&M University, 1995.

Scudday, James F. "Two Recent Records of Gray Wolves in West Texas." *Journal of Mammalogy* 53, no. 3 (1972): 598.

———. "The Vertebrate Fauna of Capote Canyon." In *Capote Falls: A Natural Area Survey*, edited by the Office of Research, Lyndon B. Johnson School of Public Affairs, 174–200. Austin: University of Texas, 1973.

Seebach, John D. *Late Prehistory along the Rimrock.* Papers of the Trans-Pecos Archeological Program No. 3. Alpine, Tex.: Center for Big Bend Studies, Sul Ross State University, 2011.

Seebach, John D., III. "El Despoblado: Folsom and Late Paleoindian Occupation of Trans-Pecos, Texas." PhD diss., Southern Methodist University, 2011.

Self, Stephen. "The Effects and Consequences of Very Large Explosive Volcanic Eruptions." *Philosophical Transactions of the Royal Society* 364, no. 1845 (2006): 2073–97.

Serra, Mari Carmen, and Carlos A. Lazcano. "The Drink Mescal: Its Origin and Ritual Uses." In *Pre-Columbian Foodways: Interdisciplinary Approaches to Food, Culture, and Markets in Ancient Mesoamerica*, edited by John Staller and Michael Carrasco, 137–56. New York: Springer, 2010.

Shipman, O. L., and Jack Shipman. "The Savage Saga: A Vivid Story of the Settlement and Development of Presidio County." *Voice of the Mexican Border* (1938): 6–15.

Shkuda, Aaron. *The Lofts of SoHo: Gentrification, Art, and Industry in New York, 1950–1980.* Chicago: University of Chicago Press, 2016.

Shurley, Mike. "The Dowes of Pinto Canyon." *Junior Historian* 28, no. 5 (March 1968): 12–24.

Smith, Ralph A. "Mexican and Anglo-Saxon Traffic in Scalps, Slaves, and Livestock, 1835–1841." *West Texas Historical Association Year Book* 36 (October 1960): 98–115.

———. "Poor Mexico, So Far from God and So Close to the Tejanos." *West Texas Historical Association Year Book* 44 (1986): 78–105.

Smith, Thomas T. *The Old Army in Texas: A Research Guide to the US Army in Nineteenth-Century Texas.* Austin: Texas State Historical Association, 2000.

Smithers, W. D. *Chronicles of the Big Bend: A Photographic Memoir of Life on the Border*. Austin, Tex.: Madrona Press, 1976.

Stegner, Wallace. *Where the Bluebird Sings to the Lemonade Springs: Living and Writing in the West*. New York: Penguin Books, 1992.

Suro, Roberto. "Drug Traffickers Are Reopening Old Routes in Texas Badlands." *New York Times*, 7 February 1992.

Texas Agricultural Extension Service. *Integrated Toxic Plant Management*. College Station: Texas A&M University System, 2011.

Texas Parks and Wildlife Department. "Chinati Mountains State Natural Area Public Use Plan." Draft 4, June 2004.

Thompson, Cecilia. *History of Marfa and Presidio County, Texas, 1535–1946*. 2 vols. Austin, Tex.: Nortex Press, 1985.

Traweek, Stella. *The Production and Marketing of Mohair in Texas*. Bureau of Business Research Monograph No. 12. Austin: University of Texas, 1949.

Tucker, Spencer. *The Encyclopedia of the Mexican-American War: A Political, Social, and Military History*. Santa Barbara, Calif.: ABC-CLIO, 2013.

Tunnell, Curtis. *Wax, Men, and Money: A Historical and Archeological Study of Candelilla Wax Camps along the Rio Grande Border of Texas*. Office of the State Archeologist Report 32. Austin: Texas Historical Commission, 1981.

Tunnell, Curtis, and Enrique Madrid. "Making and Taking Sotol in Chihuahua and Texas." *Proceedings of the 3rd Symposium on Resources of the Chihuahuan Desert: US and Mexico*, 145–62. Alpine, Tex.: Chihuahuan Desert Research Institute, 1990.

Tyler, Ronnie C. *The Big Bend: A History of the Last Texas Frontier*. Washington, DC: National Park Service, 1975.

Urbanczyk, Kevin. "Geologic History of Pinto Canyon Area." Public lecture, Casa Perez open house, Pinto Canyon, Tex., 6 September 2014.

Utley, Robert M. *Lone Star Lawmen*. New York: Penguin Group, 2007.

Vallentine, John. *Grazing Management*. New York: Academic Press, 1992.

Van Pelt, Arnold F. *An Annotated Inventory of the Insects of Big Bend National Park, Texas*. Big Bend National Park, Tex.: Big Bend Natural History Association, 1999.

Vernon, J. R. "The 1920–21 Deflation: The Role of Aggregate Supply." *Economic Inquiry* 29 no. 3 (1991): 572–80.

Wade, Maria F. *The Native Americans of the Texas Edwards Plateau, 1582–1799*. Austin: University of Texas Press, 2003.

Walton, Gary, and Hugh Rockoff. *History of the American Economy*. 12th ed. Mason, Ohio: South-Western, 2014.

Ward, James Randolph. "The Texas Rangers, 1919–1935: A Study in Law Enforcement." PhD diss., Texas Christian University, 1972.

Watson, Jeff. *The Golden Eagle*. London: T&AD Poyser, 2010.

West Texas Council of Governments. "Presidio County Land Use Program." 1972.

Williams, Oscar Waldo. *O. W. William's Stories from the Big Bend*. El Paso: Texas Western College Press, 1965.

Wilson, John Andrew. "Geochronology of the Trans-Pecos Texas Volcanic Field." *New Mexico Geological Society Guidebook – 31st Field Conference, Trans-Pecos Region*, edited by Patricia Wood Dickerson, Jerry M. Hoffer, and J. F. Callender, 205–11. Socorro: New Mexico Geological Society, 1980.

Wilson, Laura. "Donald Judd in Marfa." *Chinati Foundation Newsletter* 21 (October 2016): 12–23.

Woolford, Sam, ed. "Notes and Documents: The Burr G. Duval Diary." *Southwestern Historical Quarterly* 65 (July 1961): 487–511.

Wright, Paul. "Build It and They Will Come? Boom and Bust in Presidio." *Journal of Big Bend Studies* 20 (2008): 39–60.

Online Resources

A. J. Rod Company. "About Us." A. J. Rod Company website. Accessed 2 March 2016. http://www.ajrodco.com/about-us/.

American Merchant Marine at War. "US Merchant Marine Casualties during World War II." Accessed 1 March 2016. http://www.usmm.org/casualty.html.

Animal Health Publications. "Castrating Lambs and Kids." Infovets.com. Accessed 12 February 2016. http://www.infovets.com/books/smrm/c/c104.htm.

Bennett, Lee. "Marfa, TX." *Handbook of Texas Online*. Accessed 29 November 2016. http://www.tshaonline.org/handbook/online/articles/hjm04.

Berger, Max, and Lee Wilborn. "Education." *Handbook of Texas Online*. Accessed 26 October 2015. http://www.tshaonline.org/handbook/online/articles/khe01.

Brown, David. "1918 Flu Virus Limited the Immune System." *Washington Post*, 18 January 2007. http://www.washingtonpost.com/wp-dyn/content/article/2007/01/17/AR2007011701113.html.

Byrns, Steve. "Angora Goats: A 'Shear' Delight!" Texas Department of Agriculture and Texas Agricultural Experiment Station. November 2011. http://sanangelo.tamu.edu/files/2011/11/AngoraGoatsAShearDelight_1.pdf.

Calvert, Robert A. "Texas since World War II." *Handbook of Texas Online*. Accessed 10 February 2016. http://www.tshaonline.org/handbook/online/articles/npt02.

Centennial Museum, University of Texas at El Paso. "Desert Marigold (*Baileya multiradiata*)." Chihuahuan Desert Plants. September 2002. http://museum2.utep.edu/chih/gardens/plants/baileya.htm.

Crown X Safaris Ranch. "Crown X Safaris: A West Texas Game Ranch." Accessed 25 April 2016. http://www.cxsafaris.com/.

Derbes, Brett J. "Second Texas Cavalry." *Handbook of Texas Online*. Accessed 6 July 2016. http://www.tshaonline.org/handbook/online/articles/qks08.

Draper, Robert. "Soldiers of Misfortune." *Texas Monthly*, August 1997. http://www.texasmonthly.com/articles/soldiers-of-misfortune-2/.

"El Paso, TX [Union Depot]." The Great American Stations. Accessed 12 July 2016. http://www.greatamericanstations.com/Stations/ELP.

Encyclopædia Britannica Online. "Rheumatism." Accessed 8 April 2015. http://www.britannica.com/EBchecked/topic/501241/rheumatism.

Environmental Systems Research Institute (ESRI). Arcmap Version 10.1 [software]. Redlands, Calif., 1999–2012.

Epourstory410. "Our Cousin Armando." epourstory [Wordpress website], posted 25 August 2014. https://epourstory.wordpress.com/2014/08/25/our-cousin-armando/.

Espensen-Sturges, Jaime. "XERF." *Handbook of Texas Online*. Accessed 9 March 2016. http://www.tshaonline.org/handbook/online/articles/ebx01.

Flores, Rosie, and Peggy McCracken. "Judge's Work Praised, Even as Rulings Caused Controversy." *Pecos Enterprise*, 17 January 2001. http://www.pecos.net/news/arch2001/011701p.htm.

Hanks, Patrick, ed. "Ayala." In *Dictionary of American Family Names*. New York: Oxford University Press, 2003. Online version, 2006. Accessed 3 August 2017. http://www.oxfordreference.com/view/10.1093/acref/9780195081374.001.0001/acref-9780195081374-e-2224?rskey=80kG72&result=2201.

Hentz, Tucker F. "Geology." *Handbook of Texas Online*. Accessed 13 December 2016. http://www.tshaonline.org/handbook/online/articles/swgqz.

"History of Blackwell School." Blackwell School Alliance. Accessed 24 July 2015. https://www.theblackwellschool.org/history.

Incon Corporation. "Early History." Perlite.info, 2011. Accessed 11 April 2016. http://www.perlite.info/hbk/0031443.html.

Judd Foundation. "Donald Judd Biography." Judd Foundation.org. Accessed 30 April 2016. https://juddfoundation.org/artist/biography/.

———. "Marfa, Texas." Judd Foundation.org. Accessed 30 April 2016. http://www.juddfoundation.org/marfa.

Kerr, K. Austin. "Prohibition." *Handbook of Texas Online*. Accessed 21 March 2016. http://www.tshaonline.org/handbook/online/articles/vap01.

Koestler, Fred L. "Bracero Program." *Handbook of Texas Online*. Accessed 26 April 2016. http://www.tshaonline.org/handbook/online/articles/omb01.

———. "Operation Wetback." *Handbook of Texas Online*. Accessed 26 April 2016. http://www.tshaonline.org/handbook/online/articles/pq001.

Landon School. "John F. 'Jeff' Fort III '59 Business Leader." Landon School website. Accessed 2 March 2015. http://www.landon.net/page.cfm?p=4145.

Madigan, Charles M. "A Libel Case with a 6-Cent Verdict." *Chicago Tribune*, 8 June 1997. http://articles.chicagotribune.com/1997–06–08/news/9706300080_1_henry-ford-chicago-tribune-anarchist.

Maestre, Fernando T., Manuel Delgado-Baquerizo, Thomas C. Jeffries, David J. Eldridge, Victoria Ochoa, Beatriz Gozalo, José Luis Quero, Miguel García-Gómez, Antonio Gallardo, Werner Ulrich, et al. "Increasing Aridity Reduces Soil Microbial Diversity and Abundance in Global Drylands." *Proceedings of the National Academy of Sciences* 112 no. 51 (2015): 15684–89. http://www.pnas.org/content/112/51/15684.

Maremont, Mark, and Laurie P. Cohen. "Tyco Spent Millions for Benefit of Kozlowski, Its Former CEO." *Wall Street Journal*, 7 August 2002. http://www.wsj.com/articles/SB1028674808717845320.

Maupin, Janice. "Sheep and Goat Ranching—Texas Style." Oral Business History Project. Graduate School of Business, University of Texas, 1972. Accessed 12 February 2016. http://www.thebhc.org/sit Maestre, Fernando T., Manuel Delgado-Baquerizo, Thomas C. Jeffries, David J. Eldridge, Victoria Ochoa, Beatriz Gozalo, José Luis Quero, Miguel García-Gómez, Antonio Gallardo, Werner Ulrich, et al. "Increasing Aridity Reduces Soil Microbial Diversity and Abundance in Global Drylands." *Proceedings of the National Academy of Sciences* 112 no. 51 (2015): 15684–89. http://www.pnas.org/content/112/51/15684.

Maestre, Fernando T., Manuel Delgado-Baquerizo, Thomas C. Jeffries, David J. Eldridge, Victoria Ochoa, Beatriz Gozalo, José Luis Quero, Miguel García-Gómez, Antonio Gallardo, Werner Ulrich, et al. "Increasing Aridity Reduces Soil Microbial Diversity and Abundance in Global Drylands." *Proceedings of the National Academy of Sciences* 112 no. 51 (2015): 15684–89. http://www.pnas.org/content/112/51/15684.

McCroskey, Mona Lange. "The Most Efficient Fiber Producers on Earth: Angora Goat Ranching in Yavapai County, Arizona, 1880–1945." Prescott Corral of Westerners International. Accessed 9 February 2016. http://www.prescottcorral.org/TT3/V2AngoraGoatRanching.htm.

"Moon Phases 1934." Calendar 12.com. Accessed 18 February 2013. http://www.calendar-12.com/moon_phases/1934.

Murphy, Kate. "Let's Go Wild: No. 6 Pinto Canyon, Chinati Mountains." *Texas Monthly*, October 2014. http://www.texasmonthly.com/list/lets-go-wild/no-6-pinto-canyon/.

Murphy, Philip G. "The Drought of 1934: A Report of the Federal Government's Assistance to Agriculture." Report presented to the President's Drought Committee, 15 July 1935. https://fraser.stlouisfed.org/docs/publications/books/drought_1934_aaa.pdf.

Newspapers.com. Date and subject search [Tompkins and Chinati / 1969–1971]. Accessed 21 April 2016. www.newspapers.com.

Odencrantz, William, Steven T. Nutter, and Josie M. Gonzalez. "Immigration Reform and Control Act of 1986: Obligations of Employers and Unions." *Berkeley Journal of Employment and Labor Law* 10, no. 1 (1988): 92–115. Accessed 29 April 2016. http://scholarship.law.berkeley.edu/bjell/v0110/iss1/7.

Oldershausen, Sasha von. "The Tragic Story of a Texas Teen and the Marines Who Killed Him for No Reason." Fusion. Accessed 19 July 2017. http://fusion.kinja.com/the-tragic-story-of-a-texas-teen-and-the-marines-who-ki-1795441102.

Olmstead, Alan L., and Paul W. Rhode, "Cotton, Cottonseed, Shorn Wool, and Tobacco: Acreage, Production, Price, and Cotton Stocks, 1790–1999 [Annual]." In *Historical Statistics of the United States, Earliest Times to the Present: Millennial Edition*, edited by Susan B. Carter, Scott Sigmund Gartner, Michael R. Haines, Alan L. Olmstead, Richard Sutch, and Gavin Wright, table Da755–765. New York: Cambridge University Press, 2006. Accessed 1 September 2015. https://hsus.cambridge.org/HSUSWeb/toc/treeTablePath.do?id=Da755–765 2.

Patoski, Joe Nick. "What Would Donald Judd Do?" *Texas Monthly*, July 2001. http://www.joenickp.com/texas/donaldjudd.html.

Pennington, Wayne D., and Scott D. Davis. "Notable Earthquakes Shake Texas on Occasion." *Texas Almanac.com*. Accessed 19 February 2015. https://texasalmanac.com/topics/media/notable-earthquakes-shake-texas-occasion.

Premier One Supplies. Shearing Instructions. Accessed 12 February 2016. https://www.premier1supplies.com/img/instruction/41.pdf.

Rock, Rosalind Z. "San Francisco de La Junta Pueblo." *Handbook of Texas Online*. Accessed 5 May 2016. http://www.tshaonline.org/handbook/online/articles/uqs45.

"San Esteban Lake." *Handbook of Texas Online*. Accessed 30 April 2015. http://www.tshaonline.org/handbook/online/articles/ros05.

Schoenian, Susan. "Sheep 201: A Beginners Guide to Raising Sheep." Updated October 2012. http://www.sheep101.info/201/ewerepro.html.

Sides, Anne Goodwin. "Donald Judd Found Perfect Canvas in Texas Town." *Weekend Edition/NPR*, 31 January 2009. http://www.npr.org/2009/01/31/99130809/donald-judd-found-perfect-canvas-in-texas-town.

Smith, Julia Cauble. "Indio TX (Presidio County)." *Handbook of Texas Online*. Accessed May 14, 2015. http://www.tshaonline.org/handbook/online/articles/hri03.

———. "Ochoa, TX." *Handbook of Texas Online*. Accessed 15 January 2016. http://www.tshaonline.org/handbook/online/articles/ht003.

———. "Pilares, TX." *Handbook of Texas Online*. Accessed 7 December 2016. http://www.tshaonline.org/handbook/online/articles/hrp87.

Taubenberger, J. K., and D. M. Morens. "1918 Influenza: The Mother of All Pandemics." *Emerging Infectious Diseases* 12, no. 1 (2006). Accessed 26 May 2015. http://dx.doi.org/10.3201/eid1209.050979.

Texas beyond History Web Team. "Sotol." *Texas beyond History*, University of Texas at Austin. Accessed 29 January 2016. http://www.texasbeyondhistory.net/ethnobot/images/sotol.html.

———. "Trans-Pecos Mountains and Basins." *Texas beyond History*, University of Texas at Austin. Accessed 23 February 2015. http://www.texasbeyondhistory.net/trans-p/index.html.

———. "Wax Camps." *Texas beyond History*, University of Texas at Austin. Accessed 30 March 2016. http://www.texasbeyondhistory.net/waxcamps/index.html.

Texas Parks and Wildlife Department. "Chinati Mountains State Natural Area." Accessed 15 August 2017. https://tpwd.texas.gov/state-parks/chinati-mountains.

Timmons, W. H. "Rodríguez-Sanchez Expedition." *Handbook of Texas Online*. Accessed 12 December 2016. http://www.tshaonline.org/handbook/online/articles/upr01.

Tyco International Ltd. "John Fort Assumes Primary Executive Responsibility for Tyco." Press release, 3 June 2002. Investors.tyco.com. http://investors.tyco.com/phoenix.zhtml?c=112348&p=irol-newsArticle&ID=301812 (no longer available).

Vallance, Tom. "John Drew Barrymore." *The Independent* (UK), 1 December 2004. http://www.independent.co.uk/news/obituaries/john-drew-barrymore-728653.html.

Vernon, Walter N. "Lydia Patterson Institute." *Handbook of Texas Online*. Accessed 24 July 2015. https://www.tshaonline.org/handbook/online/articles/kb121.

Wade, Dale A., Donald W. Hawthorne, Gary L. Nunley, and Milton Caroline. "History and Status of Predator Control in Texas." *Proceedings of the Eleventh Vertebrate Pest Conference*, paper 42 (1984), 122–31. Accessed 11 March 2016. http://digitalcommons.unl.edu/vpc11/42.

Washington State Department of Health. "Foot and Mouth Disease." June 2007. https://agr.wa.gov/FP/Pubs/docs/183-FMDBrochure.pdf.

Weatherby, James Sim. Find a Grave. Accessed 9 October 2015. www.findagrave.com/cgi-bin/fg.cgi?page=gr&GRid=63771638&ref=acom.

Western Whitetail.com. "Carmen Mountain Whitetail." Accessed 25 April 2016. http://www.westernwhitetail.com/whitetail/carmen-mountain-whitetail/.

Wikipedia contributors. "Aconcagua." Wikipedia. Accessed 22 September 2017. https://en.wikipedia.org/w/index.php?title=Aconcagua&oldid=801736089.

Wikipedia contributors. "Atherosclerosis." Wikipedia. Accessed 14 March 2016. https://en.wikipedia.org/w/index.php?title=Atherosclerosis&oldid=709897206.

Wikipedia contributors. "Ayala (surname)." Wikipedia. Accessed 3 August 2017. https://en.wikipedia.org/w/index.php?title=Ayala_(surname)& oldid=774386030.

Wikipedia contributors. "Ayala-Aiara." Wikipedia. Accessed 3 August 2017. http://fr.wikipedia.org/w/index.php?title=Ayala-Aiara&oldid=1372 59096.

Wikipedia contributors. "Catholic Funeral." Wikipedia. Accessed 15 April 2016. https://en.wikipedia.org/w/index.php?title=Roman_Catholic_funeral&oldid=713224710.

Wikipedia contributors. "Denali." Wikipedia. Accessed 22 September 2017. https://en.wikipedia.org/w/index.php?title=Denali&oldid=800368323.

Wikipedia contributors. "Dennis Kozlowski." Wikipedia. Accessed 25 May 2016. https://en.wikipedia.org/w/index.php?title=Dennis_Kozlowski &oldid=710367587.

Wikipedia contributors. "Giant (1956 film)." Wikipedia. Accessed 4 March 2016. https://en.wikipedia.org/w/index.php?title=Giant_(1956_film)& oldid=708028466.

Wikipedia contributors. "Hurricane Celia." Wikipedia. Accessed 25 April 2016. https://en.wikipedia.org/w/index.php?title=Hurricane_Celia& oldid=687172432.

Wikipedia contributors. "John Drew Barrymore." Wikipedia. Accessed 3 March2016.https://en.wikipedia.org/w/index.php?title=John_Drew_Barrymore&oldid=707426768.

Wikipedia contributors. "José Inés Salazar." Wikipedia. Accessed 22 December 2015. https://en.wikipedia.org/w/index.php?title=Jos%C3%A9_In%C3%A9s_Salazar&oldid=675460198.

Wikipedia contributors. "List of Mountain Peaks of Texas." Wikipedia. Accessed 11 February 2015. http://en.wikipedia.org/w/index.php?title =List_of_mountain_peaks_of_Texas&oldid=629332097.

Wikipedia contributors. "Lydia Patterson Institute." Wikipedia. Accessed 24 July 2015. https://en.wikipedia.org/w/index.php?title=Lydia_Patterson _Institute&oldid=601667704.

Wikipedia contributors. "Mount Washington (New Hampshire)." Wikipedia. Accessed 22 September 2017. https://en.wikipedia.org/w/index .php?title=Mount_Washington_(New_Hampshire)&oldid=799042716.

Wikipedia contributors. "Pancreatic Cancer." Wikipedia. Accessed 22 March 2016. https://en.wikipedia.org/w/index.php?title=Pancreatic_cancer& oldid=709652842.

Wikipedia contributors. "Perlite." Wikipedia. Accessed 11 April 2016. https:// en.wikipedia.org/w/index.php?title=Perlite&oldid=703068988.

Wikipedia contributors. "Piper PA-18 Super Cub." Wikipedia. Accessed 27 January 2016. https://en.wikipedia.org/w/index.php?title=Piper_PA-18_Super_Cub&oldid=701596491.

Wikipedia contributors. "Poliomyelitis." Wikipedia. Accessed 8 July 2015. https://en.wikipedia.org/w/index.php?title=Poliomyelitis&oldid =669813888.

Wikipedia contributors. "Supervolcano." Wikipedia. Accessed 13 February 2015. http://en.wikipedia.org/w/index.php?title=Supervolcano&oldid =646290843.

Wikipedia contributors. "Warner & Swasey Company." Wikipedia. Accessed 3 March 2016. https://en.wikipedia.org/w/index.php?title=Warner_%26_ Swasey_Company&oldid=676497587.

Williamson, Samuel H. "Seven Ways to Compute the Relative Value of a US Dollar Amount, 1774 to present." MeasuringWorth, 2017. Accessed 26 July 2017. www.measuringworth.com/uscompare/.

World Health Organization (WHO). "Global Solar UV Index: A Practical Guide." Accessed 6 March 2015. http://www.who.int/uv/publications/ en/UVIGuide.pdf.

Index